Pericyclic Reactions
Volume II

This is Volume 35 of
ORGANIC CHEMISTRY
A series of monographs
Editors: ALFRED T. BLOMQUIST and HARRY H. WASSERMAN

A complete list of the books in this series appears at the end of the volume.

Pericyclic Reactions
Volume II

Edited by

Alan P. Marchand
Roland E. Lehr

Department of Chemistry
University of Oklahoma
Norman, Oklahoma

Academic Press   New York   San Francisco   London   1977
A Subsidiary of Harcourt Brace Jovanovich, Publishers

COPYRIGHT © 1977, BY ACADEMIC PRESS, INC.
ALL RIGHTS RESERVED.
NO PART OF THIS PUBLICATION MAY BE REPRODUCED OR
TRANSMITTED IN ANY FORM OR BY ANY MEANS, ELECTRONIC
OR MECHANICAL, INCLUDING PHOTOCOPY, RECORDING, OR ANY
INFORMATION STORAGE AND RETRIEVAL SYSTEM, WITHOUT
PERMISSION IN WRITING FROM THE PUBLISHER.

ACADEMIC PRESS, INC.
111 Fifth Avenue, New York, New York 10003

United Kingdom Edition published by
ACADEMIC PRESS, INC. (LONDON) LTD.
24/28 Oval Road, London NW1

Library of Congress Cataloging in Publication Data

Main entry under title:

Pericyclic reactions.

(Organic chemistry series ; vol. 35)
Includes bibliographies and index.
1. Cyclic compounds. 2. Chemical reactions.
I. Marchand, Alan P. II. Lehr, Roland E.
QD331.P46   547'.5   75-40611
ISBN 0-12-470502-2

PRINTED IN THE UNITED STATES OF AMERICA

Contents

List of Contributors vii
Preface ix
Contents of Volume I xi

1 Carbocations
T. S. SORENSEN AND A. RAUK

    I. General Introduction 1
   II. Electrocyclic Reactions 3
  III. Cycloaddition and Cycloreversion Reactions 34
  IV. Sigmatropic Reactions 45
   V. Pericyclic, Longicyclic, and Laticyclic Stabilization 64
      References 71

2 Pericyclic Reactions of Cumulenes
LÉON GHOSEZ AND MARTIN J. O'DONNELL

     I. Introduction 79
    II. Electronic Structure of Cumulenes 80
   III. [2 + 2] Cycloadditions of Ketenes to Alkenes 85
   IV. Ketene Dimerization 104
    V. Ketene–Allene Cycloadditions 104
   VI. [2 + 2] Cycloadditions of Allenes 109
  VII. [2 + 2] Cycloadditions of Ketenimines 117
 VIII. [2 + 2] Cycloadditions of Keteniminium Salts 118
   IX. [2 + 4] Cycloadditions of Cumulenes 126
    X. Cycloadditions of Vinylcumulenes 130
      References 135

3 Cheletropic Reactions
WILLIAM L. MOCK

    I. Introduction 141
   II. Sulfolene Reactions 142
  III. Cheletropy in Other Systems 161
  IV. Conclusion 174
      References 174

4 Applications of Frontier Molecular Orbital Theory to Pericyclic Reactions
K. N. HOUK

      I. Introduction 182
     II. The Development of the Frontier Molecular Orbital Method . . 182
    III. The Frontier Orbital Approximation 184
    IV. Practical Applications of Frontier Molecular Orbital Theory . . 199
     V. Applications to Thermal Intermolecular Cycloadditions . . . 208
    VI. Applications to Thermal Intramolecular Reactions 240
   VII. Photochemical Pericyclic Reactions 258
  VIII. Conclusion 266
      References 266

5 Thermally Forbidden Reactions
JOHN E. BALDWIN

    I. Introduction 273
   II. Theory 275
  III. Examples 280
  IV. Conclusions 297
      References 297

Author Index 303
Subject Index 319

List of Contributors

Numbers in parentheses indicate the pages on which the authors' contributions begin.

John E. Baldwin (273), Department of Chemistry, University of Oregon, Eugene, Oregon

Léon Ghosez (79), Laboratoire de Chimie Organique de Synthèse, Université Catholique de Louvain, Louvain-La-Neuve, Belgium

K. N. Houk (181), Department of Chemistry, Louisiana State University, Baton Rouge, Louisiana

William L. Mock (141), Department of Chemistry, University of Illinois at Chicago Circle, Chicago, Illinois

Martin J. O'Donnell* (79), Laboratoire de Chimie Organique de Synthèse, Université Catholique de Louvain, Louvain-La-Neuve, Belgium

A. Rauk (1), Department of Chemistry, The University of Calgary, Calgary, Alberta, Canada

T. S. Sorensen (1), Department of Chemistry, The University of Calgary, Calgary, Alberta, Canada

* Present address: Department of Chemistry, Indiana University–Purdue University at Indianapolis, Indianapolis, Indiana.

Preface

Probably no single development has had so profound an effect on the field of organic chemistry in the past ten years than that of orbital symmetry and its related rules and their application to pericyclic reactions. Yet, to date, no contributed volume devoted entirely to pericyclic reactions has appeared. This two-volume treatise fills that void.

The contributions included in the volumes bear testimony to the wide range of topics justifiably contained in such a treatise. Thus, chapters dealing with theoretical approaches to pericyclic reactions are included, as well as reviews of pericyclic reactions of reactive intermediates and of particular reaction types. An introductory chapter examines some of the experimental approaches used to establish the authenticity of an apparent pericyclic reaction.

While the volumes will be of greatest interest to those actively engaged in research in an area specifically reviewed in one of the chapters, they should also be of value to those chemists wishing to familiarize themselves with modern developments in this general area.

We are most fortunate in having secured contributions from experts in their respective fields. We asked each contributor to provide critical discussions of recent developments in his area, with particular emphasis to be placed on those which relate to pericyclic reactions. We are pleased with the responses we received and with the resulting contributions. Whatever merit these volumes possess is due to the excellence of these contributions.

Alan P. Marchand
Roland E. Lehr

Contents of Volume I

Operational Criteria for Evaluation of Concertedness in Potential Pericyclic Reactions
ROLAND E. LEHR AND ALAN P. MARCHAND

The Möbius–Hückel Treatment of Organic Systems and Reactions and MO Following as a Technique in Organic Chemistry
HOWARD E. ZIMMERMAN

Some Pericyclic Reactions of Carbenes
W. M. JONES AND UDO H. BRINKER

Pericyclic Reactions of Carbanions
STUART W. STALEY

Author Index–Subject Index

I

Carbocation

T. S. Sorensen and A. Rauk

I.	General Introduction	1
II.	Electrocyclic Reactions	3
	A. Introduction	3
	B. The $k = 2$ Systems	3
	C. The $k = 4$ Systems	21
	D. Polyenylic Cations $k = 6$ or Larger	33
III.	Cycloaddition and Cycloreversion Reactions	34
	A. Introduction	34
	B. Rules for Cycloaddition and Cycloreversion Reactions	35
	C. [2 + 2] Cycloaddition or Cycloreversion Processes in Cationic Systems	36
	D. [$_\pi$2 + $_\pi$4] Cycloadditions	41
	E. [$_\pi$4 + $_\pi$6] Cycloadditions	45
IV.	Sigmatropic Reactions	45
	A. General Considerations	45
	B. Linear vs. Bridged Structures	46
	C. [1,2]-Sigmatropic Rearrangements	52
	D. [1,3]-Sigmatropic Rearrangements	54
	E. [1,4]-Sigmatropic Rearrangements	59
	F. [1,5]-Sigmatropic Rearrangements	62
	G. [1,6]-Sigmatropic Rearrangements	63
	H. [3,4]-Sigmatropic Rearrangements	64
V.	Pericyclic, Longicyclic, and Laticyclic Stabilization	64
	References	71

I. General Introduction

In organizing this chapter on carbocations we have kept the original divisions and descriptions of Woodward and Hoffmann. Occasionally, a classification of some pericyclic reaction will seem rather arbitrary, but this is relatively unimportant because in a generalized sense the orbital symmetry arguments merge. We have in general used the HOMO–LUMO orbital descriptions, and carbocations appear particularly well-behaved in this regard.

We have restricted our coverage to spin-paired (closed shell) carbocations and have thus excluded cation radicals. Problems arise in considering heteroatom stabilized species, i.e., a carbonyl group can be considered an oxyanion-substituted carbocation $\diagup\!\!\!\!\diagdown\!\!\mathrm{C}^+\!\!-\!\mathrm{O}^-$. We do not cover neutral carbonyl compounds but have included reactions which probably involve the protonated carbonyl function $\diagup\!\!\!\!\diagdown\!\!\mathrm{C}^+\!\!-\!\mathrm{OH} \leftrightarrow \diagup\!\!\!\!\diagdown\!\!\mathrm{C}\!\!=\!\!\mathrm{O}^+\mathrm{H}$.

There are several features of carbocation chemistry which are unique. First, the electron-deficient center allows very rapid bond rearrangement processes. An everyday analogy is to consider a street of houses with one house empty. It is relatively easy to permute the families since each movement results in a new vacancy. The analogy for neutral systems is no vacancy, any net movement involves first removing the occupants of one house to some higher level (the street?). Although the rearrangements in carbocations are characteristically rapid, this very rapidity can lead to problems since even orbital-symmetry-disallowed processes may be rapid, the relative "allowed–disallowed" comparison being between rapid and very rapid rates. Second, carbocations are studied as both reactive intermediates and as observable species in strong acid solution. Most literature cases involve the former, and, here, one has a problem with "concertedness," i.e., many rearrangements are concerted with a leaving group departure. These may still be subject to orbital symmetry control, the two modes of disrotatory opening in incipient cyclopropyl cations being a classical example. In this chapter we will have occasion to discuss both transient and observable carbocation rearrangements. Examples of photochemical carbocation transformations are few, most of the work involving relatively stable species like protonated ketones. Successful experiments have to be done on observable carbocations where any competing thermal rearrangements are slow (particularly the reverse reaction), and these criteria certainly limit the possibilities.

We have added a concluding section in this chapter which deals not with carbocation rearrangements but with orbital symmetry interactions which are "extra" stabilizing or destabilizing compared to some reference state. Most chemists are familiar with delocalization through space, e.g., the myriad of homoconjugated structure proposals. "Through-bond" interactions are much less obvious, but they may be just as important. In fact, second-order perturbation treatments are one area in which continuing experimental and theoretical work is occurring. Other interesting areas involve the modification of the relatively well-known pericyclic processes, for example, by adding electron-withdrawing or electron-donating substituents, providing for homoconjugation, etc. At the theoretical level we shall probably find more and more detailed mapping of transition-state surfaces. In some ways, and unavoidably, these distract from the simple pedagogical picture provided us by the HOMO–LUMO description.

II. Electrocyclic Reactions

A. INTRODUCTION

Electrocyclic transformations, as defined by Woodward and Hoffmann (1965), are either the formation or breaking of a single bond between the termini of a linear system, containing k electrons in the case of closure or $k - 2$ in the case of opening.

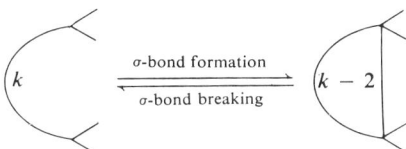

This σ-bond formation in a planar π-electron system involves the eventual twisting of both termini and can be either disrotatory or conrotatory. The simple rules for deriving these are well known and the allowed carbocation transformations for the lower homologs are shown below.

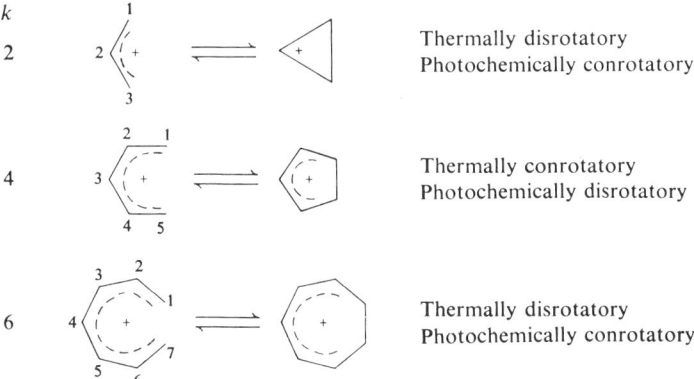

Many variations of these basic systems are possible and this discussion will not be restricted solely to the stereochemical aspects of electrocyclic transformations.

B. THE $k = 2$ SYSTEMS

1. Allyl Cation–Cyclopropyl Cation Energy Considerations

The title system has been very ably investigated by a number of workers. In contrast to the longer chain homologs, the open π member of this pair (the allyl cation) is normally the more stable and studies have therefore involved the ring opening of the cyclopropyl cation. The gross energy equation in all of these systems is relatively simple if one considers only the π-system energies and ring strain as in this case.

$$\text{Allyl cation resonance energy} + \pi \text{ bond} \rightleftharpoons \sigma \text{ bond} + \text{ring strain}$$

There have been many calculations on the allyl cation ⇌ cyclopropyl cation transformation, both with respect to the stereochemistry of the process, including detailed investigations of the transition state surface, and on the energy difference of the two ground states (Kutzelnigg, 1967; Hoffmann, 1968; Clark and Smale, 1969; Clark and Armstrong, 1969; Peyerimhoff and Buenker, 1969; Olsen et al., 1971; Pearson, 1971; Dewar and Kirschner, 1971; Liberles et al., 1972; Tee and Yates, 1972; Radom et al., 1973a,b; Yamaguchi and Fueno, 1973; Coffey and Jug, 1974; Merlet et al., 1974). Adding stabilizing groups to the allyl termini will of course accentuate the difference. However, adding cation stabilizing groups to the central atom should minimize the difference. The recent *ab initio* calculations of Radom et al. (1973b) bear this out and are shown in Table I.

In support of these calculations, several aminocyclopropyl cations do not open to allyl systems (Jongejan et al., 1972). Further, in preparing enamines of 2-chloro- and bromocyclohexanone, Cantacuzène and Tordeux (1971) reported that both compounds **1** and **2** were formed, the latter perhaps from an electrocyclic closure of the corresponding 2-aminoallyl system. However,

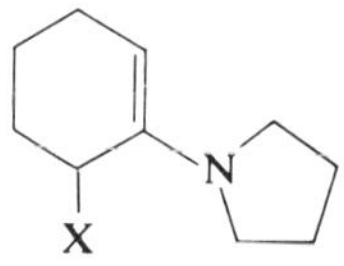

(1)   (2)
X = Cl or Br

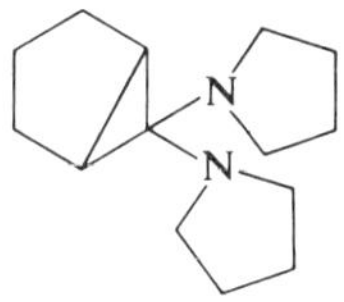

the halogen in **1** can be displaced by an alcohol without forming a cyclopropyl system. The extreme example of a stabilizing substituent is found in cyclopropanone (**3**) and the corresponding 2-oxyallyl open form (**4**), where there is no doubt that the closed form is the more stable.

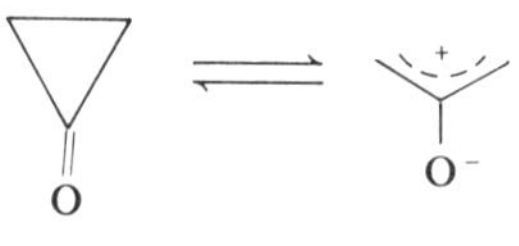

(3)   (4)

The cyclopropyl–allyl cation energy separation is relevant in considering possible mechanisms for the interconversion of cis–trans (E–Z) substituents in allyl cations. This could occur by the following mechanism. From Table I,

TABLE I Calculated Relative Energies of Allyl, Cyclopropyl, and Perpendicular Allyl Cations

Substituent X	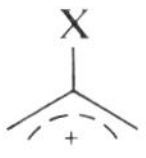 allyl	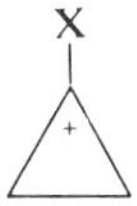 cyclopropyl	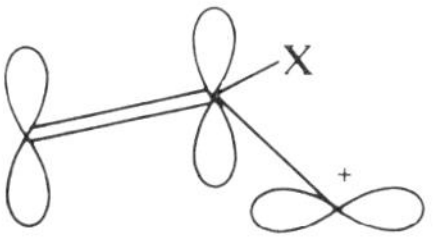 perpendicular allyl
H	0	39.2[a]	34.8
CH$_3$	0	18.8	32.8
NH$_2$	0	−33.5	41.6
OH	0	−9.8	35.6
F	0	14.9	39.4

[a] In kcal/mole, all taken relative to the allyl member = 0.

this mechanism is calculated to be preferable to the simple rotation process involving a perpendicular allyl cation; however, still other mechanisms are possible and there are at present no established examples of the isomerization involving cyclopropyl cations.

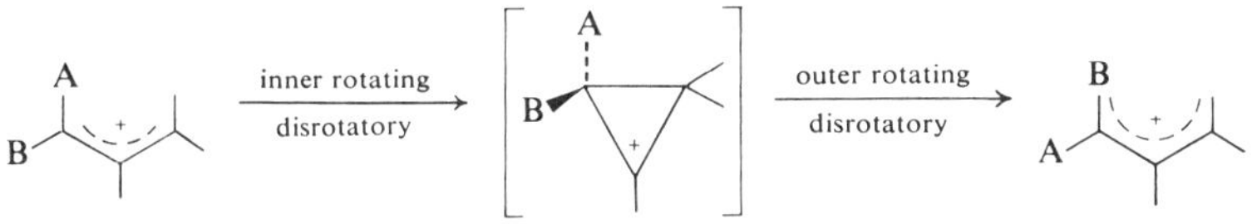

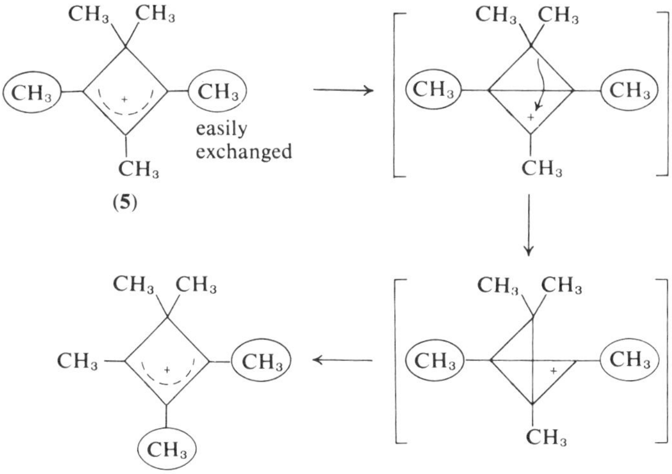

Closure of an allyl cation to an intermediate cyclopropyl cation would also provide a mechanism for more deep-seated allyl cation rearrangements. Koptyug et al. (1971) have reported, from H–D exchange studies, that the central methyl group in the pentamethylcyclobutenyl cation (5) interchanges with a terminal methyl group, and suggests a mechanism involving 1,2-migration in the cyclopropyl cation. It is possible that the cyclopropyl cation is relatively more stable in this instance and that such isomerizations would not occur as readily in open allyl systems.

The conrotatory closure of an allyl cation should be photochemically possible but no examples are known. One would not expect to see the cyclopropyl cations in simple systems since the thermal disrotatory reverse cannot easily be stopped. However, overall allyl cation isomerizations might be observed.

2. Stereochemical Aspects

A number of theoretical studies, semiempirical and *ab initio*, have been published (see previous list of references). Although several authors urge caution in using only frontier orbitals in making correlations, every calculation to date has predicted the disrotatory thermal opening. The chief interest among these studies is the degree of opening at the transition state. Here again, there is near unanimity; the calculations show either no activation barrier (the cyclopropyl cation is already over the barrier starting from a neutral precursor) or only a small one. For simple systems, the cyclopropyl cation is probably not on the transition state surface since the opening is concerted with the leaving group departure.

The initial evidence for the disrotatory opening was provided by kinetic studies; the most extensive studies on simple systems are those of De Puy *et al.* (1966), Cristol *et al.* (1965), Schleyer *et al.* (1966), Schleyer (1967), Schöllkopf (1968), Parham and Yong (1970), Hausser and Grubber (1972), and

TABLE II Relative Acetolysis Rates for Cyclopropyl Tosylates

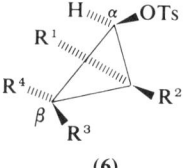

(6)

Compound	R¹	R²	R³	R⁴	$k_{\text{relative}}(150°)$
6a	H	H	H	H	1
6b	CH₃	H	H	CH₃	18,000
6c	H	CH₃	CH₃	H	4
6d	CH₃	CH₃	CH₃	H	80
6e	CH₃	CH₃	CH₃	CH₃	5,500
6f	CH₃	H	CH₃	H	260
6g	CH₃	CH₃	H	H	330

Hausser and Uchic (1972). Using methyl substituents, Schleyer and Schöll-Kopf *et al.* observed the results shown in Table II.

One notes that β-methyl substituents increase the solvolysis rate; this is evidence for charge development at these centers in the transition state, in agreement with theory. The other notable feature is the large rate difference between **6b** and **6c**. This dramatic difference is due to a further feature of cyclopropane ring openings, which has no parallel in the higher homologous series. The explanation of this was due originally to De Puy (Cristol *et al.*, 1965), who noted that two disrotatory processes are possible, an "inner" rotation, and an "outer," namely,

If one actually had formed the cyclopropyl cation, one would expect only the most favorable of the two disrotatory processes to occur since both are "allowed." To explain these large rate differences, De Puy suggested that the solvolysis was concerted with ring opening and did not involve a planar cyclopropyl cation intermediate. Substituents on the same side of the three-membered ring as the leaving group rotate inward, whereas those opposite rotate outward. Several semiempirical MO calculations, which include the leaving group (Clark and Smale, 1969; Woodward and Hoffmann, 1965), and the Principle of Least Motion (Tee and Yates, 1972) all attest to this. If one uses the electrons of the $C(\beta)-C(\beta)$ bond to do a backside displacement on the leaving group, then this result can be predicted in a simple qualitative way, namely,

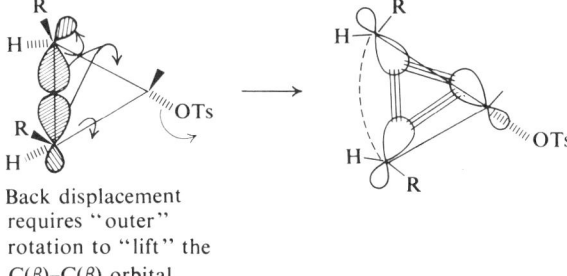

Back displacement requires "outer" rotation to "lift" the $C(\beta)-C(\beta)$ orbital

The rate difference between **6b** and **6c** is now perfectly understandable. Inner rotation of *cis*-methyl groups (to the leaving group) involves large steric interactions, while the situation in **6b** probably involves some steric acceleration. A very similar situation is found in the solvolysis of the cyclopropyl chlorides **7** and **8** using propyl substituents (Parham and Yong, 1970).

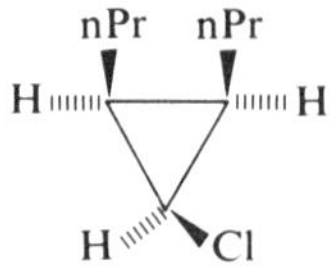

(7) $k < 2.2 \times 10^{-8}$ sec^{-1}

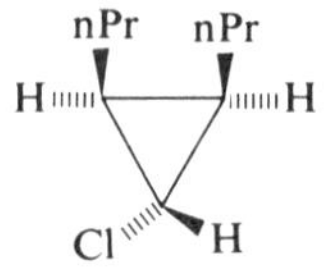

(8) $k = 2.2 \times 10^{-5}$ sec^{-1}

Using phenyl groups as "markers," however, Hausser (Hausser and Uchic, 1972; Hausser and Grubber, 1972) finds evidence for only one kind of disrotatory process. The evidence again is kinetic and his data are shown in Table III.

Although compound **9e** is solvolyzing much slower than **9f** or **9g**, it is still faster than the monophenyl compounds and **9d**. This result is substantially different from that shown in Table II and this leads Hausser to believe that both **9e** and **9g** open via the "outer" rotation. This contradiction of De Puy's rule is understandable since an "inner" rotation in **9e** would lead to intolerable crowding of the phenyl groups. Nevertheless, it would still be desirable to verify this conclusion utilizing conditions under which the ion can

TABLE III Relative Acetolysis Rates of Phenyl-Substituted Cyclopropyl Chlorides

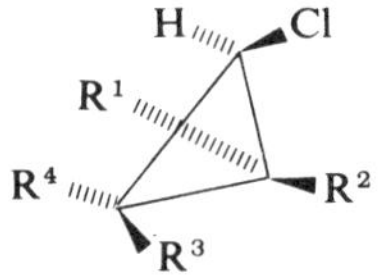

(9)

Compound	R^1	R^2	R^3	R^4	k_{relative}
9a	H	H	H	H	10^{-4}
9b	H	Ph	H	H	1.0
9c	Ph	H	H	H	4
9d	Ph	Ph	H	H	6
9e	H	Ph	Ph	H	145
9f	Ph	H	Ph	H	10,000
9g	Ph	H	H	Ph	465,000

be directly observed, as has been done in the corresponding system **9** (R = methyl).

In all these kinetic studies, the product is generally an allylic species. Nucleophile capture at one end of the allyl cation destroys one center. However, from the (1*E*, 3*E*)-disubstituted cation **10**, one expects exclusively an *E*-substituted double bond, from (1*Z*, 3*Z*)-**12**, exclusively a *Z*-substituted double bond and from (1*E*, 3*Z*)-**11**, a probable mixture. In practice, the neutral allyl

(**10**) 1*E*,3*E* *E*

(**11**) 1*E*,3*Z* *E* *Z*

(**12**) 1*Z*,3*Z* *Z*

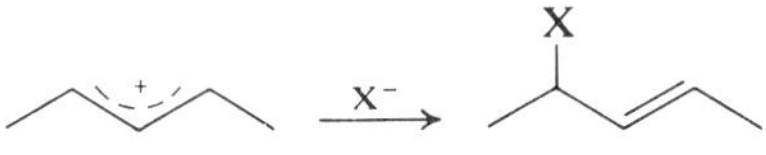

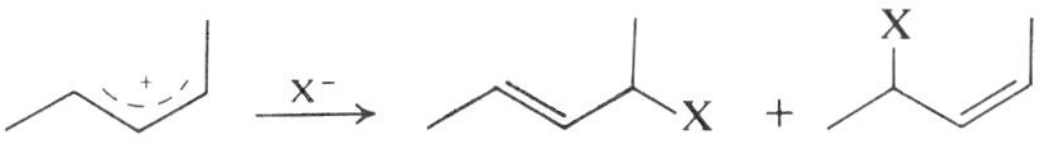

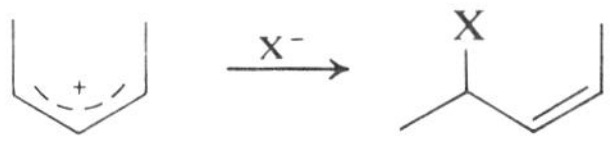

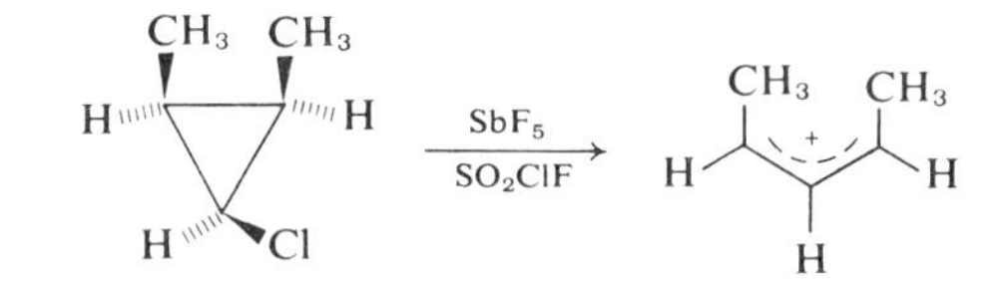

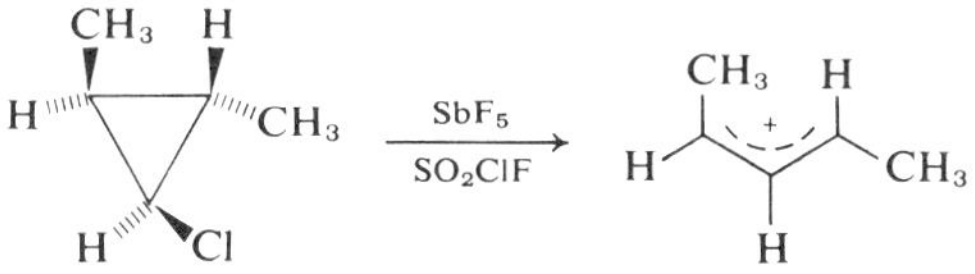

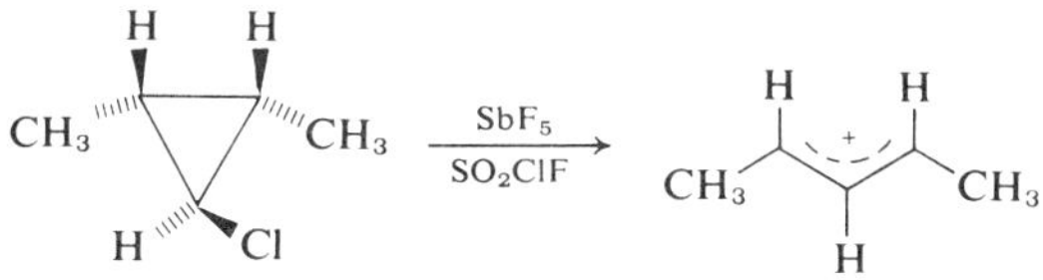

products in the simple cases often isomerize under the reaction conditions to give only the thermodynamic product. This situation can be avoided by directly observing the allyl cations formed. In a beautifully simple experiment, Schleyer *et al.* (1969), showed that the three isomeric chlorides (**13, 14, 15**) do

(**13**)

(**14**)

(**15**)

indeed give the expected allyl cations. T. S. Sorensen and E. Wagner (unpublished results, 1970) have observed that the 2,2,3-trimethyl analogs **16** and

(16)

(17)

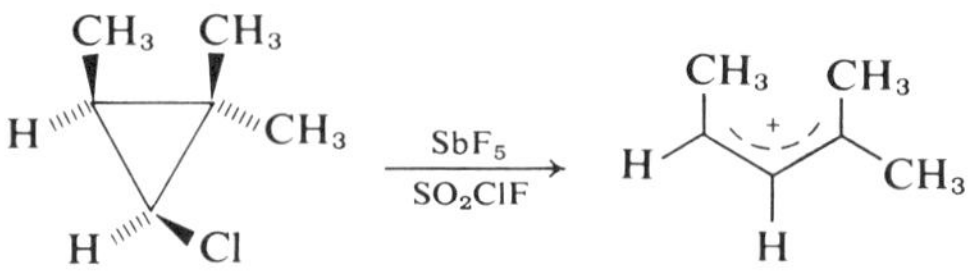

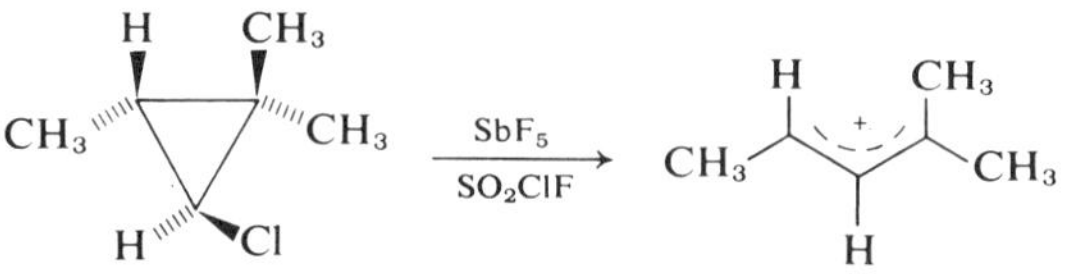

17 are similarly specific. The 1-methyl analogs **18, 19**, and **20** behave reasonably well, as shown by Bollinger *et al.* (1970).

(18) 90% **(22)** 10% **(23)**

(19) 75% **(20)** 25% **(21)** 41%

mixture

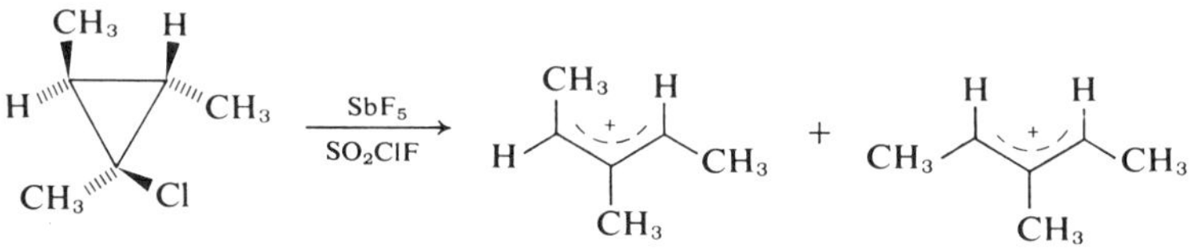

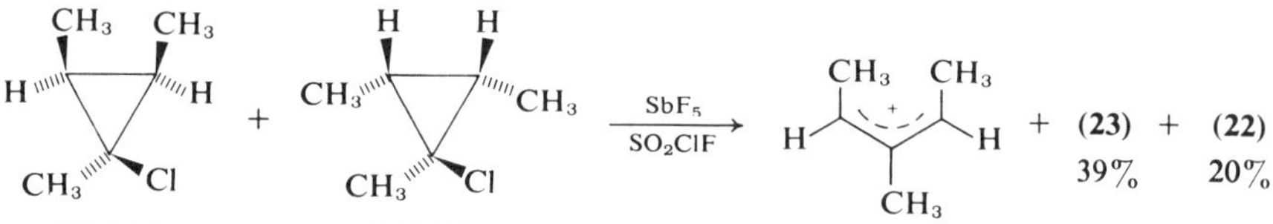

In bicyclic systems, the distinction between inner and outer rotations is even more dramatic. For cis-fused bicyclo[*n*.1.0]alkanes, this is illustrated as follows:

(24)

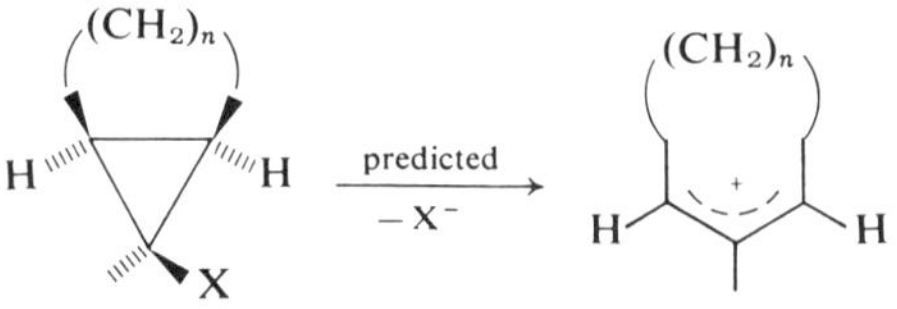

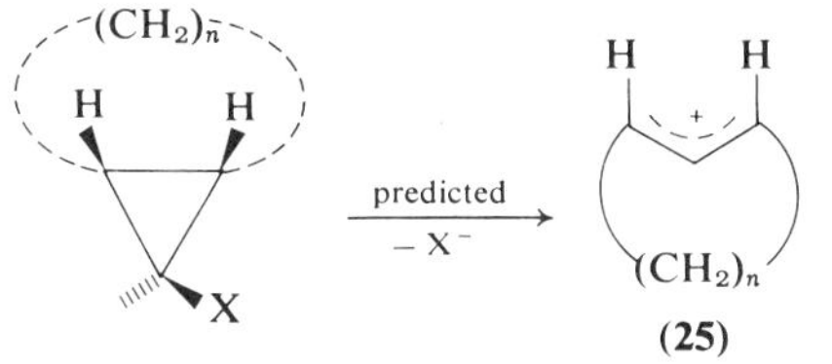

For small rings, $n = 1, 2, 3, 4$, the cis ion (24) is more favorable; for $n = 1, 2, 3$, a cis ion is a fairly absolute requirement. For larger rings, $n = 5, 6$, the di-trans-(or E, E)-fusion is better. The kinetic results of Schöllkopf et al. (1967; Schöllkopf, 1968) illustrate this and are shown in Table IV.

The products observed for $n = 5$ and 6 (exo-tosylate) were mainly the cis double bond allylic acetates; thus, isomerization must have occurred at some later stage. The $n = 2$ case has been studied by both Tufariello et al. (1972) and by Fellenberger et al. (1972). The solvolysis rate of the exo isomer, although slow, is much faster than predicted and this leads to the conclusion that a prior inversion has occurred, the actual measured rate and activation energy being that of the inversion process.

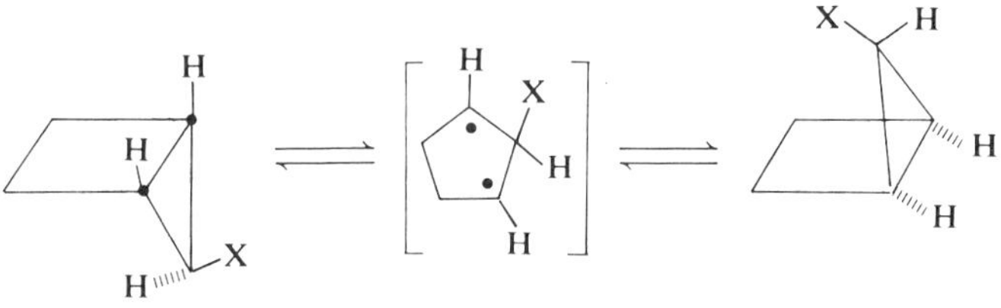

Table IV Relative Acetolysis Rates for Bicyclo[n.1.0]alkyl Tosylates[a]

	Rate	
n	endo	exo
3	25,000	≪0.01
4	62	1.7
5	3.1	2,500
6	3.5	10,000

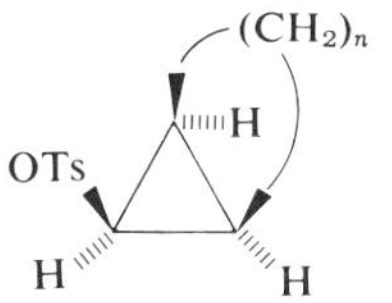

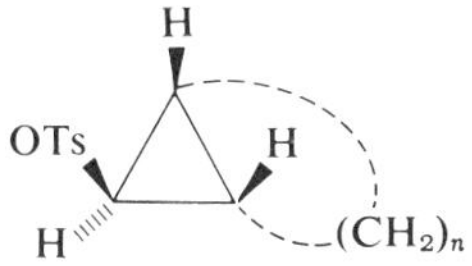

[a] Relative to cyclopropyl tosylate = 1.0.

The reactions of the bicyclic systems can be synthetically useful. Whitham and Wright (1967) have shown that the hydrolysis of *exo*-8-bromobicyclo-[5.1.0]octane gives the expected *trans*-cyclooctenol. Reese and Shaw (1970b) have since studied both the exo and endo isomers. The trans alcohol is formed from the exo isomer, while the endo isomer much more slowly forms the *cis*-cyclooctenol. These reactions are done in the presence of Ag^+ ion, which catalyzes the ring opening and allows one to use milder hydrolysis conditions. Reese and Shaw (1970a) have also studied the corresponding *exo*- and *endo*-nonanes, which give comparable results to the bicyclooctane series.

The cyclopropyl cation opening has also been used for a structure proof. By carrying out a degradation to a cyclopropyl tosylate and noting the subsequent solvolysis rate, it was possible to confirm the stereochemistry of a tricyclic Diels–Alder adduct (Monti and Bertrand, 1973).

Although a solvolysis reaction is the usual way to generate the cations (concertedly), other methods have been used. Rodewald and Lewis (1971) have studied the anodic oxidation of the cyclopropanecarboxylic acids (**26, 27**). Both give the same trans double-bond product (**28**), perhaps indicating a free cyclopropyl cation intermediate. However in other cases, different ratios were obtained. Pasto *et al.* (1973) have studied the electrophilic addition

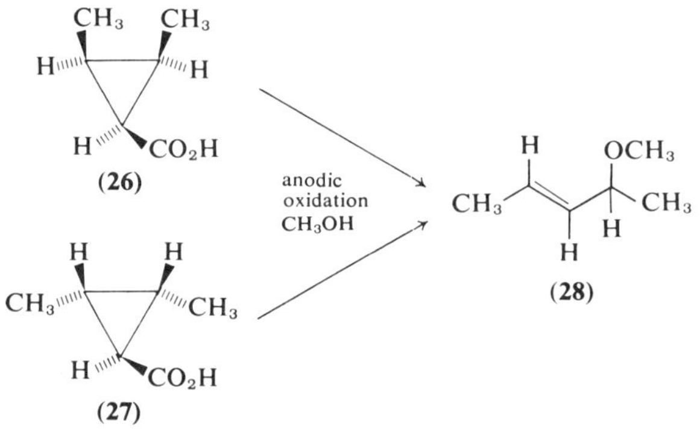

of chlorosulfonyl isocyanate to several isobutenylidenecyclopropanes and observed products due to stereospecific cyclopropane opening. Kirmse and Scheidt (1970) have studied the fragmentation of [*n*.1.0]bicyclic diazonium cyclopropanes. In this case, the unfavorable isomers (*n*, small for exo; *n*, large for endo) apparently isomerize to the reactive isomer via a diazoalkane intermediate. Wilt and Malloy (1970) have described a reaction where a cyclopropane ring becomes activated because of phenyl participation, the stereospecific process here being a combination of participation geometry and orbital symmetry geometry.

While the electrocyclic ring opening is concerted in most instances, there are several examples which appear to involve a "free" cyclopropyl cation. Gajewski and Oberdier (1972) have studied the rearrangement of the spiro cations (29) (a or b = methyl). Both give the same product (30), which argues for a planar intermediate (31), opening in that disrotatory mode dictated by

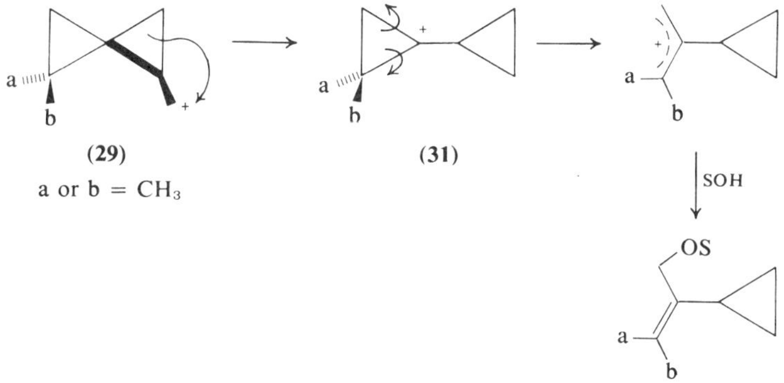

(29) a or b = CH$_3$

(31)

(30)

a = H ⎫ from both
b = CH$_3$ ⎭ precursors

steric interactions. Schöllkopf et al. (1970) have reported the solvolysis of a series of cyclopropyl chlorides containing a stabilizing phenylthio substituent, e.g., 32. They find a combination of direct substitution of the cyclopropane ring, and ring opening. The direct substitution gives an identical product from 32a or 32b. Also, the ring-opened product is a single stereoisomer from both. These results are difficult to rationalize and a "free" pyramidal cation is suggested as the intermediate, i.e., 33a ⇌ 33b, although this structure seems rather unlikely to us.

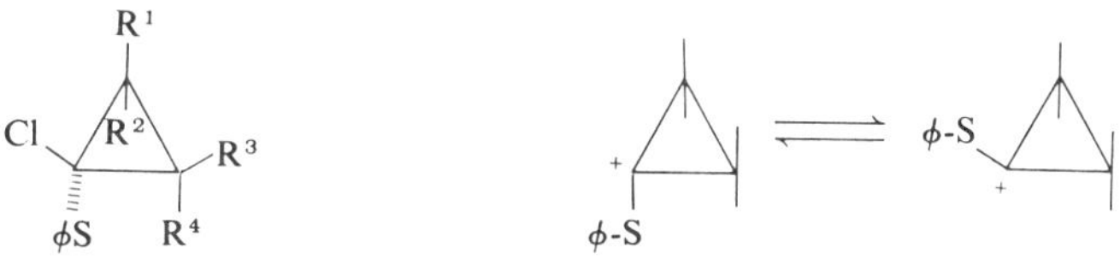

(32a) R^1 = R^3 = CH$_3$, R^2 = R^4 = H
(32b) R^1 = R^3 = H, R^2 = R^4 = CH$_3$

(33a)   (33b)

There are several examples where the initially formed cyclopropyl cation intermediate rearranges to either a more reactive or a more stable structure, which then opens. (See the previous discussion for similar rearrangements starting with allyl cations.)

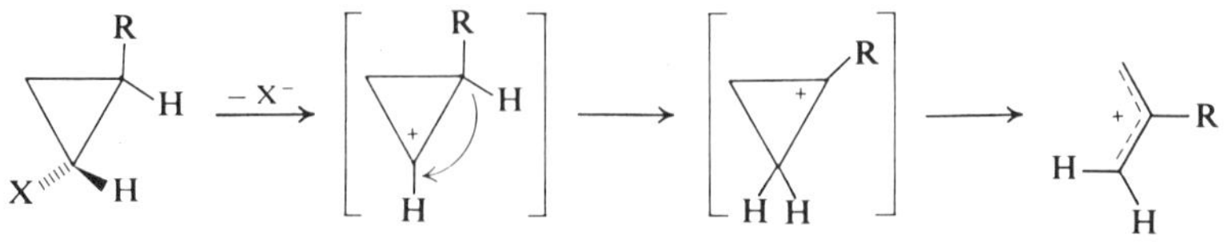

Ledlie (1972) has observed the following reaction starting with **34**.

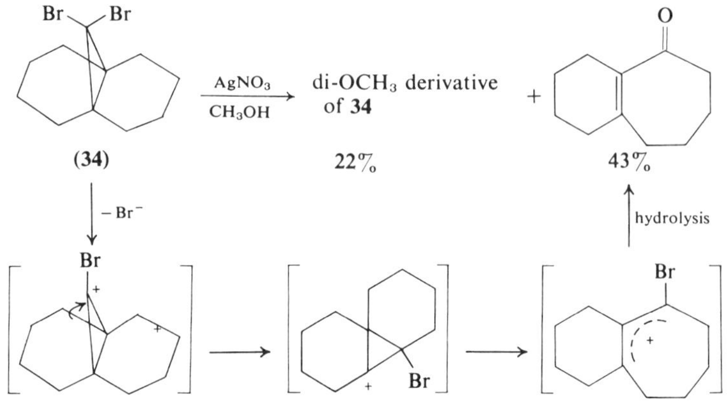

The initially formed cyclopropyl cation **35** cannot easily open since, regardless of the rotation, a di-trans-fused seven-membered allyl cation ring would result. Reese and Strebles (1972) have investigated an even simpler bicyclic system,

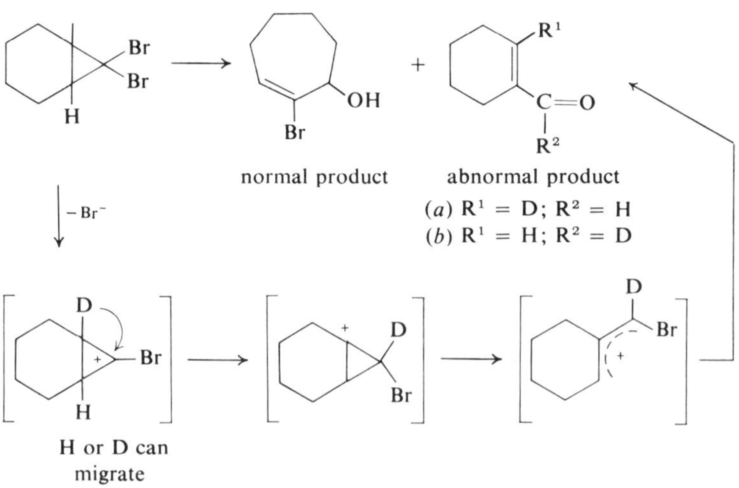

normal product

abnormal product
(a) $R^1 = D$; $R^2 = H$
(b) $R^1 = H$; $R^2 = D$

H or D can migrate

using a deuterium label as a mechanistic probe. The abnormal product in this case can be explained by the mechanism shown. One-half of the label should appear in the aldehyde group and one-half on the double bond, as found. The reaction may involve "free" cyclopropyl cations, but a concerted process is certainly not ruled out. Parham *et al.* (1970) have also observed similar rearrangements.

Information is also available on the stereochemistry of the nucleophile replacement in Ag$^+$-cation catalyzed reactions (Reese and Shaw, 1970a). A single diastereomer, assigned the stereochemistry shown in **37**, is formed from the Ag$^+$-methanol treatment of **36**. It follows that **Y** will be the bromide ion to leave since this results in the more favorable di-*trans*-allyl cation. The

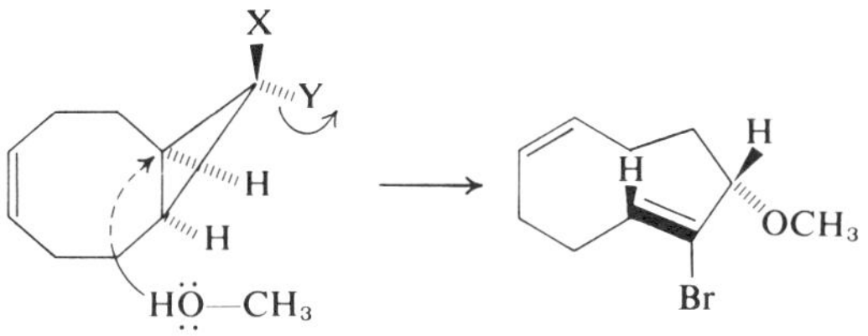

(36)   (37)

X = Y = Br

methoxy group must therefore have attacked the allyl cation from the same side as **Y** departed, with resulting inversion at this center. The nucleophile is usually pictured as a latecomer in the overall reaction, simply quenching the allyl cation. However, the whole process may well be concerted, the nucleophile attacking the incipient allyl cation as the leaving group departs.

3. Dihalocyclopropanes

Synthetically, it is somewhat more convenient and reliable to prepare 1,1-dihalocyclopropanes instead of cyclopropanols or monohalocyclopropanes. That these substances behave identically to the monohalo derivatives has been clearly shown.

The two disrotatory modes are no longer distinguishable if the two halogen atoms are identical, because either one can leave and one will expect the most easily formed product. Parham and Yong (1970) have shown that dichloro derivatives solvolyze somewhat slower than the monochloro analogs and that the *cis*-2,3-di-*n*-propyl and *trans*-2,3-di-*n*-propyl isomers solvolyze at similar rates. Bollinger *et al.* (1970), in observable ion studies, have shown the following.

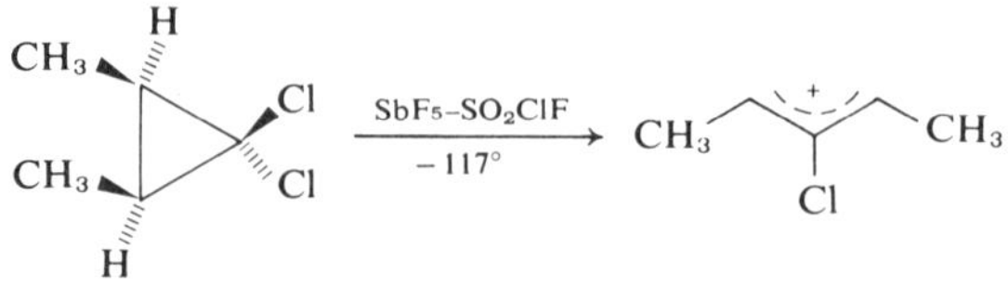

(38) → **(40)** SbF$_5$–SO$_2$ClF, −117°

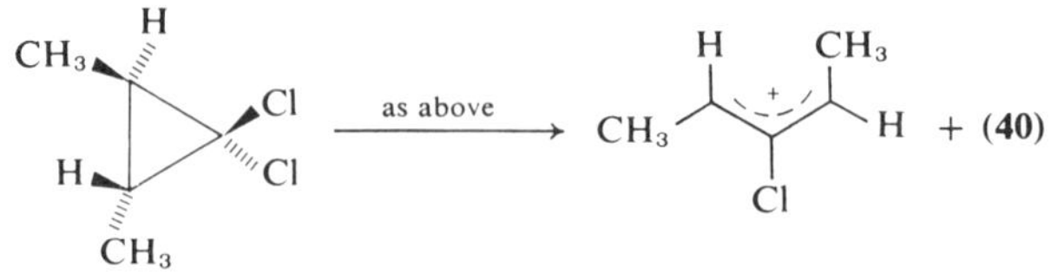

(39) → **(41)** 77–99% + **(40)** (as above)

Synthetically, the electrocyclic opening of dihalocyclopropanes has been used to increase ring size by one or even two carbons. In many cases, the stereochemical aspects are uninteresting since most adducts involve addition to five- and six-membered *cis*-cyclic olefins; this leads to the di-*cis*(or *Z,Z*)-cyclic allyl cation and on quenching, ultimately to another *cis*-olefin. Several *trans*-cyclic olefins have been subjected to this ring-expansion reaction and do indeed yield the expected *cis*- and *trans*-cycloalkene mixture (Parham, 1968; Graefe and Mühlstädt, 1970; Casanova and Waegell, 1972). However, the most useful synthetic procedure is in the stereospecific synthesis of *trans*-cycloalkenes from $n-1$ medium ring *cis*-olefins (Whitham and Wright, 1967; Parham, 1968; Reese and Shaw, 1970a,b; Baird and Reese, 1969, 1970; Duffin and Sutherland, 1970). The allyl cations from dihalocyclopropanes

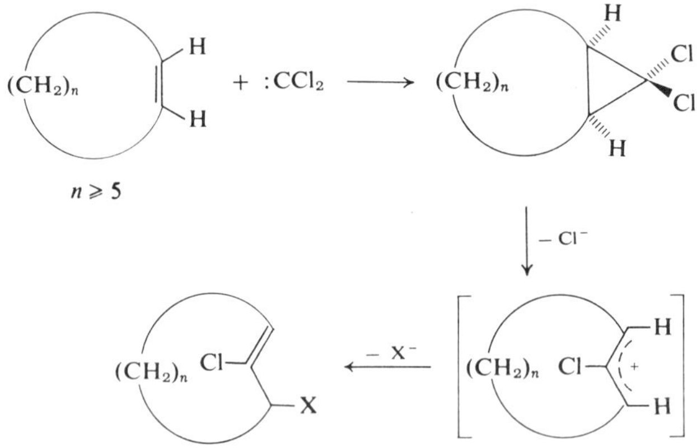

occasionally react further; for example, both cleavage (Seyferth *et al.*, 1970) and cyclization (Dehmlow and Schönefeld, 1971) can occur, as shown for compounds **42** and **43**, respectively.

4. Heteroatom Analogs of the Cyclopropyl Cation

a. Nitrenium Ions. The nitrogen heterocycle corresponding to cyclopropane is aziridine (**44**). Leaving groups can be easily attached to the nitrogen, e.g., *N*-chloroaziridines, and the corresponding cation (**45**), from heterolytic cleavage, is termed a nitrenium ion. In contrast to carbon chemistry, this

ion has been much less studied and, in fact, one often has difficulty in proving whether the cleavage is heterolytic (giving ions) or homolytic (radicals). In the aziridine case, one appears to get heterolytic cleavage, and Gassman *et al.* (1970) have studied the stereochemistry of the aziridine nitrenium ion ring opening. The kinetic results are shown in Table V. The accelerating effect of methyl substituents is compared to that observed for the cyclopropyl system, although the former is a hydrolysis reaction and the latter acetolysis. The agreement in relative rates is impressive and provides evidence for an incipient nitrenium ion intermediate. The inner and outer disrotatory modes have not been observed because the aziridine with $R^1 = R^2 = CH_3$, $R^3 = R^4 = H$ isomerizes to the $R^3 = R^4 = CH_3$, $R^1 = R^2 = H$ isomer, before hydrolysis occurs.

The initial product of ring opening is expected to be the azaallyl cation (**46**). The initial capture product is an imine and this further hydrolyzes to give the

TABLE V Relative Hydrolysis Rate of Methyl-Substituted *N*-Chloroaziridines and Comparison to Cyclopropyl Tosylate Acetolysis[a]

		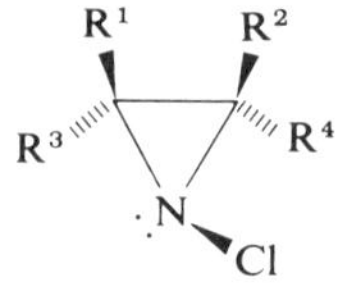 Relative rate at 60°C[a]	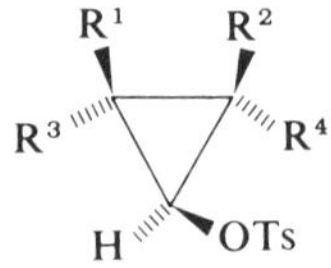 Relative rate at 100°C[a]
	$R^1 = R^2 = R^3 = R^4 = H$	1	1
$R^1 = CH_3$	$R^2 = R^3 = R^4 = H$	15	6
$R^3 = CH_3$	$R^1 = R^2 = R^4 = H$	210	138
$R^1 = R^4 = CH_3$	$R^2 = R^3 = H$	1,500	490
$R^3 = R^4 = CH_3$	$R^1 = R^2 = H$	150,000	41,000

[a] Relative rates are internal to each series.

two carbonyl compounds and ammonium chloride. Very recently, Gassman *et al.* (1975) have been able to isolate an imine in impure form.

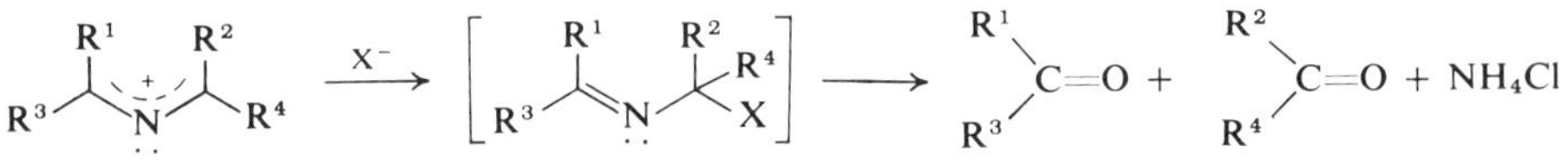

(46)

Weiss (1971) has predicted the disrotatory opening from CNDO MO calculations. This semiempirical method commonly favors cyclic structures and, in this instance, the calculations are erroneous in predicting the closed ion to be more stable than the open form.

The azaallyl cation situation is not strictly analogous to the allyl cation. The latter is formally sp^2 hybridized throughout, while the azaallyl cation, $C_2H_4N^+$, can be written as all sp^2 hybridized (as in **46**) or with sp hybridization of the nitrogen (**47**). One predicts **46** to be formed initially on electrocyclic opening but **47** is probably much more stable and may well be the

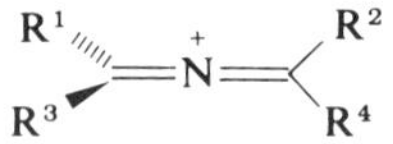

(47)

species captured. This would destroy any stereochemistry present in the allyl form and would make product studies and observable ion studies of the stereochemistry difficult.

b. α-Haloepoxides. McDonald and co-workers (McDonald and Tabor, 1967; McDonald and Steppel, 1969) have shown that the thermal rearrangement of α-chloroepoxides very likely proceeds through cationic intermediates.

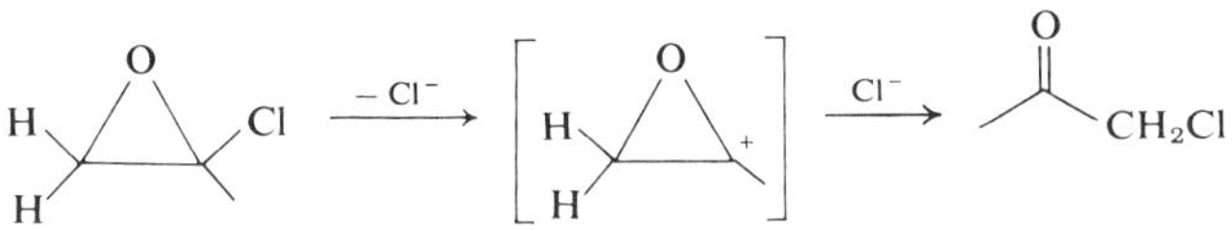

This reaction can be considered an electrocyclic ring opening. The stereochemistry of the oxygen lone pairs is, of course, indeterminate and the dis-

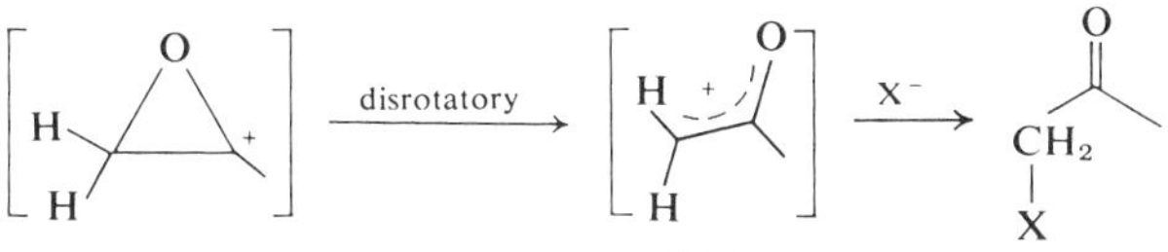

(**48**)

rotatory and conrotatory modes are not distinguishable. All that one might hope to distinguish would be "inner" and "outer" disrotatory modes, e.g.,

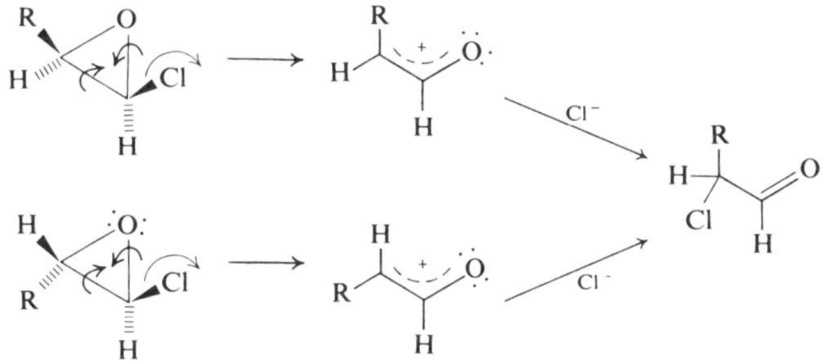

The product will be the same in both cases but the reaction rates might differ.

The energy of the planar 1-oxyallyl cation structure (**48**) has been calculated (A. Rauk, unpublished calculations, 1973) using a limited basis set *ab initio* method and has been found to be 79 kcal/mole less stable than the 90° twisted form. Since an electrocyclic opening in the Woodward–Hoffmann theory requires a rotation to give the planar allyl system, it would appear that some other route leaving this center unrotated may well be more favorable in the chloroepoxide rearrangement.

5 Homoallyl Cation Systems

The homoallyl (**49**), cyclobutyl (**50**), and cyclopropylcarbinyl (**51**) cations have been studied experimentally by many workers and the energy surface

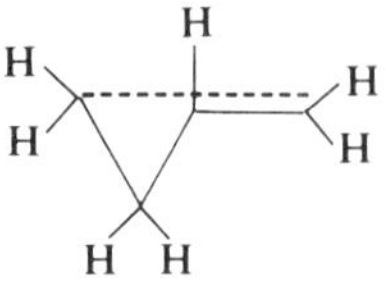

(49) 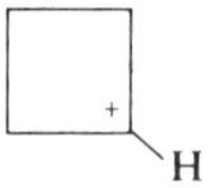   (50) 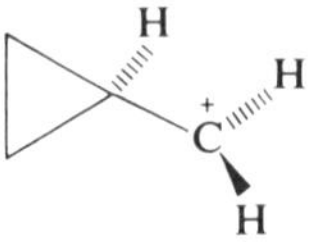   (51)

interconverting these has been investigated theoretically (see discussion later in this chapter). The homoallyl–cyclobutyl cation interconversion can be considered as an electrocyclic reaction. The best analogy to the cyclopropyl

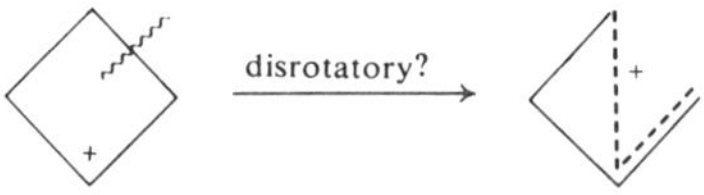

(50)   disrotatory?   (49)

case is one suggested by De Puy (1968), based on work by Wiberg and Fenoglio (1963). This makes use of the further feature of cyclopropane openings, i.e., the existence of two separate disrotatory modes. The isomeric tosylates (52, 53) solvolyze in acetic acid with a rate difference of 10^6, the endo isomer (52) being faster, suggesting that the inner rotation in 52 is easier than the inner rotation in 53. It is somewhat difficult to visualize the

(52)   (53)

opening without models and the comparison is clearer if one looks at the corresponding cyclopropane system, the bicyclo[n.1.0]alkanes discussed earlier in this section (imagine the dotted line in 52 and 53 to be the bond and the CH_2 removed). For small rings (see Table IV), the endo isomers react much faster than the exo isomers.

Berson et al. (1969) have considered, but discounted, the possibility of a bishomoallyl cation (54) closure to a cyclopentyl cation (55), in rationalizing the stereoselectivity of the nucleophilic attack in a tricyclic system. One can picture a simple case as follows.

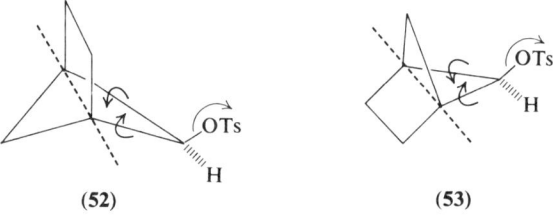

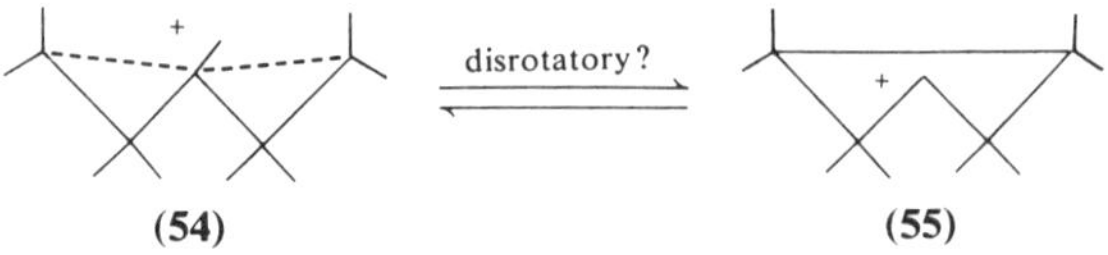

(54)   (55)

Whereas the allyl cation is more stable than the cyclopropyl cation, and the homoallyl–cyclobutyl cations are roughly isoenergetic, the cyclopentyl cation (the closed form) will be more stable than the bishomoallylic cation.

C. The $k = 4$ Systems

1. Pentadienyl Cation–Cyclopentenyl Cation

The gross energy equation in the simple pentadienyl (**56**)–cyclopentenyl cation (**57**) case is as follows.

(**56**)    (**57**)

pentadienyl cation resonance energy + π bond $\underset{?}{\rightleftharpoons}$ σ bond + allyl cation resonance energy

There is expected to be little or no strain in the closed form, the cyclopentenyl cation, and this inverts the relative energies compared to the $k = 2$ case, the cyclopentenyl cation being much more stable than the pentadienyl cation. No accurate experimental estimate of this energy difference in the simple system is known, but the difference is sizable since no reversibility in this reaction has ever been established (even with the pentadienyl cation as a high energy intermediate). With some ring systems, the pentadienyl cation can be the more stable; for example, the benzenium cation (**58**) (a pentadienyl cation) is more stable than the corresponding closed ion, the bicyclo[3.1.0]hexenyl cation (**59**). If the ring strain in **59** is roughly taken as 40 kcal/mole, then the simple pentadienyl–cyclopentenyl cation energy difference probably lies between 15 and 35 kcal/mole.

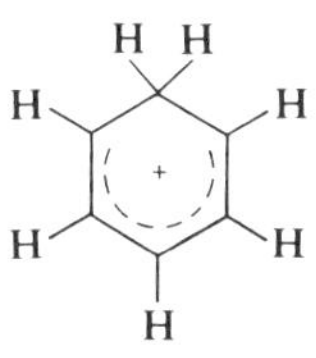

(**58**) 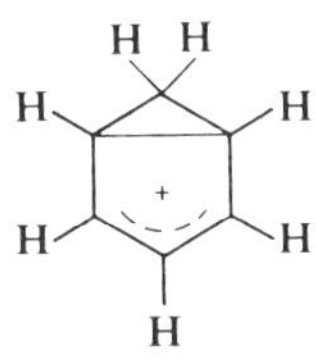   (**59**)

Pentadienyl cations, as solvolysis intermediates, have been little studied and most of the work on the stereochemistry of the ring closure comes from observable ion studies. Observable linear pentadienyl cations were first prepared in 1964 (Deno and Pittman, 1964; Sorensen, 1964) and the thermal cyclization of these to cyclopentenyl cations was noted immediately. The early examples were all complicated by further rearrangements of the (presumed) initially formed cyclopentenyl cation; for example, the obvious pentadienyl cation for stereochemical studies is **60**, but this cyclizes and rearranges to **61** (Sorensen,

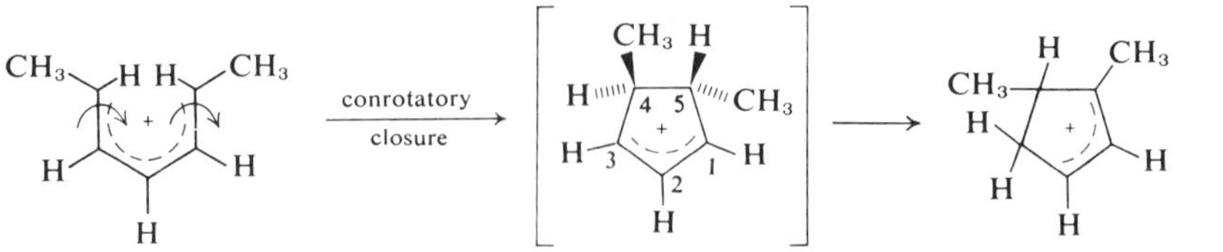

(60) → conrotatory closure → [not observed] → (61) first observed product

1964, 1965a). One has to maintain the original stereochemistry at C(4) and C(5) in order to test the mode of closure.

Another problem arises in this system which has no parallel in the $k = 2$ case. The stable form of pentadienyl cations has the $2E,3E$ backbone (or W configuration). The cyclization of C(1) to C(5) requires a $2Z,3Z$ (or U) con-

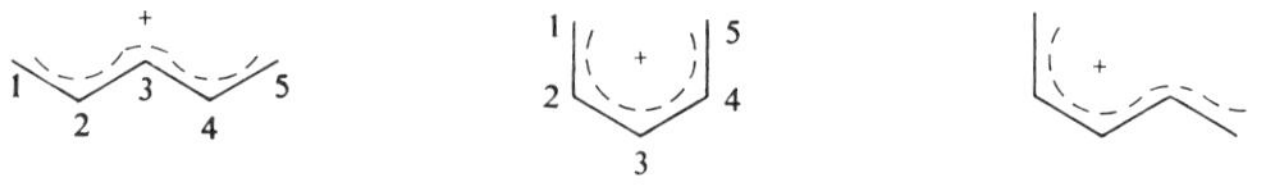

"sickle"

formation which is probably not planar. The third conformation possible is the $2E,3Z$ (or "sickle" conformation). The three conformations are interconverted by rotations about the C(2)–C(3) [or C(3)–C(4)] partial double bond. These rotations are rapid since attempts to prepare the $2E,3Z$ ("sickle") conformer lead only to the $2E,3E$ (W) form even at 173°K (T. S. Sorensen and E. Wagner, unpublished results, 1971).

The first stereochemical study of the pentadienyl cation cyclization was that of Campbell *et al.* (1969), who used a pentamethylpentadienyl cation series. Addition of the geometrically pure alcohols **62a,b,c** to strong acids led in the optimum cases to nearly stereospecific formation of the predicted (conrotatory closure) cyclopentenyl cations. The presumed intermediates in each

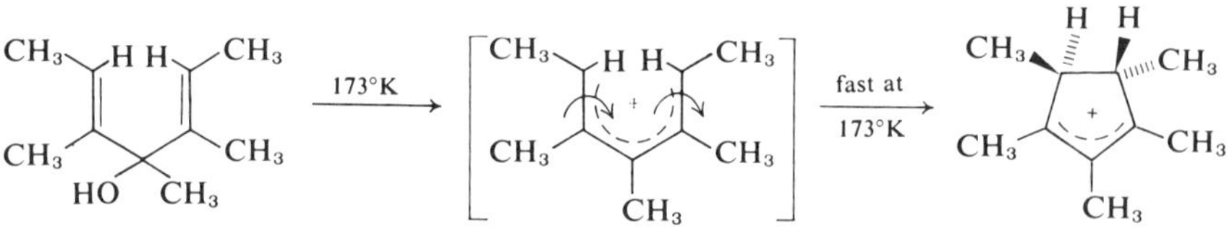

(62a) → 173°K → (63a) → fast at 173°K → (64) 94%

(**62b**) (**63b**) (**65**) 94%

(**62c**) (**63c**) (**64**) 82%

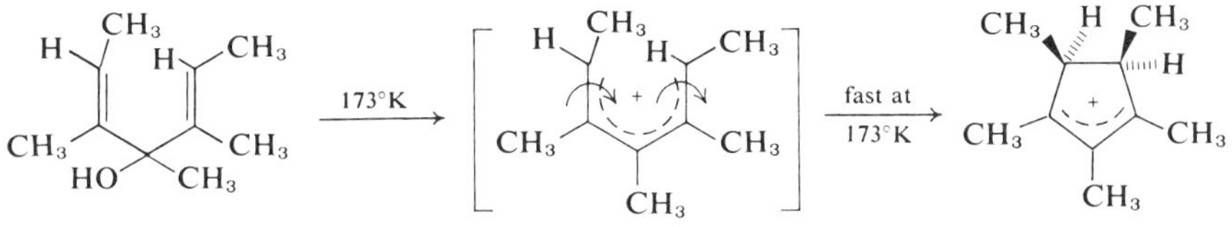

case are the corresponding pentadienyl cations **63a,b,c**, which are written in the 2*Z*,3*Z* conformation for convenience only. These observations do not absolutely require a discrete pentadienyl cation intermediate, but, following this work, Chiu and Sorensen (1973) studied an analogous tetramethyl series where the pentadienyl cations (**65a**, **65b**) could be observed. The third pentadienyl isomer (analogous to **63c**) could not be directly observed. The pentadienyl cations (**65a**, **65b**) cyclize in about 1 hr at 188°K. The kinetics of

(**65a**) (**66**) 91%

(**65b**) (**67**) 84%

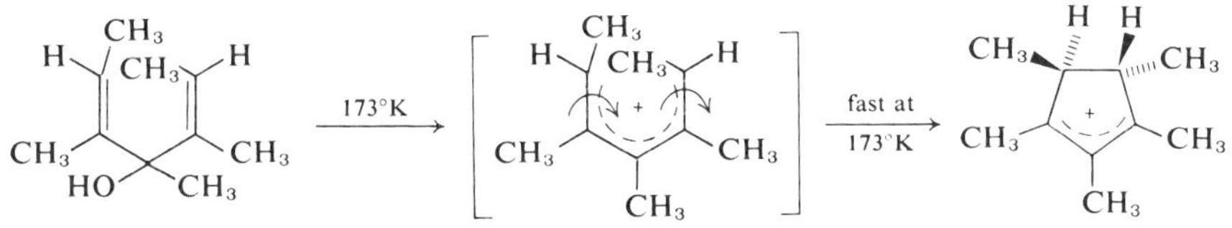

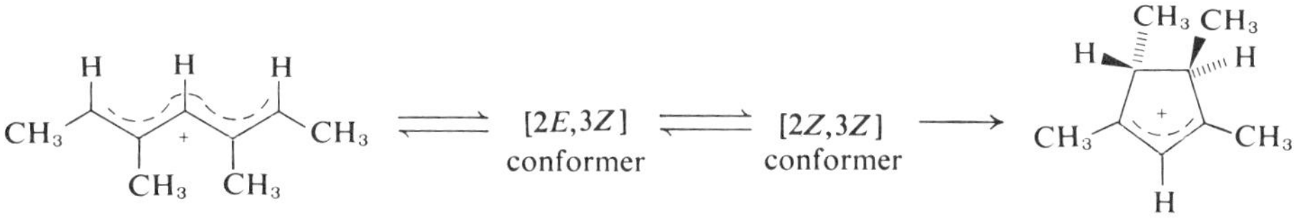

pentadienyl cation cyclizations have been extensively studied (Bladek and Sorensen, 1972) and it was predicted that removal of the central methyl group in **63a,b,c** would slow the cyclization rate, as found.

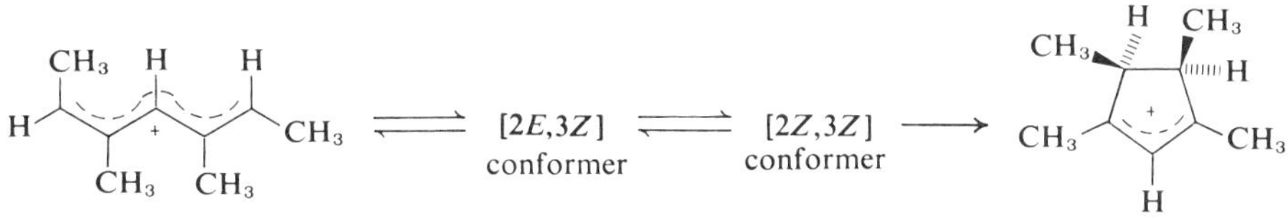

These pentadienyl cation cyclizations are never 100% stereospecific and it is believed that loss of geometric purity probably occurs at the pentadienyl cation stage. Calculations of the bond rotation barriers in the parent pentadienyl cation (MINDO, NNDO in brackets) (Baird, 1972) give the following results (kcal/mole) for the 2E,3E conformer.

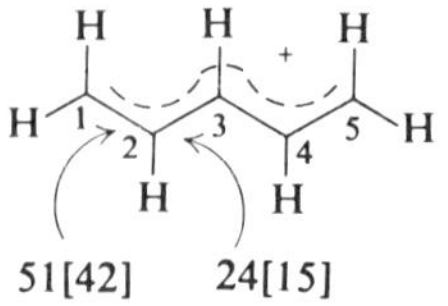

$$\underset{51[42]\quad\quad 24[15]}{\text{[structure]}}$$

One of course requires rotation about the C(2)–C(3) [C(3)–C(4)] bond to get the U form but wishes no rotation about C(1)–C(2) [or C(4)–C(5)] in the geometrically labeled pentadienyl cations. That one probably does get some C(1)–C(2) bond rotation is indicated by the fact that a lower degree of stereospecificity is observed in the tetramethyl cases **65a,b**, where the pentadienyl cations live longer [and hence have more time to isomerize by way of C(1)–C(2) bond rotation], compared to **63a,b,c**.

The photochemical cyclization of pentadienyl cations was first observed by Winstein and co-workers (Childs and Winstein, 1968; Childs *et al.*, 1968),

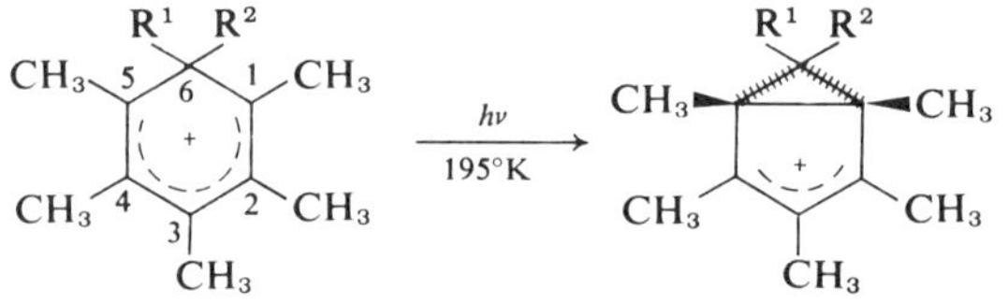

(68a) R^1 = R^2 = H
(68b) R^1 = H, R^2 = CH$_3$
(68c) R^1 = R^2 = CH$_3$

(69a) R^1 = R^2 = H
(69b) R^1 = H, R^2 = CH$_3$
(69c) R^1 = R^2 = CH$_3$

using polymethylbenzenium cations **68a,b,c**. Benzenium cations may not be typical pentadienyl cations if there is significant overlap of the terminal C(1)–C(5) orbitals or "through-bond" interaction at the C(6) methylene group. However, the observed photocyclization can be adequately described as an electrocyclic pentadienyl–cyclopentenyl transformation, as opposed, for example, to a [$_\sigma 2_a + {}_\pi 2_a$] cycloaddition (Childs and Parrington, 1970). The allowed photochemical mode is disrotatory, which gives the observed cis-fused cyclopropane ring junction. The thermal kinetic stability of the bicyclo[3.1.0]hexenyl cations **69a,b,c** results because the allowed conrotatory opening would give a *trans*-benzenium ion (**70**). However, the "forbidden" reversion to the *cis*-benzenium cation does eventually occur.

The bicyclo[3.1.0]hexenyl cations are very interesting in their own right and exhibit a degeneracy in which the cyclopropane ring moves around the periphery of the five-membered cyclopentenyl structure (Childs and Winstein, 1968; Isaev *et al.*, 1969; Childs and Parrington, 1970). This is discussed under sigmatropic rearrangements. Similar rearrangements have also been observed for protonated cyclohexadienones (Hart *et al.*, 1969; Parrington and Childs, 1970).

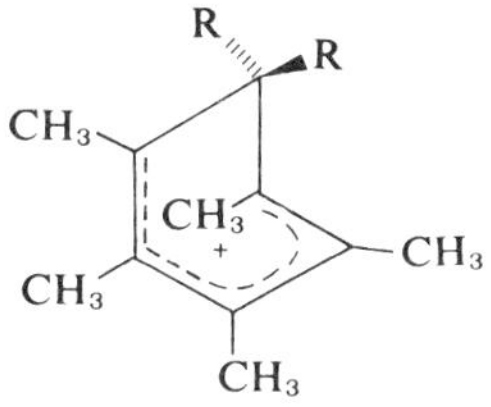

(70)

Other photocyclizations have been reported. On irradiation in 5% sulfuric acid, the tropylium cation (71) yields the bicyclic alcohol (72) and the ether (73) (van Tamelen *et al.*, 1968, 1971). This can be rationalized by assuming a

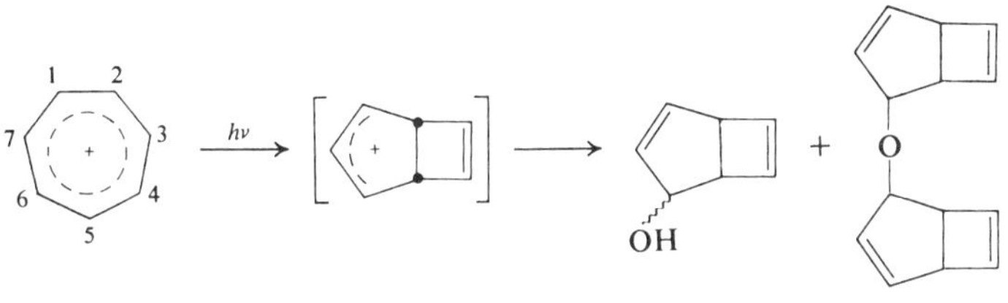

(71)   (74)   (72)   (73)

cyclization from C(1) to C(5) (a pentadienyl subunit), which would need to occur in a disrotatory mode in order to get the cis ring junction in 74 (a *trans* fusion is nearly impossible). In stronger acids, the cation (74) is not captured by a nucleophile but rearranges further to a 7-norbornadienyl cation (Childs and Taguchi, 1970; Hogeveen and Gaasbeek, 1970). The homotropylium cation (75) is also cyclized photolytically (Hogeveen and Gaasbeek, 1970) and the photoproduct was recently identified as 76 (Christensen *et al.*, 1974). The species (76) is not the expected primary photoproduct (77) but its formation

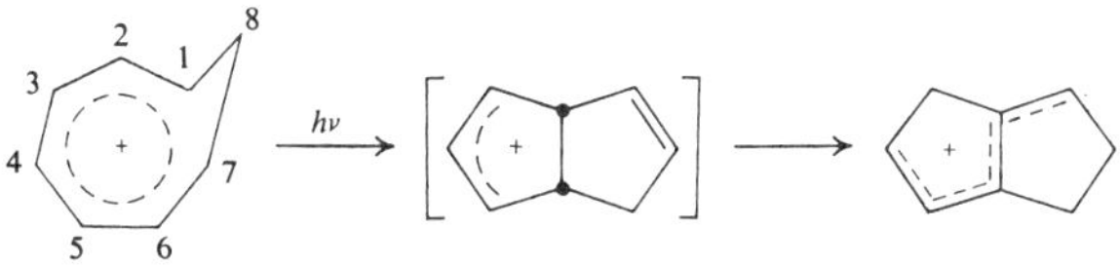

(75)   (77)   (76)

from 77 can be rationalized. The formation of 77 again involves the closure of a pentadienyl subunit of the homotropylium cation.

The thermal opening of the bicyclic allyl cation (78), to give the heptachlorotropylium ion (79), has been reported by Kusuda et al. (1971). The

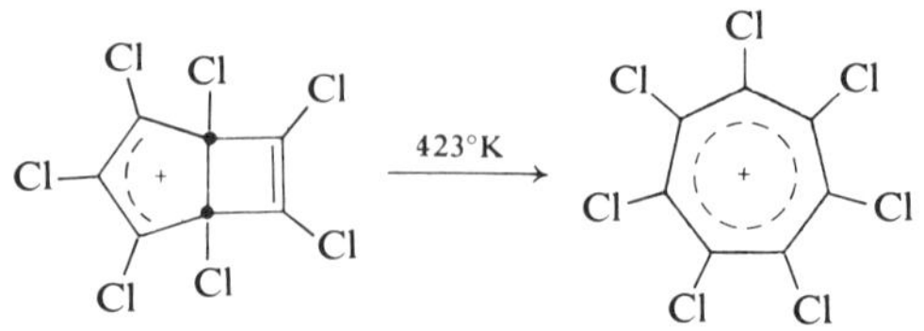

ring junction in 78 is very likely cis so that the needed disrotatory opening to give 79 is not symmetry allowed. Since rate comparisons to similar allowed processes cannot be made, little can be said about the importance of orbital symmetry in this case.

2. Protonated Ketone Analogs

A protonated ketone can be regarded in a formal sense as a hydroxycarbonium ion. There is little question, however, that the bulk of the charge

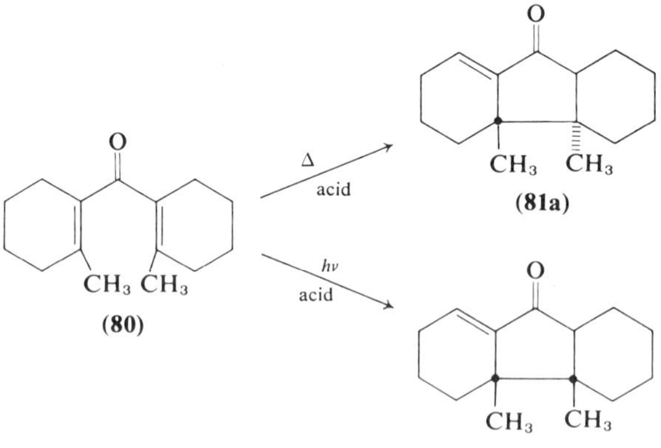

resides on oxygen. Nevertheless, protonated dienones (hydroxydienylic cations) appear to behave in an orbital symmetry way like the "pure" hydrocarbon species (pentadienyl cations). The cross-conjugated dienone (80) is cyclized in acid, both thermally and photochemically, to yield the different cyclopentenones (81a, 81b), corresponding in stereochemistry to a conrotatory thermal process and a disrotatory photochemical process, respectively (Kurland, 1967; Lehr, 1968). The ring system dienone is required because the proton removal step will destroy a stereochemical center in a simple model such as the following.

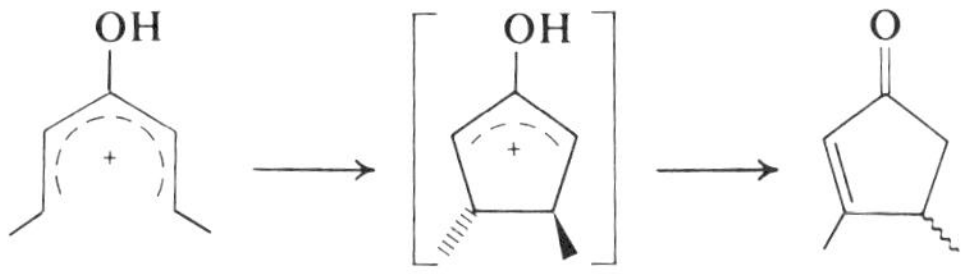

This reaction is a useful cyclopentenone synthesis known as the Nazarov cyclization. Alkyl migrations can occur at the hydroxyallyl cation stage, as evidenced by the following rearrangement, ketone (**82**) yielding in part the cyclized ketone (**83**) (Ohloff *et al.*, 1971).

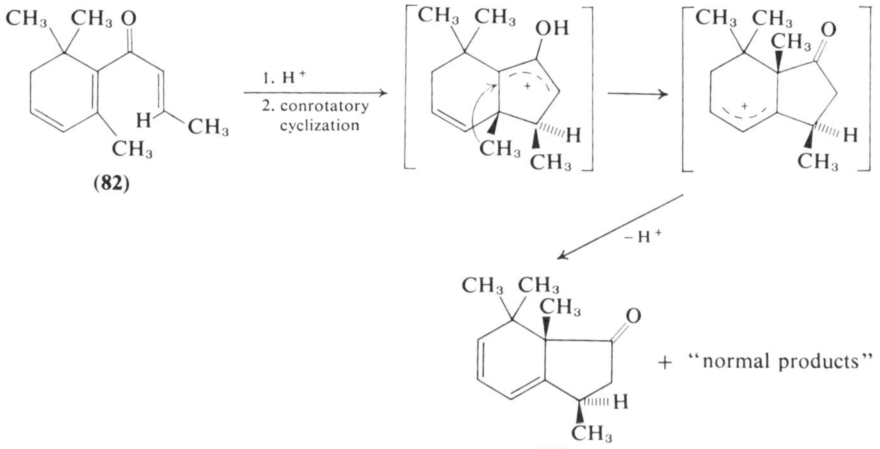

Shoppee and Cooke (1972) have investigated the acid-catalyzed cyclization of *E,E*-1,5-diphenylpentadien-3-one (**84**) which affords the conrotatory product (**85**). In this case the stereochemistry is maintained at C(4) and C(5) because there is nucleophile capture rather than proton loss. A dibromodienone cyclization has also been studied (Shoppee and Lack, 1969).

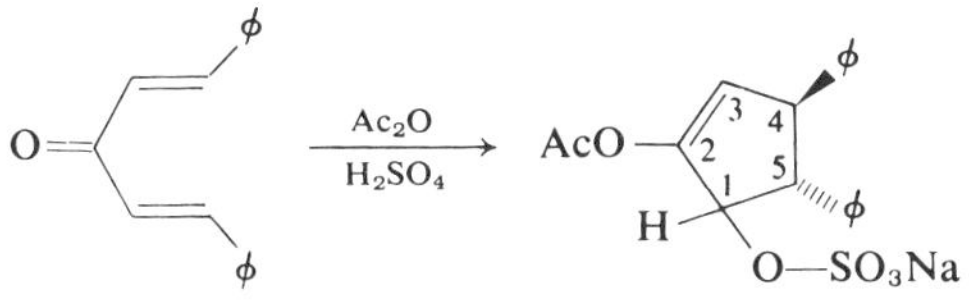

Although cyclization of the protonated penta-2,4-dien-1-one system can also yield cyclopentenones, the stereochemistry of the closure is not readily discernible, and this reaction is therefore of marginal interest to us, e.g.,

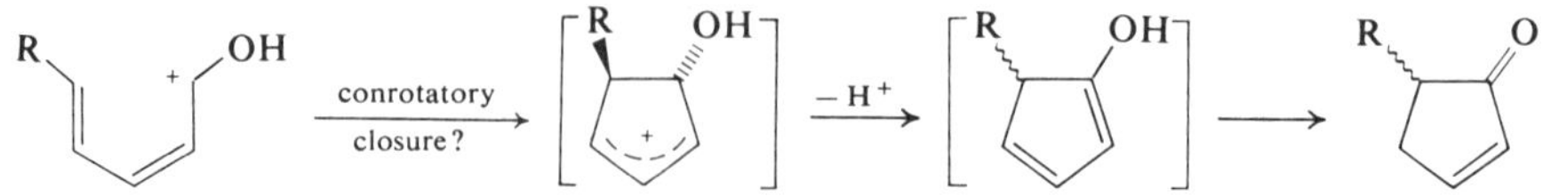

OH geometry not easy to fix

stereochemistry lost in any case

Protonated ketone analogs of the benzenonium ion are also photocyclized by the allowed disrotatory closure. The conjugated and cross-conjugated cyclohexadienones (**86, 87**), on photolysis in acid, both give the same photoproduct (**88**) (Hart et al., 1969; Parrington and Childs, 1970). In neither case

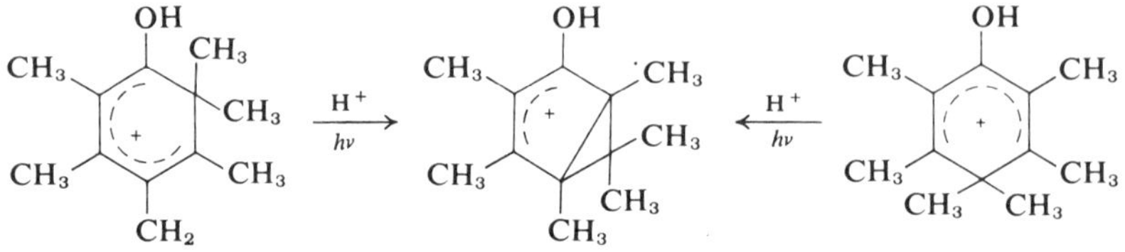

is **88** the expected primary photoproduct but the formation of **88** is thermodynamically favored (OH on the formally charged center) and can occur by way of an allowed [1,4]-sigmatropic shift of the cyclopropane ring in the initially formed intermediates.

3. Heterocyclic Analogs

The electrocyclic closure of the 3-azapentadienyl cation system has been proposed (Okamoto and Shudo, 1973) to rationalize the formation of **89** from **90**.

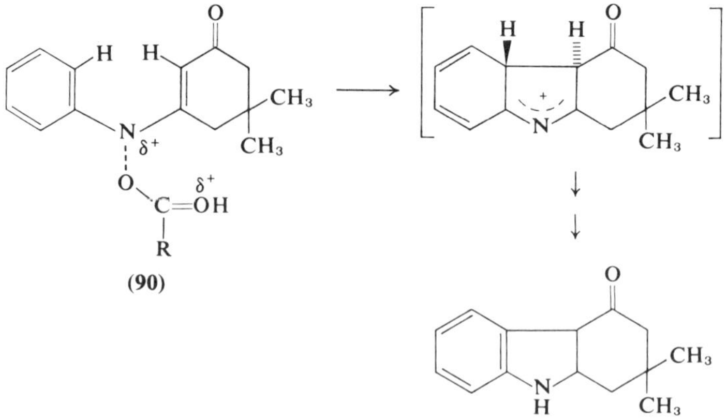

The photocyclization of a novel boron analog (**91** to **92**) has been reported (Clark *et al.*, 1971).

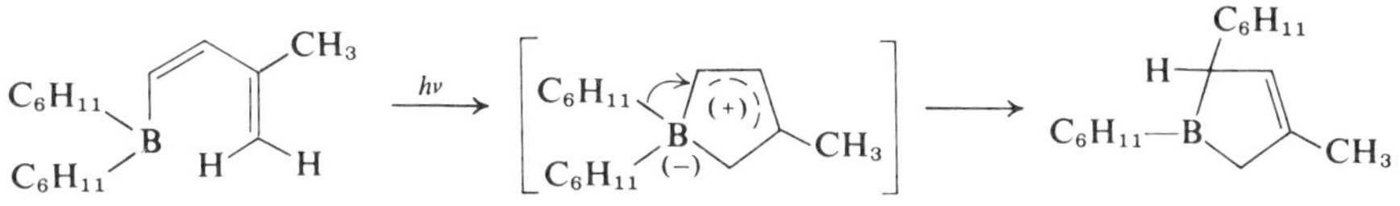

4. Homodienyl Systems

There are two homosystems to consider, cations **93** and **94**. These have been written in the closed form appropriate for cyclization but would probably be most stable in an extended geometry. The electrocyclic cyclization products differ in the two cases; a cyclohexenyl cation is formed from **93** and a bicyclo[3.1.0]hexyl cation is formed from **94**.

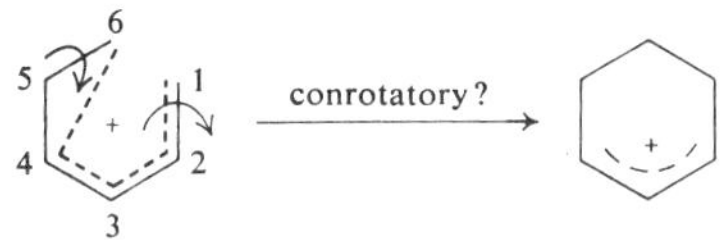

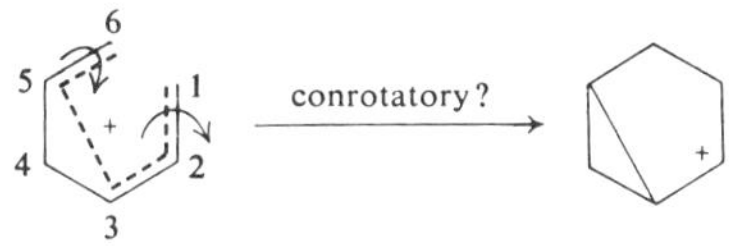

a. The 4,6-Homo System. The localized resonance forms of cation **93** are the structures **95** and **96** and one could approach the homodienyl system using a dienol related to structure **95** or a cyclopropylenol related to **96**.

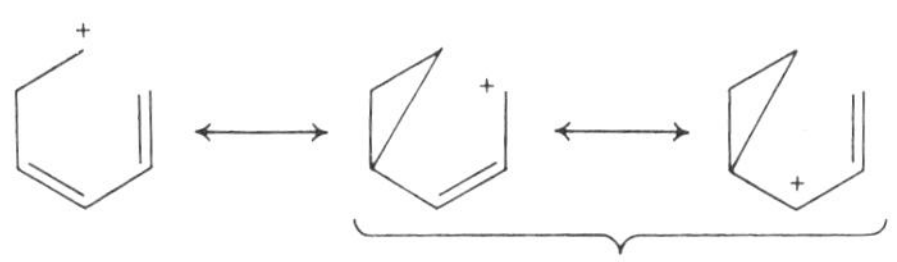

1-Cyclopropylallyl cations (i.e., **96**) have been extensively studied and do indeed rearrange to cyclohexenyl cations. However, it has been conclusively shown (Rajeswari and Sorensen, 1971, 1973) that the mechanism does not involve an electrocyclic closure of **93**, even where this would result in a much more stable product than is in fact formed. This shows up clearly in α-methylcyclopropyl systems, e.g., the cyclization of the general structure **97**

gives **98** as the product, whereas an electrocyclic closure would have given the more stable **99** (methyl at the charged allyl center). The reaction giving the observed product **98** will be discussed further under cycloaddition processes.

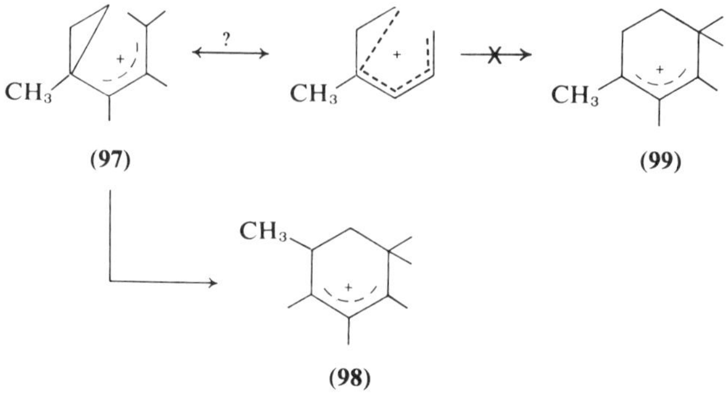

A single exception to this mode of reaction occurred in a β,β-dimethylcyclopropyl case, where cation **100** did give some of cation **101** as product (Rajeswari, 1971). The two β,β-methyl groups might enhance the homodienylic

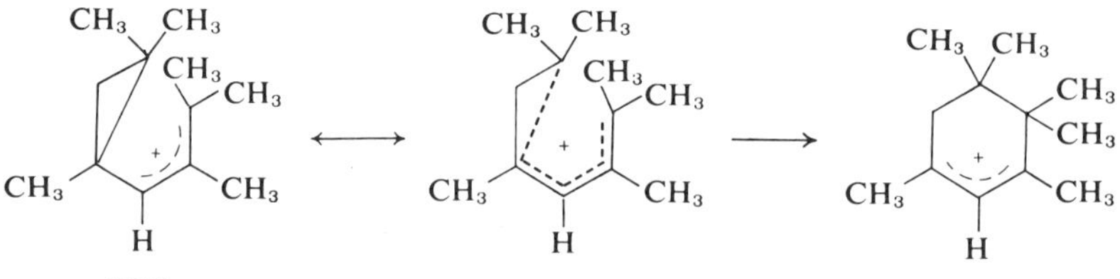

character of the cation, i.e., **93**. No stereochemical investigation of this closure has been made.

b. *The 3,5-Homo System.* The obvious starting point for preparing this cation are the easily prepared alcohols of structure **102**. A number of methyl-substituted examples have been treated with strong acid but in no case was it

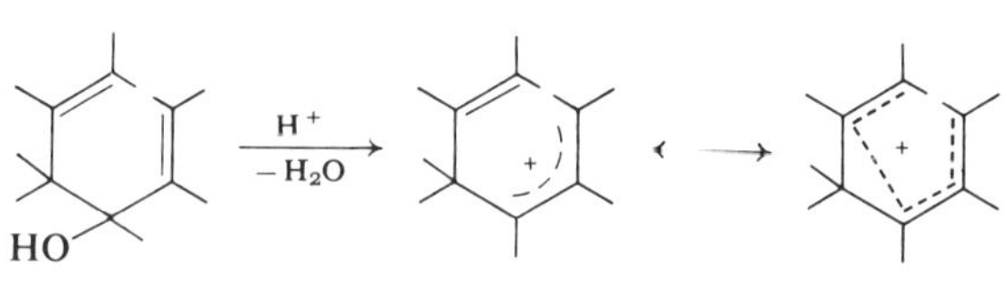

possible to observe a bicyclo[3.1.0]hexyl cation (Christensen, 1971). However, cyclohexenyl cations are a major product of the reaction and these appear to be derived from the initially formed bicyclo[3.1.0]hexyl cations by a simple 1,2-alkyl or hydrogen shift:

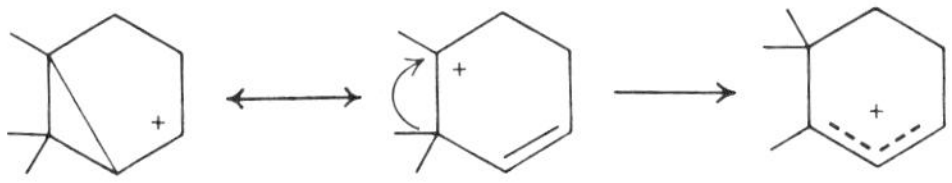

No stereochemical studies have been attempted as yet and, in fact, the reaction may be better studied under solvolysis conditions, where one might well capture the initial bicyclic cation and hence study the stereochemistry of this neutral product.

The homoconjugated systems (**93, 94**) are only two examples of the concept of through space cationic interactions,* pioneered by Winstein and co-workers in numerous studies. In this connection, cations **93** and **94** are interesting because their cyclization (electrophilic attack) would be expected regardless of whether there is significant homoconjugation or not. Any stereospecificity in the cyclization might well be the test of whether orbital overlap (and hence symmetry control) is important.

5. *1-Phenylallyl Cation Systems*

The electrocyclic cyclization of 1-phenylallyl cations (**103**) is a reaction generally discussed in the literature only in terms of electrophilic attack.

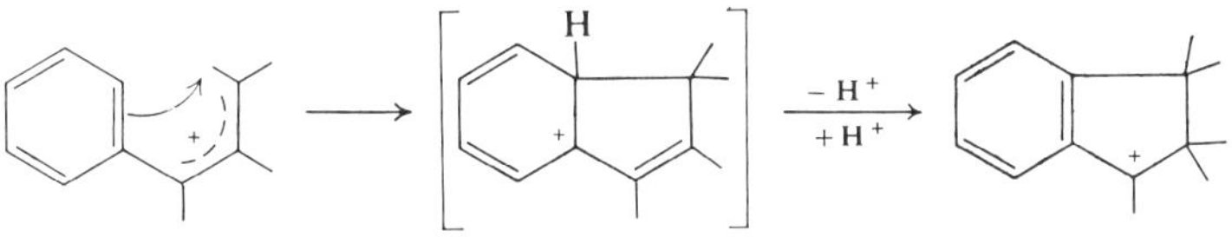

(**103**)　　　　　　　　　　　　　　　　　(**104**)

Many examples are known and the rearranged indanyl cation has invariably been the end product (Deno *et al.*, 1965; Farnum *et al.*, 1971; Pittman and Miller, 1973). Rearomatization via deprotonation destroys the stereochemical center at the bridgehead; to stop this, the obvious answer is to put an alkyl group at the bridgehead position. Thus, for example, the closure of the mesitylallyl cation (**105**) occurs cleanly to yield the single observable trienylic cation (**106**), in which the stereochemistry of the two methyl groups is retained

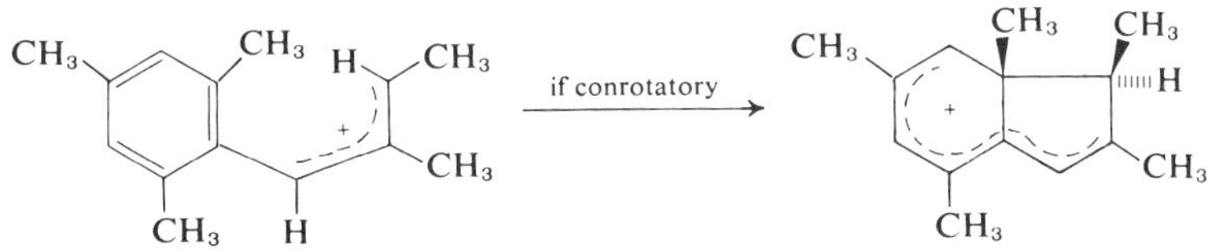

(**105**)　　　　　　　　　　　　　　　　　(**106**)

* See Section V for a discussion of the importance of through-bond interactions, a more recently investigated phenomenon.

intact (T. S. Sorensen, unpublished results, 1974). The geometry of the methyl groups in cation **106** has not yet been determined but the HOMO coefficients (simple Hückel) at the two termini are opposite in sign. One therefore predicts a conrotatory closure. The gross energy equation for the cyclization of **105** to **106** is

benzene resonance + π bond + allyl (benzene) resonance $\rightleftharpoons$ σ bond + trienylic cation resonance

It is significant that, even with loss of the benzene resonance, this electrocyclic closure is still spontaneous.

6. Diphenylmethyl Systems

In principle, the cyclization of the diphenylmethyl cation **107** could yield two geometric isomers (**108a** or **108b**) depending on whether the C–H fragments of the aromatic system move disrotatory or conrotatory in changing hybridization from sp^2 to sp^3. The gross energy equation for this is

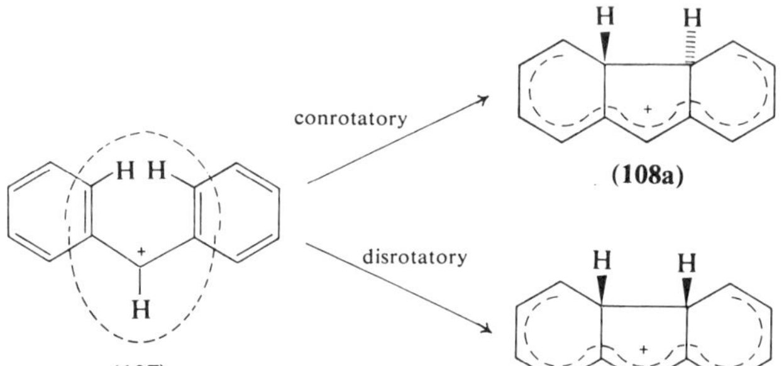

2 × benzene resonance + π bond $\rightleftharpoons$ σ bond + pentaenylic cation resonance

In this case, the cyclization of **107** to **108** is not thermodynamically favorable. However, in the normal circumstance, the polyenylic cation **108** (as a high energy intermediate) would suffer rapid rearomatization to give the fluorenyl system, with concomitant loss of all stereochemical information about the bridgehead positions in **108**. The overall transformation of **107** to a fluorenyl cation is probably favorable on energetic grounds; certainly the thermal cyclization of triphenylcarbinol in acid to give 9-phenylfluorene is well known, although this does not necessarily reflect the relative cation stabilities. The photocyclization of the triphenylmethyl (trityl) cation has also been observed (van Tamelen *et al.*, 1968; van Tamelen and Cole, 1971), the product again being 9-phenylfluorene.

The only stereochemical information on the electrocyclic cyclization of **107** comes from mass spectrometry results. The cyclized ion appears to have the cis-bridged structure **108b** since two hydrogens appear to be lost in a single step (Johnstone and Ward, 1968).

A heterocyclic example, the 3,3'-dithienylcarbonium ion, is reportedly cyclized in chlorosulfonic acid, persumably via an intermediate analogous to cation **108**. Again, one only observes the rearomatized cyclic system (Östman and Sjöbert, 1970).

D. POLYENYLIC CATIONS $k = 6$ OR LARGER

Simple linear heptatrienylic cations ($k = 6$) do not undergo a 1,7-cyclization to cycloheptadienyl cations, but instead prefer a 1,5-ring closure. Thus, cation **109** does not give **110** but instead yields the 1,5-closed product **111**

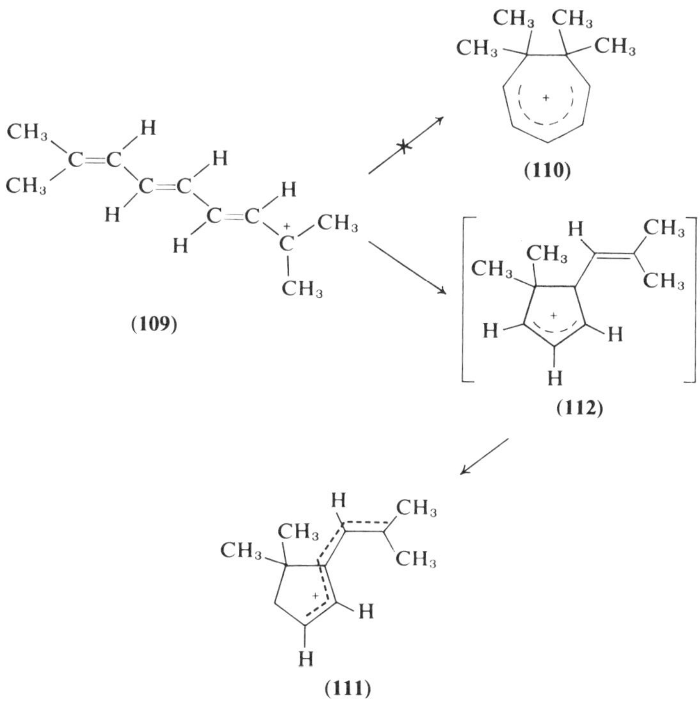

(Sorensen, 1965b). Similarly, the tetramethylnonatetraenylic analog to **109** cyclizes via 1,5-closure, no 1,7- or 1,9-closed product being observed (Sorensen, 1965b). A 1,9-closure appears sterically difficult, but a 1,7-closure might reasonably have occurred in either the trienylic (or tetraenylic) cations via the nonplanar, completely folded conformation (113). In any case, the 1,5-closure appears sterically the most favorable and higher polyenylic cations can always be written with a U subunit, i.e., conformer 114 for a trienylic cation. As

(113) 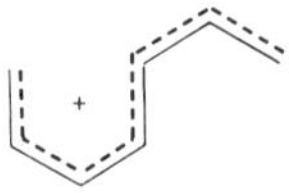   (114)

expected, with unsymmetrical polyenylic cations, several cyclic 1,5-products are formed (T. S. Sorensen and E. Wagner, unpublished results, 1970). The stereochemistry of the closure in heptatrienylic and nonatetraenylic cations remains undetermined since the initially formed product, i.e., cation 112, rearranges to a fully conjugated ion (111) with loss of stereochemical information. Simple Hückel calculations, of the type used to predict the conrotatory dienylic closure, predict a conrotatory 1,5-closure for heptatrienylic cations. However, the nonatetraenylic cation 1,5-closure mode is not predicted since C(5) in the HOMO of this cation is a nodal point.

Several photochemical transformations involving the overall ring contraction of protonated cycloheptadienones seem to require an electrocyclic opening to give a hydroxytrienylic cation (Hine and Childs, 1971; Noyori et al., 1972), which in one case then closes thermally in a 1,5 manner, e.g., the transformation of 115 to 116.

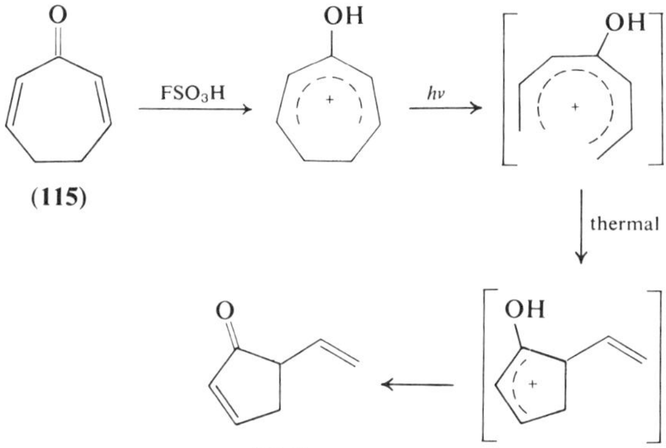

(115)

(116)

III. Cycloaddition and Cycloreversion Reactions

A. INTRODUCTION

Cycloaddition reactions, as exemplified by the Diels–Alder reaction, require no definition. Hoffmann and Woodward (1965) have provided us with orbital symmetry correlation diagrams, but predating this was the concept of aromatic ($4n + 2$) and nonaromatic ($4n$) transition states, the former favorable, the latter unfavorable. While theoreticians seem relatively agreed on the mode of electrocyclic transformations, the same is not true for cycloadditions. The question, of course, is whether both new bonds are formed simultaneously or sequentially.

In considering cationic systems, this problem is likely to be accentuated since the "electrophilic" attack of a cation on a C–C double or single bond is well known. For example, the [$_\pi 2 + _\pi 2$] addition of two ethylenes in a sequential manner would involve transition states where there is either charge separation or diradical character. In contrast, the possible sequential [$_\pi 2^+ + _\pi 2$] addition of allyl cation to ethylene is isopolar.

B. RULES FOR CYCLOADDITION AND CYCLOREVERSION REACTIONS

The categorization of cycloadditions is more open-ended than for electrocyclic transformations. We will not consider "termolecular" or higher processes, i.e., [2 + 2 + 2] additions, since we know of no clearcut examples involving cations. Furthermore, the cationic center can be orthogonal to or remote from the π system undergoing cycloaddition. The former case occurs in the known [$_\pi 2 + _\pi 2$] addition of vinyl cations to acetylenes or ethylenes. In this case, the positive charge appears to be a definite inducement for the reaction. We do not consider, in this chapter, cycloadditions involving remote cationic centers. We also exclude heteroatom systems and consider only additions of carbocations to C–C single, double, or triple bonds, and to C–H bonds.

Cycloadditions can be divided into three categories, separable by looking at the overall change in the configuration of the termini of both components. The subscripts "a" and "s" were introduced to designate *suprafacial* and *antarafacial* additions, the former designating addition to the same face or side and the latter to opposite sides of a π bond. For σ bonds, the terms imply double retention or double inversion of the termini for suprafacial and retention–inversion (or vice versa) for antarafacial. For a total of $4q$ electrons (q = integer) in both fragments, the allowed thermal reaction is $m_s + n_a$ (or $m_a + n_s$), where m and n are the number of electrons in each fragment. For photochemical additions, the allowed reactions are $m_s + n_s$ or $m_a + n_a$. For the $4q + 2$ systems (aromatic), the $m_s + n_s$ (or $m_a + n_a$) cycloadditions are allowed thermally and $m_a + n_s$ (or $m_s + n_a$) photochemically. For cycloreversions, the same rules apply, although one may decide to change the label designations; thus, in ethylene + allyl cation we are losing two π bonds and forming two σ bonds, the designation being $[_\pi 2^+ + _\pi 2]$, while in the cyclopentyl cation reversion, we are breaking two σ bonds, with the designation $[_\sigma 2^+ + _\sigma 2]$. Since σ bonds are usually much stronger than π bonds, one should note that cycloadditions are in general more commonly found than cycloreversions, steric and entropy effects usually being the driving forces for the latter.

C. [2 + 2] Cycloaddition or Cycloreversion Processes in Cationic Systems

In all cases, only one of the fragments will be positively charged. This fragment is designated by "n" superscript +, where "n" is the number of delocalizable electrons in the fragment and the + indicates its cationic character (Schmidt, 1973). The simple rules for cycloadditions involving carbocations are identical to those for the corresponding neutral species having the same number of π or σ electrons. This occurs because the HOMO–LUMO symmetry is the same for ethylene or the allyl cation. That is, the number of orbitals is unimportant, only the number of π electrons. We start by considering [2 + 2] cycloadditions since cycloadditions which are formally $[_\pi 0 + _\pi 2]$ or $[_\pi 0 + _\sigma 2]$ are always allowed and the interest there is mainly in the transition state for such reactions and in the possible existence of bridged intermediates. Such additions are considered in a later section.

1. $[_\pi 2 + _\pi 2]$ Processes

Two basic types fall into this category.
(1) the addition of a vinyl cation to a neutral π system (although an ethylene is written, we imply either ethylene or acetylene π systems)

(2) the addition of an allyl cation to a neutral π system.

Both additions are subject to the same rules; $[_\pi 2_s^+ + _\pi 2_a]$ (or $[_\pi 2_a^+ + _\pi 2_s]$) for thermal reactions and $[_\pi 2_s^+ + _\pi 2_s]$ (or $[_\pi 2_a^+ + _\pi 2_a]$) for photochemical additions.

1. CARBOCATION 37

The dimerization of acetylenes in acid to give cyclobutenyl cations involves the addition of the acid to one unit, affording a vinyl cation which then reacts with excess acetylene. Thus, diphenylacetylene (**117**) in fluorosulfonic acid–sulfur dioxide at 195°K gives the cyclobutenyl cation (**118**) (Lodder *et al.*, 1970; van der Hout–Lodder *et al.*, 1972, 1973).

$$\phi-C\equiv C-\phi + H^+ \longrightarrow \left[\phi-\overset{H}{\underset{}{C}}=\overset{+}{C}-\phi\right] + \phi-C\equiv C-\phi \longrightarrow$$
(**117**)

(**118**)

Alkyl-substituted acetylenes or allenes are also dimerized by acids (Smirnov-Zamkov, 1952; Criegee and Moschel, 1959; Griesbaum *et al.*, 1965; Griesbaum, 1966; Griesbaum and Chu, 1974).

Related to vinyl cations are ketenes, which appear to owe their reactivity in cycloadditions to their "vinyl cation character." There are, however, numerous examples of [2 + 2] cycloadditions involving ketenes and we do not propose to review this area.

$$\overset{\diagdown}{\underset{\diagup}{}}C=C=O \longleftrightarrow \overset{\diagdown}{\underset{\diagup}{}}C\overset{\delta^+}{=}C\overset{\delta^-}{-}O$$

Thermal [$_\pi 2 + {}_\pi 2$] cycloadditions are allowed as a [$_\pi 2_s + {}_\pi 2_a$] process which must involve the orthogonal approach of the reacting species, i.e., as in **119**. (HOMO–LUMO interactions are depicted.) Woodward and Hoffmann (1969, 1970) have analyzed such cycloadditions and conclude that for

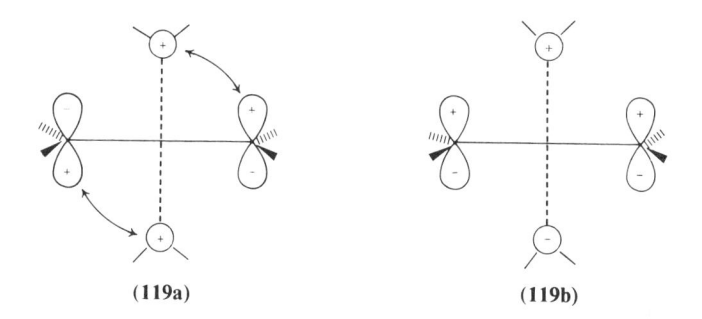

(**119a**)　　(**119b**)

normal olefins, transition state strain will essentially preclude such reactions from occurring. However, vinyl cations are regarded as especially good $_\pi 2_a$ components in this cycloaddition because the orthogonal vacant *p* orbital can effectively overlap the occupied π system of the other reactant as the two components orthogonally approach one another (structure **120**). Ketenes are also analyzed as good $_\pi 2_a$ components; in this case, the vacant p orbital of the vinyl cation is replaced by the low-lying π^* carbonyl molecular orbital. Wagner and Gompper (1971a,b) have more recently carried out EHMO and MINDO/2 calculations on the vinyl cation–ethylene, vinyl cation–acetylene and vinyl cation–allene reactions. These calculations suggest a transition state (**121**) akin to those for electrophilic attack [$_\pi 0 + {}_\pi 2$], and the question remains

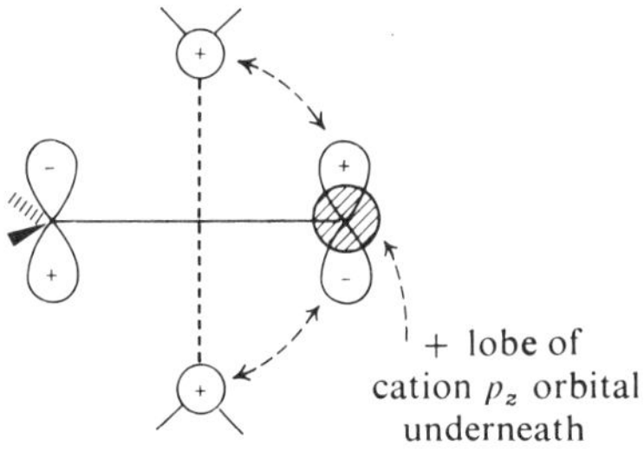

(120) 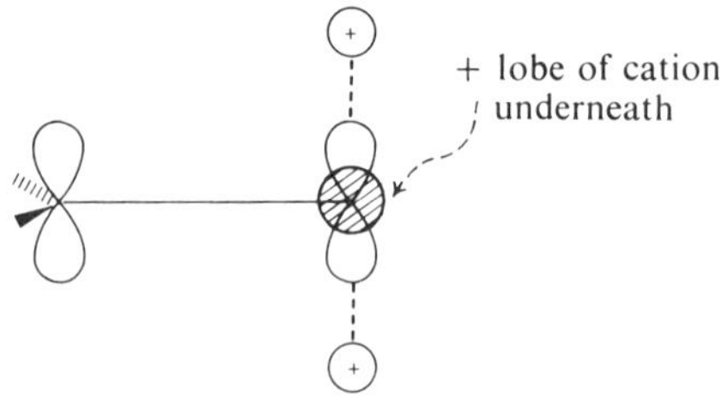   (121)

+ lobe of cation p_z orbital underneath

+ lobe of cation underneath

whether this overall cycloaddition reaction is concerted or not. Although there is considerable evidence that ketene cycloadditions are concerted, there is evidence that a keteneimine cycloaddition goes stepwise via polar intermediates (Ghosez and de Perez, 1971).

The thermal [$_\pi 2^+ + {}_\pi 2$] cycloaddition of an allyl cation to a π bond is also allowed for the [$_\pi 2_s^+ + {}_\pi 2_a$] (or [$_\pi 2_a^+ + {}_\pi 2_s$]) mode. Once again, one has the usual problem concerning the steric compression in the orthogonal transition state required for the allowed process.

Akhtar *et al.* (1972) have reported the "forbidden" [$_\pi 2_s^+ + {}_\pi 2_s$] cycloreversion of **123** to cation **122**. However, in the parent hydrocarbon system, there is no evidence that these two ions are interconvertible and, in fact,

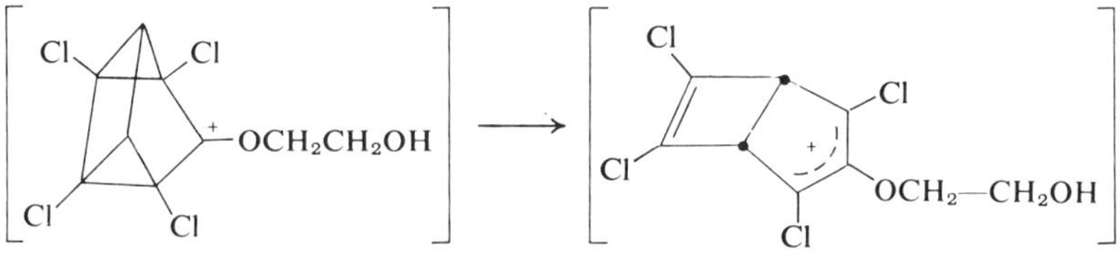

(123)   (122)

both rearrange readily to 7-norbornadienyl cations (Richey and Buckley, 1963; Story and Fahrenholtz, 1964, 1966; Lustgarten *et al.*, 1968).

Stereospecific suprafacial addition of the oxyallyl cation (as a ligand to Fe^{2+}) to substituted ethylenes has been reported (Noyori *et al.*, 1973). For an allowed cycloaddition, this then requires antarafacial addition to the oxyallyl cation termini. However, as these authors point out, stepwise stereospecific addition is also possible.

One interesting possibility for a $[_\pi 2_a^+ + _\pi 2_s]$ cycloaddition involves speculation on the possible mechanism for the biosynthetic transformation of nerol (124) to isoborneol (125) via an internal cycloaddition of the neryl cation (126). The attractive feature of such cycloadditions is that both new bonds are formed together. It has been pointed out, for example, that the

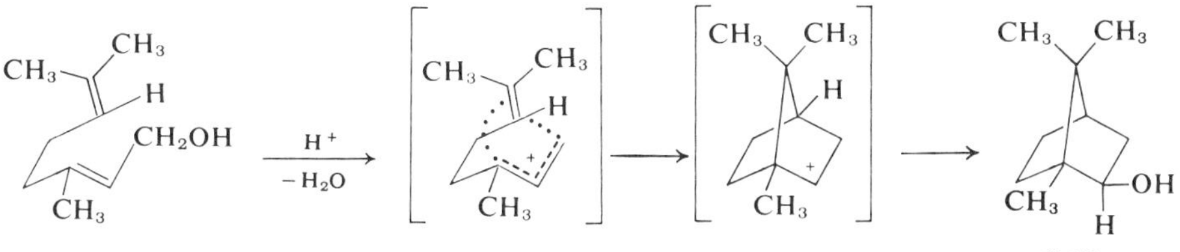

overall bicyclization of nerol is thermodynamically favorable, but starting from the already formed monocyclic state (i.e., as in stepwise addition), the bicyclization is not favorable (Gascoigne, 1958).

Photochemical $[_\pi 2^+ + _\pi 2]$ cycloadditions are likely to be extremely difficult to carry out experimentally since competing electrophilic attack of a cation on an alkene is so rapid.

2. $[_\pi 2^+ + _\sigma 2]$ *Processes*

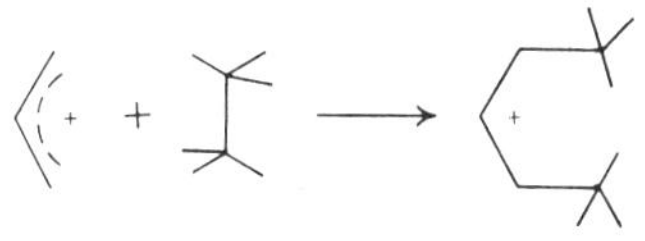

The only postulated example of this reaction involves an internal addition of an allyl cation to a cyclopropane σ bond (Rajeswari and Sorensen, 1971, 1973). A series of 1-cyclopropylallyl cations 127 was found to rearrange thermally to give cyclohexenyl cations 128, sometimes mixed with dienylic cations. The puzzle in this rearrangement was how R^1 and R^2 became separated in 128, particularly since the presence of R^1 on the allyl center would have

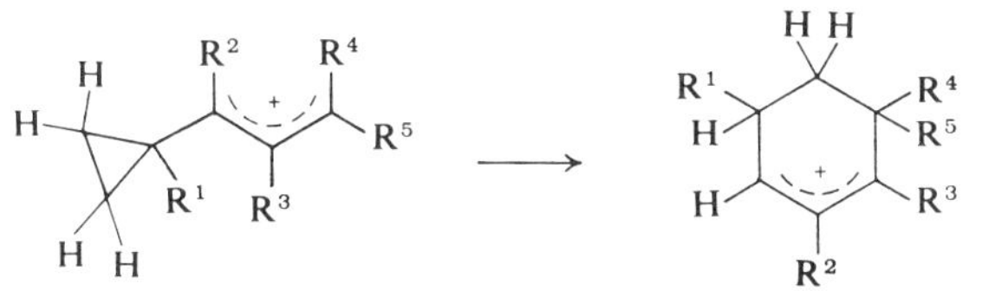

(127)
$R^1 = CH_3$
$R^2, R^3, R^4, R^5 = H$ or CH_3

(128)
$R^1 = CH_3$
$R^2, R^3, R^4, R^5 = H$ or CH_3

given a more stable cyclohexenyl cation. The overall rearrangement was rationalized by postulating a $[_\pi 2_s^+ + {_\sigma 2_a}]$ (or $[_\pi 2_a^+ + {_\sigma 2_s}]$) cycloaddition between the C_β–C_β cyclopropane bond and the allyl cation. This cycloaddition requires the cyclopropylallyl cation conformation (**129**) and leads to the intermediate (**130**), which is easily converted to a cyclohexenyl cation by a

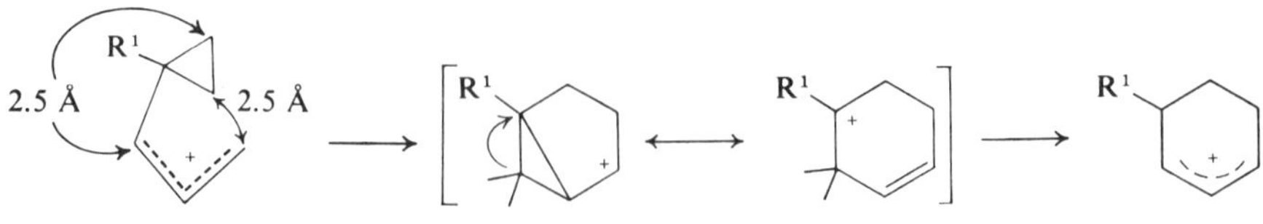

(**129**)   (**130**)

1,2-hydride or alkyl shift. The distance between the interacting centers in **129** is close enough for orbital symmetry considerations to apply and the particular bisected conformer shown in **129** automatically ensures an orthogonal approach of the two interacting groups. The only drawback in this analysis is that the cyclopropane ring and the allyl cation are conjugated and one is not therefore considering the internal cycloaddition of two "insulated" fragments. The reaction appears very general and has some similarity to the cyclopropane ring shift processes which occur in bicyclo[3.1.0]hexenyl cations; indeed, these can be looked upon as cycloaddition processes or as sigmatropic shifts (see section on sigmatropic rearrangments).

There are several interesting cation systems where cycloaddition or cycloreversion $[_\pi 2 + {_\sigma 2}]$ processes have not been observed. Cations **131** and **132** have been separately generated (as intermediates) and show no tendency to interconvert (Lambert *et al.*, 1972, 1973). Similarly cation **133** shows no sign of rearrangement to **134**, although one must admit in this case that the thermodynamics may favor **133** (Diaz *et al.*, 1971). The important thing to note is that the geometry in these cations is such that only $[_\pi 2_s + {_\pi 2_s}]$ cycloadditions would be possible, and these are of course "forbidden."

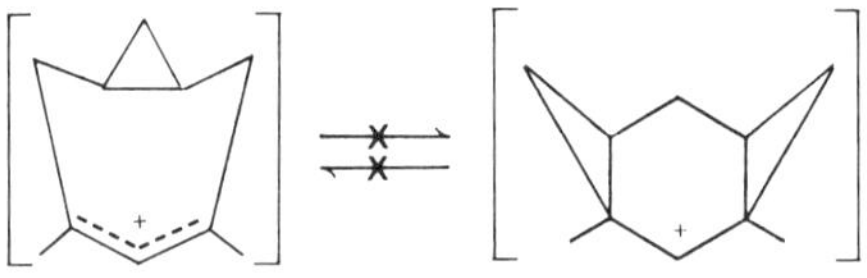

(**131**)   (**132**)

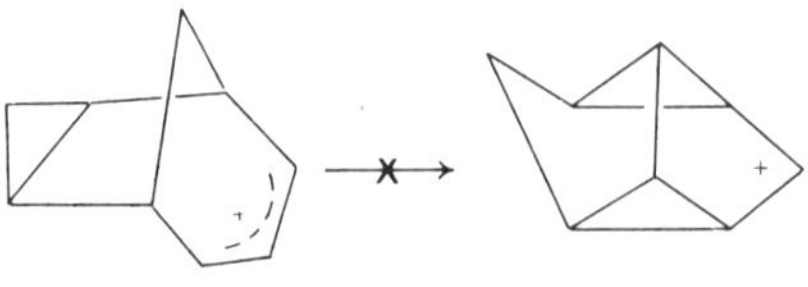

(**133**)   (**134**)

The cycloaddition (insertion) of the vinyl cation into a C–H bond has been considered theoretically (Kollmar and Smith, 1972). The calculations suggest an initial interaction of the empty p orbital of the cation with the C–H bond, implying a stepwise process for the overall cycloaddition.

There are no examples known to us of $[_\sigma 2^+ + {}_\pi 2]$ cycloadditions, i.e.,

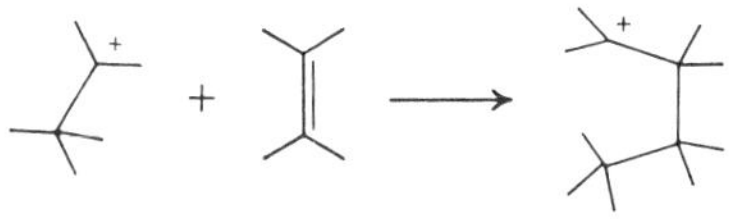

D. $[_\pi 2 + {}_\pi 4]$ Cycloadditions

Two cationic types are possible: (1) allyl cation + diene $[_\pi 2^+ + {}_\pi 4]$ and (2) pentadienyl cation + alkene $[_\pi 2 + {}_\pi 4^+]$. Both are cationic equivalents of the Diels–Alder reaction and are allowed as $[_\pi 2_s + {}_\pi 4_s]$ processes.

1. $[_\pi 2 + {}_\pi 4^+]$ Processes

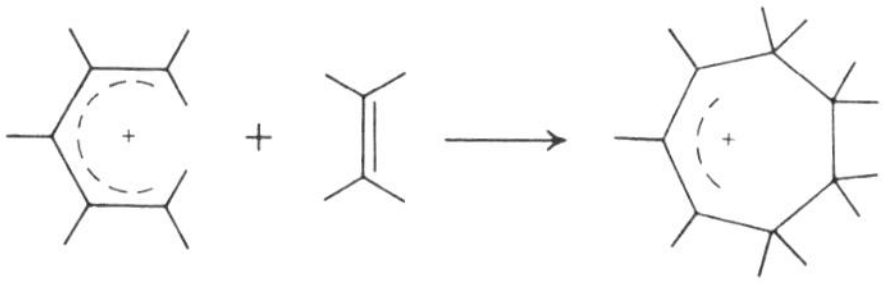

No simple examples of this cycloaddition have been reported. In addition to the usual problems of generating a reactive cation in the presence of an alkene, the pentadienyl cations are prone to undergo rapid electrocyclic closure to a cyclopentenyl cation (see previous section).

The best example of this reaction is the addition of cyclopentadiene or cycloheptatriene to the tropylium ion (Itô and Itoh, 1971). In this cycloaddition, one bond of the cyclopentadiene adds to a pentadienyl cation subunit of the tropylium ion. Both the *exo*-135 and *endo*-136 adducts are formed,

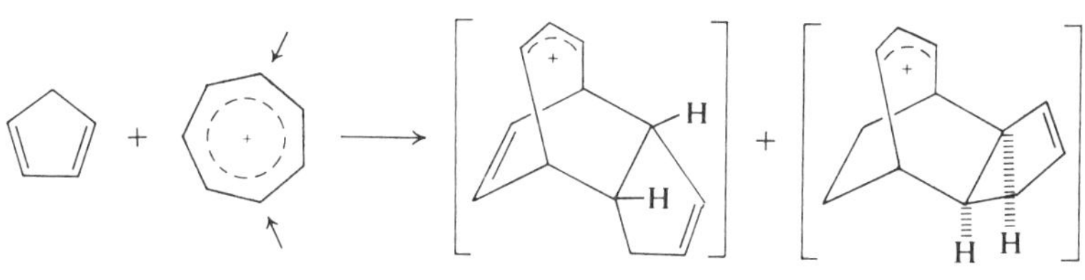

(136)   (135)

in a 1:4 ratio. The alcohol products isolated are formally derived from cations 135 and 136 and a cyclopropylcarbinyl form of these, i.e., cation 137 from 136.

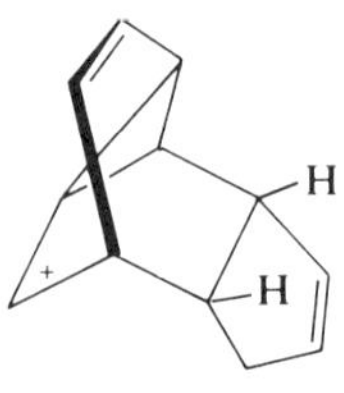

(137)

The rearrangement of perezone (138) into pipitzol (139) has been postulated as an example of a [$_\pi 2 + {}_\pi 4^+$] cycloaddition (Woodward and Hoffmann, 1969, 1970).

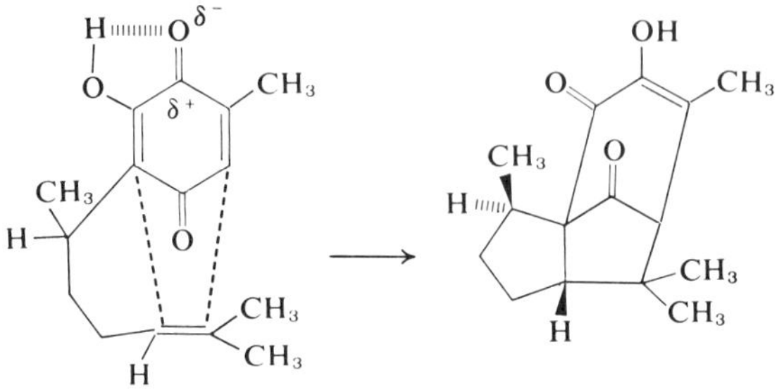

(138) → (139) (two stereoisomers)

2. [$_\pi 2^+ + {}_\pi 4$] *Processes*

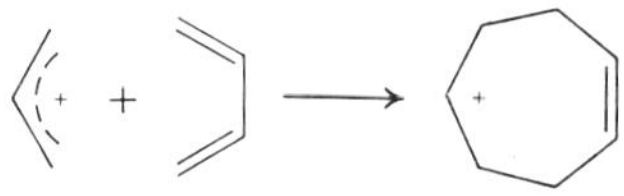

The first examples of these cycloadditions were reported by H. M. R. Hoffmann *et al.* (1968), who were able to add the 2-methallyl cation (140) to cyclopentadiene or cyclohexadiene, giving (in the case of cyclopentadiene) the two bicyclic dienes (141, 142). Somewhat later, the 2-methallyl cation was

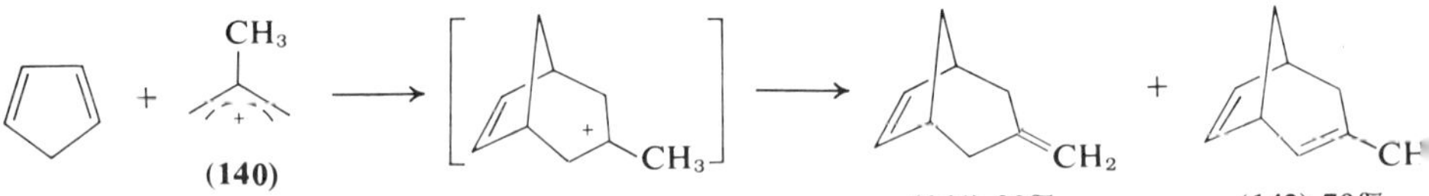

(140) (generated) *in situ*

(141) 30%

(142) 70%

added to butadiene itself, giving a cycloheptadiene (H. M. R. Hoffmann *et al.*, 1971). In carrying out these reactions, these workers have overcome formidable experimental problems. The choice of diene is quite critical and only those dienes with a cisoid configuration are really successful; that is, the cycloaddition must occur quickly because the competing electrophilic attack by the cation on the diene is also a rapid reaction.

The parent allyl cation itself has been added to cyclopentadiene; the product yield is lower and the reaction more difficult than in the methallyl cation case (H. M. R. Hoffmann *et al.*, 1971).

It is important to note that these reactions are unlikely to be sequential electrophilic attacks, i.e., attack by one end of the allyl cation on the diene followed by a second stage closure. Such a mechanism would likely have led to the preferential formation of a five-membered ring.

In the simple allyl cation–butadiene case, two transition states are possible, a boatlike (**143**) and chairlike (**144**) form. When substituted oxyallyl cations

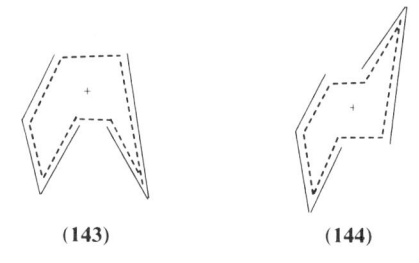

(143)   (144)

are used as reactants, the boatlike transition state (**143**) is found to be lower in energy, while with the 2-methallyl cation, the reverse was true (H. M. R. Hoffmann *et al.*, 1972b).

Subsequent efforts by H. M. R. Hoffmann and others have centered on attempts to make this cycloaddition useful in a synthetic sense for the synthesis of seven-membered rings. Thus, the 2-methoxyallyl cation has been added to furan, cyclopentadiene, and 6,6-dimethylfulvene, to give after workup the respective bicyclic ketones, e.g., **145** from furan (Greenwood and Hoffmann, 1972; Hill *et al.*, 1973). Ketones are also produced when the oxyallyl or

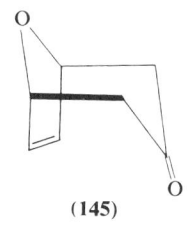

(145)

hydroxyallyl cations are added to dienes. These can be generated *in situ* from 1,3-dihaloketones and a dehalogenating agent (H. M. R. Hoffmann *et al.*, 1972a,b; Noyori *et al.*, 1971). Recently, the oxyallyl cation has been trapped with dimethylformamide to give 4-alkylidene-2-dimethylamino-1,3-dioxolanes (146), which provide a convenient source of the oxyallyl cation under mild reaction conditions (H. M. R. Hoffmann *et al.*, 1972b).

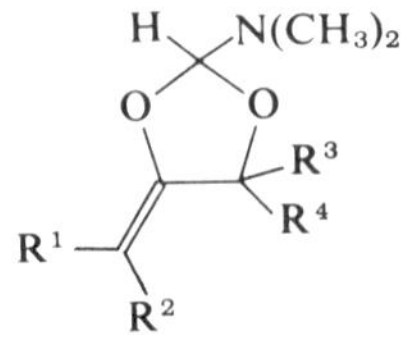

$$\underset{(146)}{\text{structure}}$$

Cyclopropanones undergo cycloaddition to dienes to give the same products that are produced from oxyallyl cation cycloadditions (Turro *et al.*, 1969; Turro, 1969). The question of whether cyclopropanone additions occur directly or via the oxyallyl cation (from electrocyclic ring opening of cyclopropanone) has been investigated (Edelson and Turro, 1970, see, however, H. M. R. Hoffmann, 1973). In additions of methyl-substituted cyclopropanones to furan, the reaction is found to be first order in both components, and, hence, the electrocyclic opening (if it occurs) cannot be rate determining. Nevertheless, the reaction rates seem related to the expected ease of electrocyclic opening of the cyclopropanones and the authors suggest that a rapid steady-state equilibrium between the cyclopropanone and the oxyallyl cation is established prior to the rate-determining cycloaddition.

3. *Other* [4 + 2] *Cases*

For cycloadditions involving one σ bond, one has in theory four types. The C–C σ bond is shown as a heavy line in each case and could be substituted for by a C–H bond (in fact, a cyclopropane C–C σ bond would be the

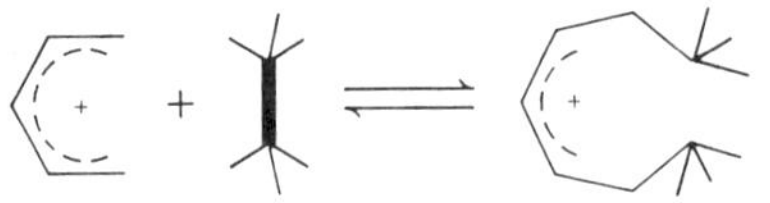

$[_\pi 4^+ + _\sigma 2]$ 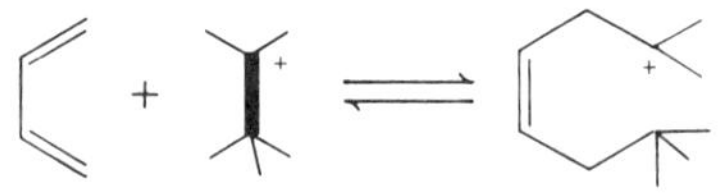 $[_\pi 4 + _\sigma 2^+]$

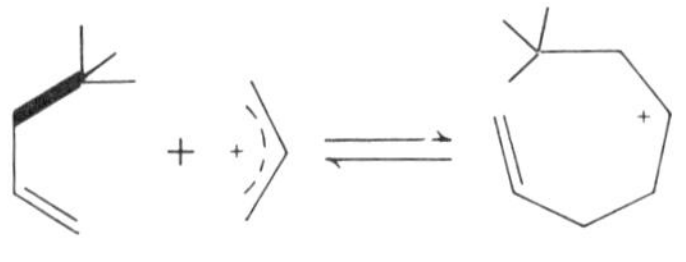

$[_{(\pi\sigma)}4 + _\pi 2^+]$ 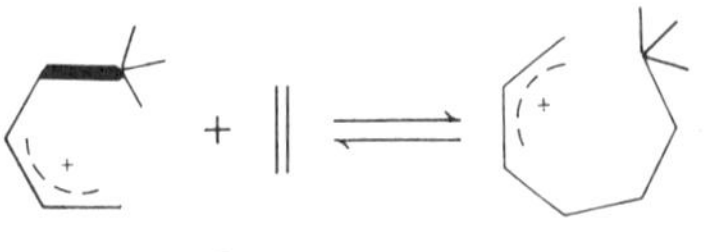 $[_{(\pi\sigma)}4^+ + _\pi 2]$

most promising in all likelihood). All are allowed as suprafacial–suprafacial additions but no examples of any of these are known to us.

E. [$_\pi 4$ + $_\pi 6$] CYCLOADDITIONS

Cationic [$_\pi 4$ + $_\pi 4^+$] cycloadditions require the orthogonal geometry for thermal reactions; not surprisingly, these cycloadditions have not been reported. There are, however, two reports of [$_\pi 4_s$ + $_\pi 6_s$] thermal cycloadditions. The addition of cycloheptatriene to the tropylium cation yields products resulting from both [$_\pi 4^+$ + $_\pi 2$] and [$_\pi 4^+$ + $_\pi 6$] cycloadditions.

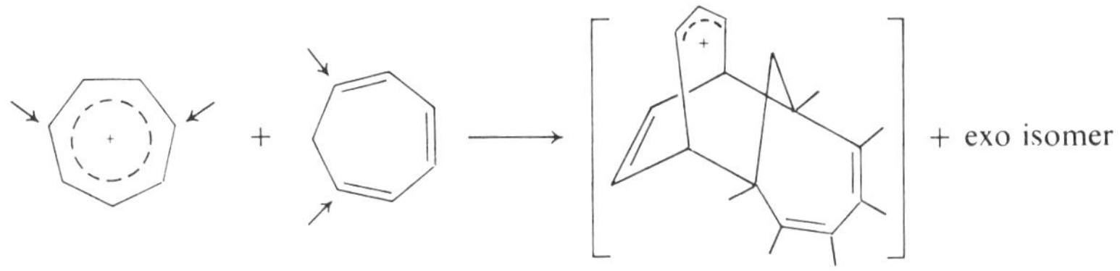

(147)

The stereoselective addition of cyclopentadiene to several alkyl-substituted ferrocenyl cations has been reported, i.e., cation **148** gives **149** (Turbitt and Watts, 1974). This reaction can be regarded as either a [$_\pi 4$ + $_\pi 2^+$] or [$_\pi 4$ + $_\pi 6^+$] cycloaddition.

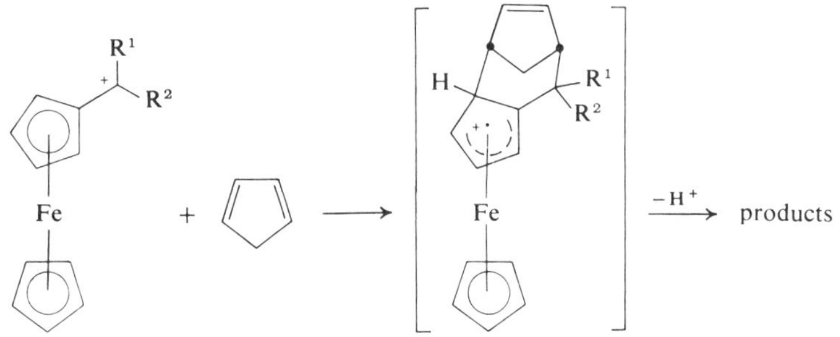

(148) **(149)**

IV. Sigmatropic Reactions

A. GENERAL CONSIDERATIONS

Woodward and Hoffmann (1970) have defined a sigmatropic change of order [i, j] as the migration of a σ bond flanked by one or more π electron systems to a new position whose termini are i-1 and j-1 atoms removed from the original bonded loci. In cations, the flanking "π system" often consists of the single, formally empty, p orbital of the cation carbon. Carbonium ions undergo hydride and alkyl abstraction reactions which, if they occur intramolecularly, must be regarded as sigmatropic rearrangements. However, they may not fit the above definition for sigmatropic change since the intervening carbon skeleton may be saturated or only partially unsaturated. In the following, we attempt as much as possible to describe the rearrangements of carbonium ions in the terminology of Woodward and Hoffmann, although at times we will have to extend our vocabulary.

B. LINEAR VS. BRIDGED STRUCTURES

Perhaps the best known and one of the most pervasive reactions of carbocations is the [1,2]-sigmatropic shift of a substituent, which usually occurs readily if it leads to a more stable cationic system. However, it is often difficult to assess experimentally whether a rapid sigmatropic rearrangement is, in fact, occurring or whether the static structure is actually "nonclassical," i.e., what one would normally suppose to be the transition state for the rearrangement. In discussing such systems, it is important to attempt to assess their instantaneous structures and, for the simpler systems at least, rigorous *ab initio* molecular orbital calculations are of considerable assistance.*

In order to attempt to rationalize the tendency of carbocations to adopt cyclic structures, one can visualize the generation of the cation by the interaction of an ethylenic fragment with an electron deficient species X, ($[_\pi 2_s + {_\pi}0_s]$ cycloaddition) as in Fig. 1. The ethylenic fragment has a π bond which is symmetrical with respect to the plane of symmetry preserved in the initial interaction along with an antisymmetrical, empty π^* orbital. Species X has a symmetrical empty p orbital and, if it is not a proton, also a filled orbital which is antisymmetrical. In the conceptualized reaction the electron acceptor (X) and the electron donor (the ethylenic fragment) form an initial π complex through interaction of the symmetrical orbitals as shown in **150**. In the process, electron density is transferred from the carbon atoms to X, thus stabilizing

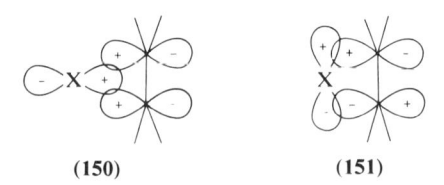

(150)   (151)

---

* An excellent review of the results of theoretical studies on the structure and stereochemistry of the smaller carbonium ions has recently appeared (Buss *et al.*, 1973).

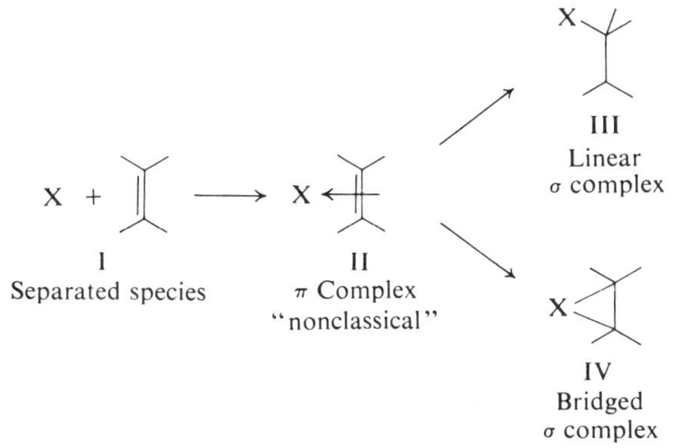

Fig. 1. Illustration of the possible stable species that may occur upon reaction of an electron-deficient species X with a C–C π bond. See text for discussion.

I Separated species
II π Complex "nonclassical"
III Linear σ complex
IV Bridged σ complex

X at the expense of the ethylenic fragment [but yielding a net stabilization in the simple one-electron orbital interaction picture (Hoffmann, 1971)]. According to our conceptualization (Fig. 1), the π complex may collapse to a σ-bonded linear complex in which the X unit is no longer electron deficient but which has a maximal deficiency at carbon, i.e., the carbonium ion site. Or, the interaction between antisymmetrical orbitals (151) may predominate, preserving the bridged structure. Which of the structures I, II, III, or IV of Fig. 1 is stabilized depends qualitatively on the severity of the electronic demand of X and the carbonium ion site as well as the availability of the pair of electrons occupying the antisymmetrical orbital of X. Various cases are summarized in Table VI.*

The classical vs. nonclassical controversy concerns itself mainly with cases 5 and 6, where the electronic demand of the species X is high, i.e., X is usually

TABLE VI Expected Structures from Fig. 1

Case	Electronic demand of X	Electronic demand of –C–	Availability of e pair on X	Expected structure (from Fig. 1)
1	Low	High	Low	I
2	Low	High	High	IV
3	Low	Low	High	IV
4	Low	Low	Low	I or III
5	High	High	Low	II
6	High	Low	Low	III

* It is assumed that if the electronic demand of X is high, the availability of the electron pair on X cannot be high.

a proton or a simpler carbonium ion and the electron pair which may participate in back donation either is not present (X = H$^+$) or is in a low lying σ bonding orbital. The detailed occurrence of cases 5 and 6 is the subject of the remainder of this chapter. Brief mention only will be made of cases 1 to 4 below.

Case 1 is the trivial case in which X is an already stabilized cation (see below), e.g., *tert*-butyl, and the ethylenic fragment is not substituted by electron releasing substituents. Case 4 is the same as case 1 except that electron releasing substituents are present on the π bond. Cases 2 and 3, in which the σ-bridged cyclic structures are favored, occur when X has nonbonded or loosely bonded electrons in p (or π) orbitals. Numerous examples of stable cyclic and bicyclic halonium ions (X = Cl$^+$, Br$^+$) are known (Olah, 1975; Olah et al., 1974a,c), as well as oxonium (X = O$^+$R) and aziridonium (X = N$^+$R$_2$) ions. Of special interest in the present context are the cases X = phenyl and X = vinyl. In such cases one has "homo" cationic species. The question of homo resonance stabilization is discussed in the last subsection. The homoallyl cation (49) (structure III, Fig. 1) has been computed to rearrange without activation to the cyclopropylcarbinyl cation (51) (structure IV, Fig. 1) which is 20 kcal/mole lower in energy (Hehre and Hiberty, 1972). The cyclopropylcarbinyl cationic system is discussed in greater detail below. Similarly, the α-phenylethyl cation (152) is computed to cyclize to the cyclic ethylenebenzenium cation (153) without activation (Hehre, 1972a). The cyclic structure (structure IV, Fig. 1) of the parent (152) as well as substituted ethylenebenzenium-type ions have been convincingly demonstrated by NMR studies under stable ion conditions (Olah and Porter, 1970, 1971). It has recently been demonstrated that benzonortricyclyl cations (154) also have the ethylenebenzenium-type structure (Olah and Liang, 1975).

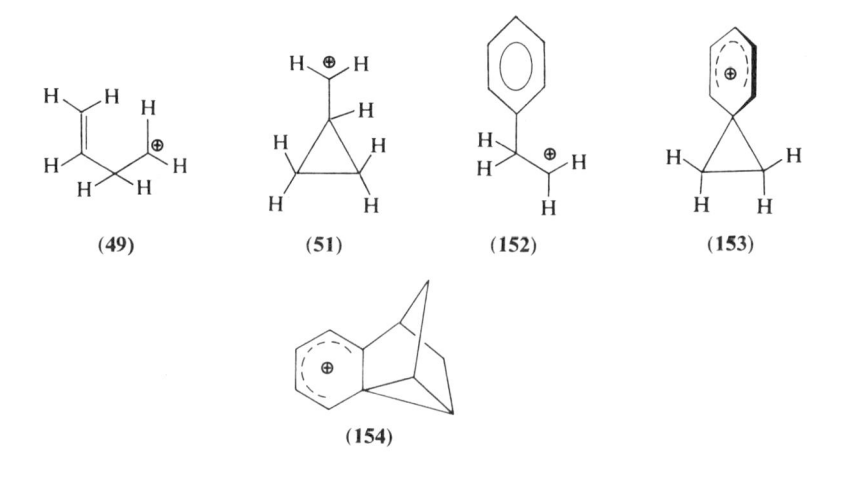

An appreciation of the electronic demand of carbonium ions may be gained by examining the geometric and electronic structure of the simplest carbonium ion, CH_3^+, which has been determined by *ab initio* computations (Kari and Csizmadia, 1969). The stable form has D_{3h} symmetry, with a C–H bond length of 1.078 Å, which is slightly shorter than that in the parent hydrocarbon, methane. The lowest *unoccupied* orbital is essentially a "$2p$" atomic orbital of carbon. The experimental heat of formation $\Delta H_f°$ (298°K) is 261 kcal/mole (Lossing and Semeliuk, 1970). By subtracting the heats of formation of $CH_3\cdot$ and CH_3^+, one may estimate the electron affinity of the empty p orbital of the carbonium ion. The reaction

$$CH_3^+ + e \rightarrow CH_3\cdot \quad \Delta H = -227 \text{ kcal/mole}$$

is exothermic by 227 kcal/mole! Clearly, a structure which can be represented in the "classical" sense as **155** is highly unstable and has an extreme tendency

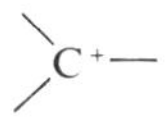

(155)

to acquire electrons from any available source. In fact, logically one must conclude that, except in the case of an isolated CH_3^+ ion, the classical, charge localized carbonium ion structure is never observed, even in the gas phase. Some electron delocalization into the formally empty p orbital always occurs. The mechanisms for the charge dispersal are discussed briefly below. Nevertheless, we will continue to write structures such as **155** and refer to these as "classical" ions if the delocalization mechanism appears not to lead to appreciable distortion in the nuclear framework toward cyclic structures, while we recognize that some distortion toward "nonclassicity" always occurs.

The simplest system which can undergo a sigmatropic reaction and for which the question of structure arises is the ethyl cation (**156, 157**). $C_2H_5^+$ has been the subject of many *ab initio* theoretical studies, most of which have

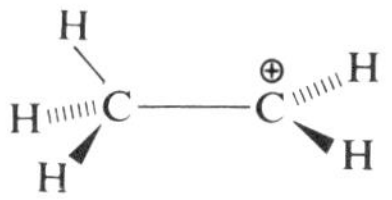

(156) 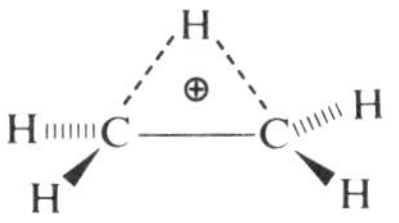   (157)

led to the prediction that a structure such as **156** describes the equilibrium nuclear configuration (Dixon and Lipscomb, 1973; see also Buss *et al.*, 1973; Lathan *et al.*, 1971; Williams *et al.*, 1970, 1971; Pfeiffer and Jewett, 1970; Sustmann *et al.*, 1969). However, very recent calculations using an extensively polarized basis set indicate that the symmetrically bridged structure (**157**) may be more stable than the classical structure (**156**) (Hariharan *et al.*, 1972). Of some significance is the fact that the same results were obtained using a more limited basis set but including electronic correlation energy (Ostlund and Bowen, 1975; Zurawski *et al.*, 1973).* From mass spectrometric studies, an upper limit of 5 kcal/mole has been set for the energy difference between the two structures (Ausloos *et al.*, 1972).

Mass spectrometric measurements indicate that ethyl cation is 42 kcal/mole more stable than methyl cation (Lossing and Semeliuk, 1970). The stabilization by the methyl group may be rationalized on the basis of orbital interactions (Hoffmann, 1971) derivable from simple perturbation theory (see, for example, Hoffmann, 1971; Salem, 1970; Dewar, 1969). The interaction between the methyl group orbitals and the empty *p* orbital of the cationic center is shown in Fig. 2. The net stabilization is accompanied by charge delocalization from one of the C–H bonding orbitals to the empty *p* orbital. The interaction, which we identify with C–H hyperconjugation, is evident in the calculated shorter bond length 1.484 Å (Lathan *et al.*, 1971) compared to 1.538 Å in ethane. Although it has been argued to the contrary on the basis of experimental data (Traylor *et al.*, 1971), computations indicate that considerable structural modification toward nonclassical or charge delocalized structures may accompany hyperconjugation (Lathan *et al.*, 1971; Radom *et al.*, 1972a). Computations also show that C–C hyperconjugation is more effective at stabilizing an adjacent carbonium ion center than is C–H hyperconjugation (Radom *et al.*, 1972b).

In light of the above considerations, it is of considerable interest that, as in the case of ethyl cation, the stable form of the 1-propyl cation has also recently been computed by *ab initio* means to be the symmetrically (methyl) bridged structure (**158**), a corner protonated cyclopropane (Hariharan *et al.*, 1974). Structure **158** is computed to be within 1 kcal/mole of a highly distorted classical form of the 1-propyl cation (**159**), but about 13 kcal/mole less stable than the 2-propyl cation (**160**). The last result is consistent with experimental results (Chong and Franklin, 1972; McAdoo *et al.*, 1972) if one assumes that structure **159** can rearrange to **158** with virtually no activation. MINDO/3 calculations yield similar results but predict that the stable form is actually an edge protonated cyclopropane (**161**) (Bischof and Dewar, 1975).

NMR and laser Raman studies (Olah *et al.*, 1970) have yielded strong

* These results suggest that the *ab initio* Hartree–Fock SCF scheme may underestimate the stability of nonclassical structures relative to classical structures. Since semiempirical schemes have long been known to grossly overestimate the stability of cyclic structures, the results from such schemes are avoided in this review. It has recently been asserted that the latest version of MINDO, i.e., MINDO/3, is free of this defect (Bischof and Dewar, 1975).

1. CARBOCATION   51

(158)   (159)   (160)   (161)

evidence that the structure of the 2-norbornyl cation may be the corner protonated nortricyclene (**162**) rather than the classical structure (**163**). The long-standing controversy continues (for some recent references, see Brown and Liu, 1975; Fong, 1974; Olah *et al.*, 1968, 1972, 1973; Olah, 1972; Goetz and Allen, 1971; Traylor *et al.*, 1971; Klopman, 1969).

In summary, we feel it is unlikely that bridged structures would be the

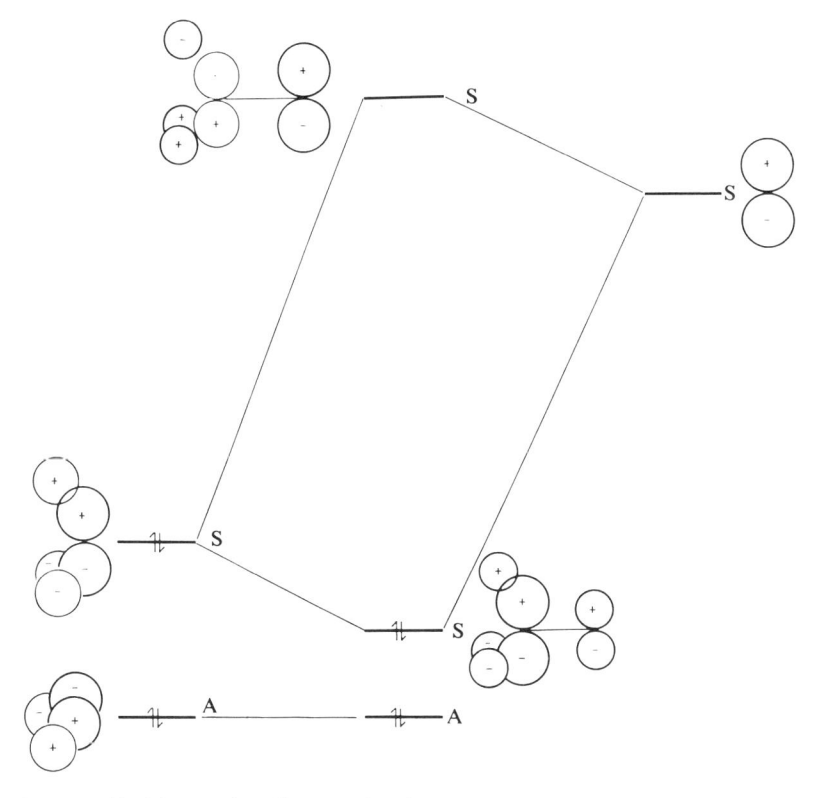

Fig. 2. Orbital interaction diagram showing stabilization by hyperconjugation. The third CH bonding orbital is of lower energy and is not shown.

most stable if other cation stabilizing groups are present. Logically, one should expect a continuum of structures between the "classical" and "nonclassical" extremes, as shown by means of the hypothetical potential energy curves in

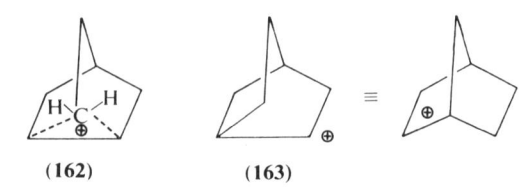

(162)   (163)

Fig. 3 for a [1,2] shift. The results of the *ab initio* MO calculations discussed above and the experimental observations of rapid 1,2-alkyl and hydride migrations suggest that in most cationic systems, the energy difference between the structural extremes is not more than a few kcal/mole if A and E (or B and D) are isoenergetic. As the electronic demand at the formally electron-deficient site decreases (upon substitution by electron releasing substituents), the tendency of the cation to reorganize toward a nonclassical structure decreases. Thus, the "classical" 2-propyl cation (160) is estimated (Hariharan *et al.*, 1974) to be 11 kcal/mole more stable than the hydrogen-bridged form and 13 kcal/mole more stable than the methyl-bridged form (158). Similarly, although there is some experimental support for the notion that the structure of the unsubstituted 2-norbornyl cation may be nonclassical (Olah *et al.*, 1968, 1970, 1972), it is agreed that the 2-methyl- (Goering *et al.*, 1974), 2-phenyl- (Farnum and Wolf, 1974), or 1,2-dimethyl-2-norbornyl (Olah *et al.*, 1971a) systems, for example, are best described by the more classical (unsymmetrical) structures. A PMR investigation of 2-aryl-2-norbornyl cations suggests that the ion "becomes nonclassical" if the aryl group is more electron demanding than phenyl (Farnum and Wolf, 1974). Similarly, evidence indicates that 2,3-dimethyl-2-butyl cation is classical, although the exchange of the hydride is rapid on the NMR time scale, even at low temperatures. Referring again to Fig. 3, if the classical structures A and E are not degenerate, but rather (say) A is more stable than E, then one would expect the transition state for the [1,2]-sigmatropic shift to resemble D more closely than the symmetrical structure C.

C. [1,2]-SIGMATROPIC REARRANGEMENTS

A [1,2]-sigmatropic (164 → 165) shift to a cationic center is a thermally allowed reaction in the Woodward–Hoffmann sense and the [1,2] shift of an alkyl group (Wagner–Meerwein shift) is predicted to proceed with retention of configuration (Woodward and Hoffmann, 1970). Theoretical computations clearly support the prediction since the migrating group should have a

1. CARBOCATION   53

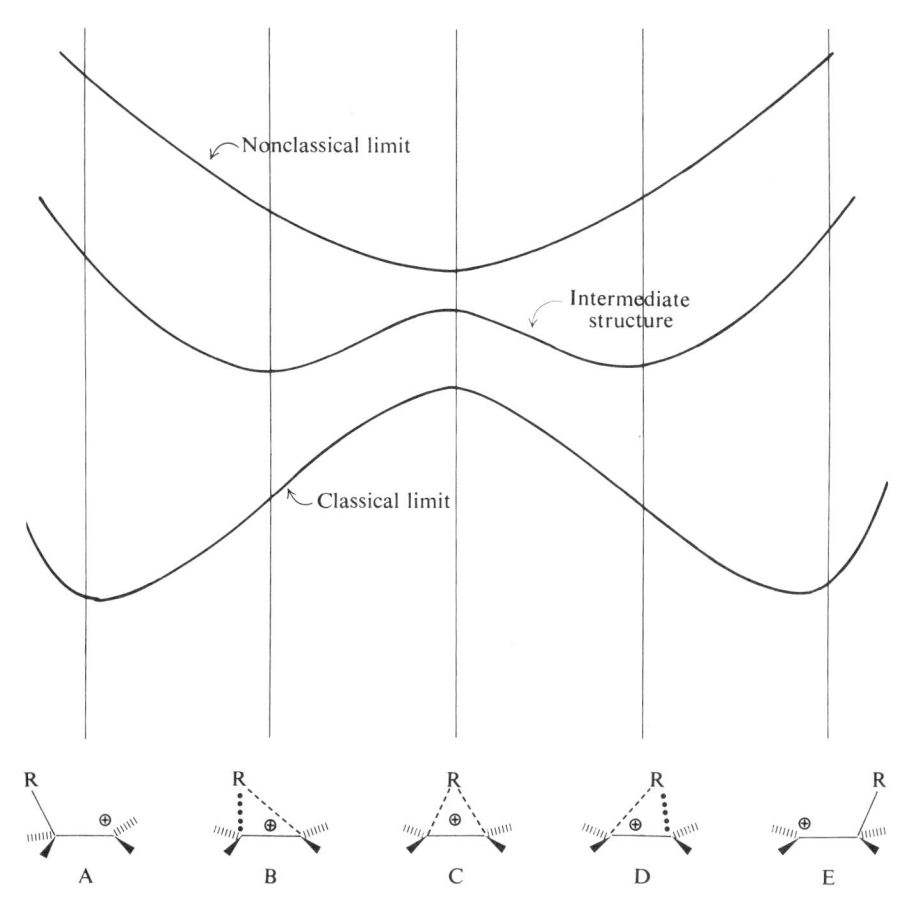

Fig. 3. Hypothetical potential surfaces for a [1,2]-sigmatropic shift showing a continuum of structural types between the "classical" and "nonclassical" limits.

pyramidal geometry comparable to **158** (Hariharan *et al.*, 1974; Radom *et al.*, 1972a). The prediction has been verified experimentally for geometrically unconstrained carbocations (Kirmse *et al.*, 1973; Kirmse and Gruber, 1973; Shono *et al.*, 1973). The rearrangement can be regarded as proceeding through the "nonclassical" or "π-complex" structure (**166**) and is aided by proper alignment of bonds adjacent to the cationic center (Nickon and Weglein, 1975). A discussion of the pericyclic stabilization in such structures is presented in the last section.

The opposite stereochemistry, inversion at the migrating carbon, has been predicted (Devaquet and Hehre, 1974) for the degenerate rearrangement in homocyclopropenyl cation (**167, 168**). Since the thermally allowed (Woodward and Hoffmann, 1970) ring opening of the cyclopropyl cation takes place within the confines of a four-membered ring, it is probably not meaningful to attempt to distinguish between structures **167** or **168** for homocyclopropenyl cation, which is known to be nonplanar from NMR evidence

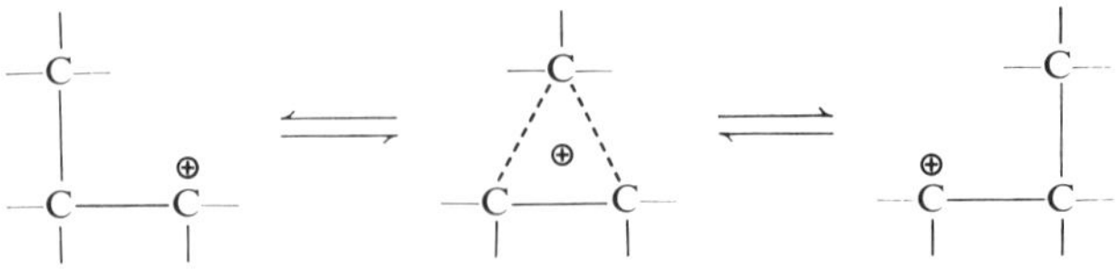

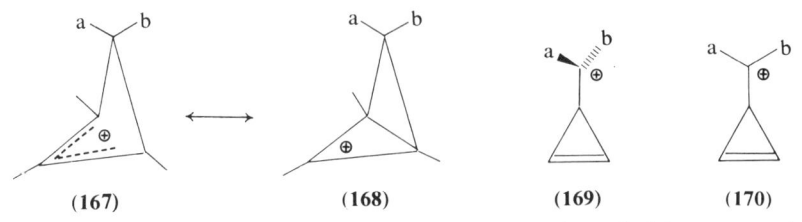

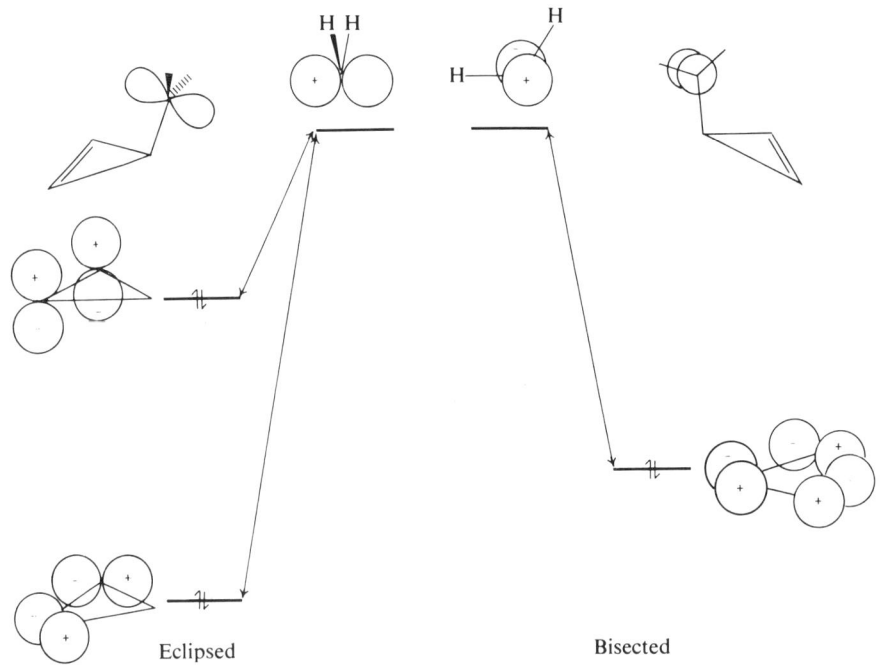

Fig. 4. Orbital interaction diagram for cyclopropenylcarbinyl cation showing how subjacent orbital control can preferentially stabilize the bisected conformation.

(Olah et al., 1974b). The transition state for the rearrangement may be regarded as a cyclopropenylcarbinyl cation (Fig. 4). As in the case of cyclopropylcarbinyl cation, the form of lower energy is the bisected structure **169** which is stabilized by the antisymmetrical Walsh orbital of the cyclopropyl ring. This form is calculated (Devaquet and Hehre, 1974) to be about 38 kcal/mole lower in energy than the eclipsed form (**170**), which would be stabilized by the HOMO (π-bonding orbital, as in structure **166**) and by the lower symmetrical Walsh orbital of the ring. This reversal in stereochemistry is an example of *subjacent orbital control* (Berson, 1972; Berson and Salem, 1972) and underlines the importance of occupied parallel *p* orbitals for the stabilization of adjacent cationic centers (see below).

D. [1,3]-Sigmatropic Rearrangements

A [1,3] shift of a hydride to a cationic center, exemplified by the reaction below, is expected to proceed through a transition state whose structure may

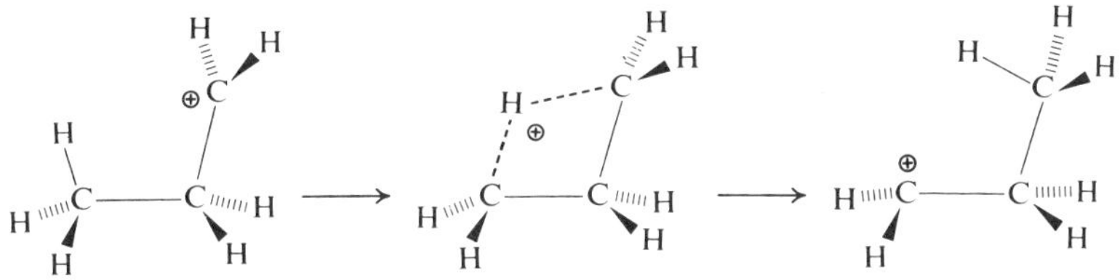

(161)

be described as an edge-protonated cyclopropane (**161**). The activation energy in the case of 1-propyl cation is estimated experimentally (Saunders et al., 1973 and references therein) and theoretically (Hariharan et al., 1974) to be approximately 5 to 6 kcal/mole. With methyl substituents, e.g., **171**, the activation energy is somewhat higher, 8.5 kcal/mole (Saunders and Stofko, 1973). The hydride shift from C(6) to C(2) of the 2-norbornyl cation system (**163**) is another example of this rearrangement.

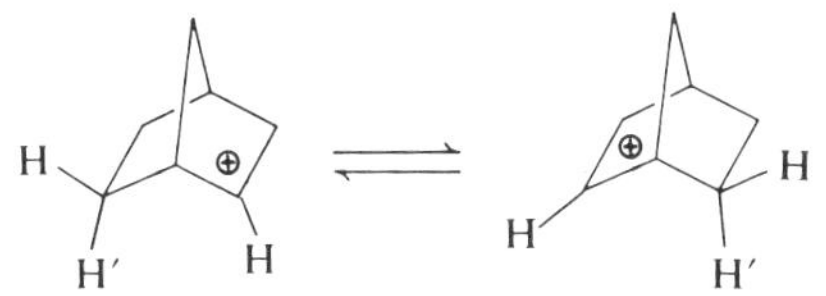

(163)

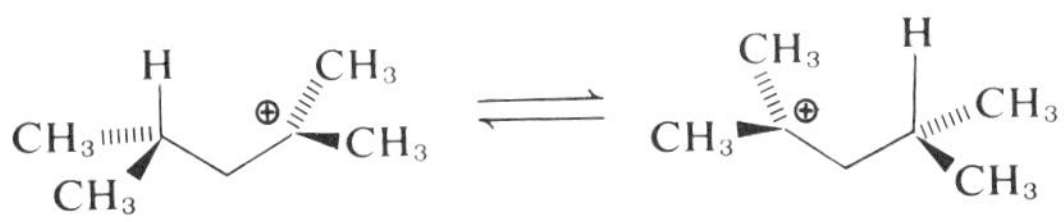

(171)

[1,3]-Sigmatropic shifts of alkyl groups have not been observed to occur in acyclic systems, presumably for steric reasons, but they can be predicted to proceed with inversion of configuration of the migrating group from considerations shown in Fig. 5. The orbitals of the transition state may be considered as arising from the interaction of the Walsh orbitals of cyclopropane and the p orbital of the carbonium ion. Although both the inversion and retention transition states are stabilized by the Walsh orbitals, extra stabilization occurs for the former through the secondary interaction with the virtual orbital which also is antisymmetrical with respect to the preserved symmetry plane. In strained cyclic systems, particularly the cyclopropylcarbinyl system (see below), the [1,3] rearrangement does occur and has the predicted stereochemistry.

The Cyclopropylcarbinyl and Allyl Systems

Considerable stabilization and geometric reorganization occurs when the cationic center is situated adjacent to a group, X, with a nonbonded or loosely bonded electron pair in a parallel p orbital. Perhaps the most extreme, but relevant, example occurs in carbonyl compounds which are often regarded as zwitterionic oxycations ($X = -O^-$) to rationalize their role in pericyclic

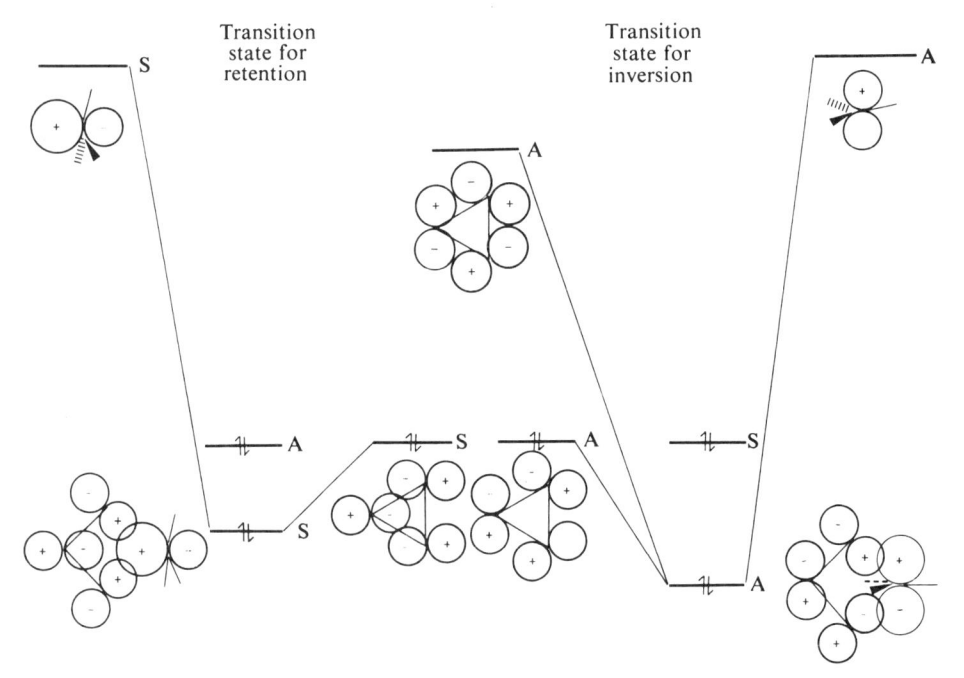

Fig. 5. Orbital interaction diagram for a [1,3]-sigmatropic alkyl shift showing how the transition state for inversion of configuration can be preferentially stabilized.

reactions. Similar stabilization obtains if X = −OR, −NR$_2$, halogen, etc. However, of particular interest in the present context are the cases X = vinyl

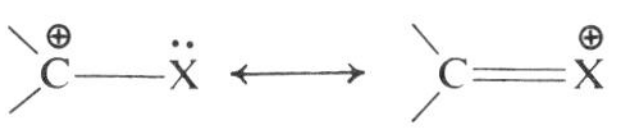

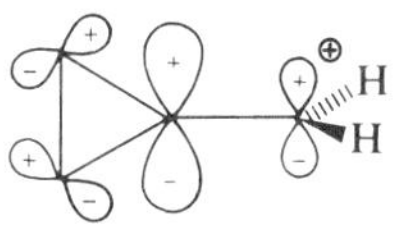

(172)

and X = cyclopropyl. The latter can interact with the empty cationic *p* orbital through the antisymmetrical ring Walsh orbital (172) (Hoffmann, 1970). In the former case, one has the allyl cation system 173 ≡ 174 or 175. Allyl cation is

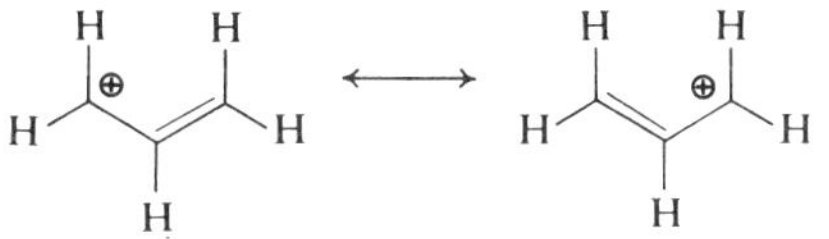

(173)　　　(174)　　　(175)

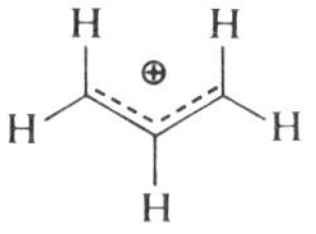

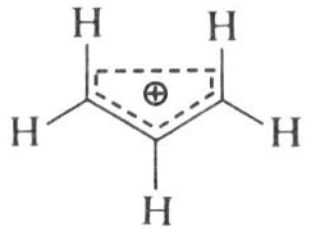

the thermodynamically stable form of formula C$_3$H$_5$$^+$. Deuterium labeling (Griffin and Jewett, 1970) and ^{13}C NMR (Olah *et al.*, 1971) experiments provide no support for structure 175, although the [1,3] interaction is often postulated under the guise of "homoaromatic stabilization" (Winstein, 1969), as in homocyclopropenyl cation (Olah *et al.*, 1971b, 1974b) and is theoretically augmented by electron donating groups at the 2-position (Carpenter, 1974).

In the case X = cyclopropyl, one has the cyclopropylcarbinyl cationic system which is known to undergo a number of intricate stereospecific rearrangements (Majerski and Schleyer, 1971) summarized in Fig. 6. The rearrangement I$_1$ ⇌ III$_1$ (cyclopropylcarbinyl ⇌ cyclobutyl) is a [1,2]-sigmatropic shift of the 4,2 σ bond between positions 1 and 2. The process I$_1$ ⇌ I$_2$ (cyclopropylcarbinyl ⇌ cyclopropylcarbinyl) is a [1,3]-sigmatropic shift of an alkyl group between positions 1 and 3. The stereochemical fate, inversion of configuration, of the migrating carbon has been demonstrated by Majerski and Schleyer (1971) in the solvolysis of hexadeuteriocyclopropylcarbinyl-methanesulfonate (176) and by Wiberg and co-workers in acid-catalyzed solvolyses of bicyclobutane (177) derivatives (Wiberg and Szeimies, 1970) and in the solvolyses of trans-fused bicyclo[6.1.0]nonyl-2-derivatives (Wiberg and Nakahira, 1971) (178). In Fig. 6, rearrangements of the type I$_1$ ⇌ II$_2$ and

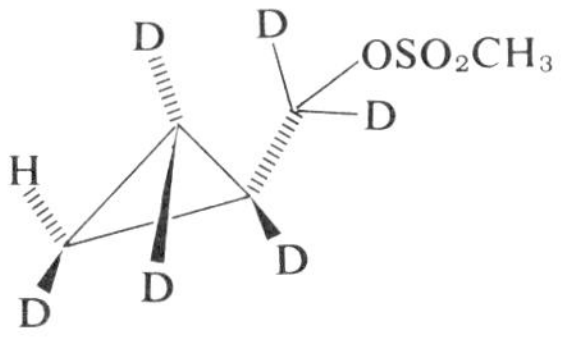

(176)　　　(177)　　　(178)

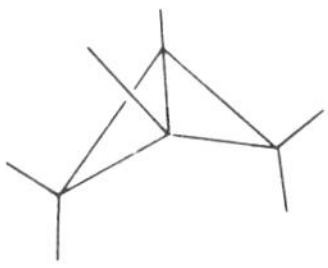

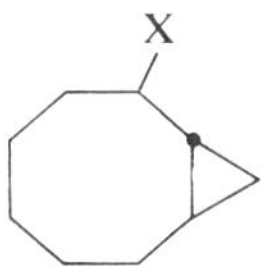

II$_2$ ⇌ III$_1$ are of the same type shown in Fig. 1 as III → II → I and may be classed as [$_\pi 0_s + _\pi 2_s$] cycloreversion reactions. The processes depicted in Fig. 6 have been previously reviewed (Majerski and Schleyer, 1971; Richey, 1972; Hanack and Schneider, 1967) and have been the subject of detailed solvolytic studies by the Winstein (Poulter and Winstein, 1972; Potter et al., 1970; Găsić et al., 1967; Whalen et al., 1967), Wiberg (Wiberg and Nakahira, 1971; Wiberg and Szeimies, 1970; Wiberg and Pfeiffer, 1970; Wiberg et al., 1970a,b), and Schleyer (Majerski and Schleyer, 1971) groups, as well as others (Gajewski and Oberdier, 1972; Dauben and Aoyagi, 1970; Lillien and Handloser, 1969a,b). Some or all of the processes shown in Fig. 6 have been invoked to rationalize observed rearrangements in more complex cationic systems (Santelli and Bertrand, 1974; Trost and Keeley, 1974; Farnum et al., 1973;

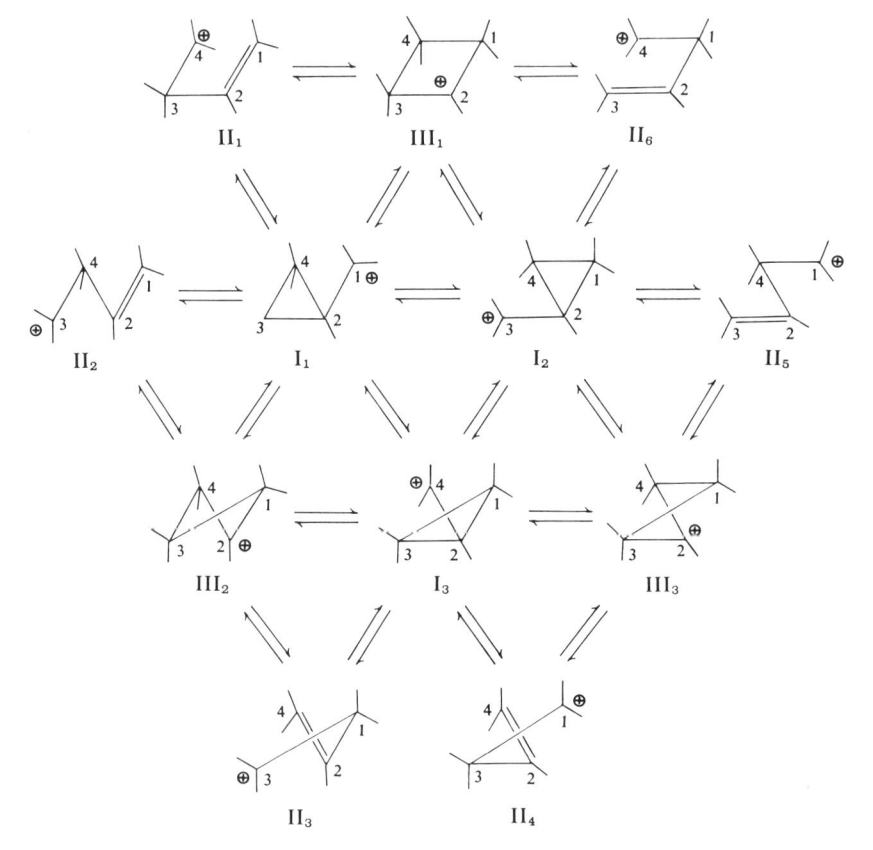

Fig. 6. The complete graph of interconversions in the cyclopropylcarbinyl–homoallyl–cyclobutyl cation system.

Paquette and Henzel, 1973; Paquette and Broadhurst, 1973; Peter-Katalinic *et al.*, 1973; Cueille and Jullien, 1972; Erman *et al.*, 1971). Several theoretical studies on the processes and structures depicted in Fig. 6 have also been carried out (Shanshal, 1972; Hehre and Hiberty, 1972; Wiberg and Szeimies, 1970; Hoffmann, 1970; Günther, 1970; Trindle and Sinanoğlu, 1969). A detailed *ab initio* study (Hehre and Hiberty, 1972) gave the result that unsubstituted homoallyl cation rearranges without activation to the more stable cyclopropylcarbinyl cation. Semiempirical calculations can yield incorrect relative energies and lead to anomalous conclusions (Trindle and Sinanoğlu, 1969).

E. [1,4]-SIGMATROPIC REARRANGEMENTS

Sigmatropic rearrangements of order [1,*j*] with *j* > 3 are also well known in cationic systems. The degenerate [1,4]-sigmatropic rearrangement in bicyclo-[3.1.0]hexenyl cations **179** ⇌ **180**, etc. is predicted by orbital symmetry considerations (Woodward and Hoffmann, 1970), to proceed thermally with

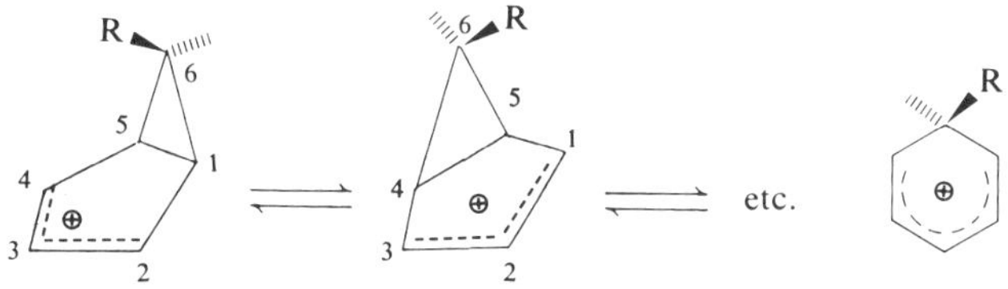

(179)   (180)   (181)

inversion of configuration at the migrating carbon atom. The prediction has been confirmed theoretically (Hehre, 1974) and was first confirmed experimentally by Childs and Winstein (1968). The activation energy for the process **179** ⇌ **180** in the parent bicyclo[3.1.0]hexenyl cation has been measured (Vogel *et al.*, 1971), to be 15 kcal/mole. Alkyl substitution at C(6) reduces the activation energy for the [1,4] shift (Koptyug *et al.*, 1970; Childs and Winstein, 1968) but also increases the tendency for the symmetry forbidden but exothermic electrocyclic reversion to benzenium ions (**181**). The photolysis of benzene in acidic media leads stereospecifically to **179**. The reaction has been postulated to proceed through benzvalene (**182**), the possible intermediacy of the benzenium ion (**181**), having been excluded on the basis that photolysis in deuterated acetic acid yields exclusively **179** (R = D) (Berson and Hasty, 1971).

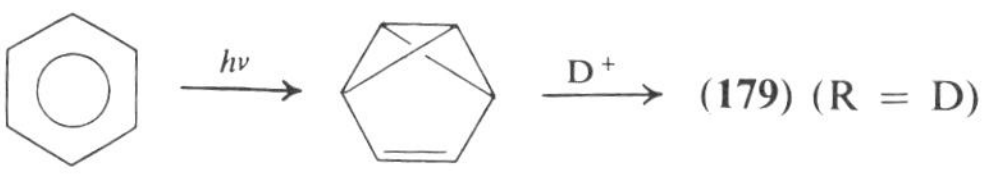

(182)

Bicyclo[3.1.0]hexenyl-type intermediates (**184**) can also be obtained by interaction of Lewis acids with bicyclo[3.1.0]hexenones (**183**) and have been shown to undergo rapid rearrangements with inversion of configuration at the migrating carbon (Hart *et al.*, 1969). Intermediates of type **185** can be derived by photolysis of 2,5-cyclohexadienones (Ogura and Matsura, 1970; Zimmerman and Crumrine, 1968) (**186** ⇌ **185**), by base-catalyzed elimination of hydrogen halides from 2-halobicyclo[3.1.0]hexan-3-ones (Zimmerman and Crumrine, 1968) (**187** → **185**) and by metal-catalyzed halogen elimination from 2,4-dihalobicyclo[3.1.0]hexan-3-ones (Brennan and Hill, 1968) (**188** → **185**). In each case, the [1,4]-sigmatropic rearrangement of **185** to the bicyclohexenone (**183**), has been shown to proceed with the predicted stereochemistry. Bicyclo[3.1.0]hexenyl cationic intermediates have also been invoked to rationalize the stereochemistry of acid-catalyzed photochemical dienone rearrangements (Filipescu and Pavlik, 1970; Schuster and Liu, 1971; Caine *et al.*, 1973). In general, acid-catalyzed thermal dienone-phenol (or -aryl ether) rearrangements involve sigmatropic shifts which are characteristic of neutral π systems, e.g., [1,3], [1,5], [3,3] (Hansen and Schmid, 1968; Miller, 1969,

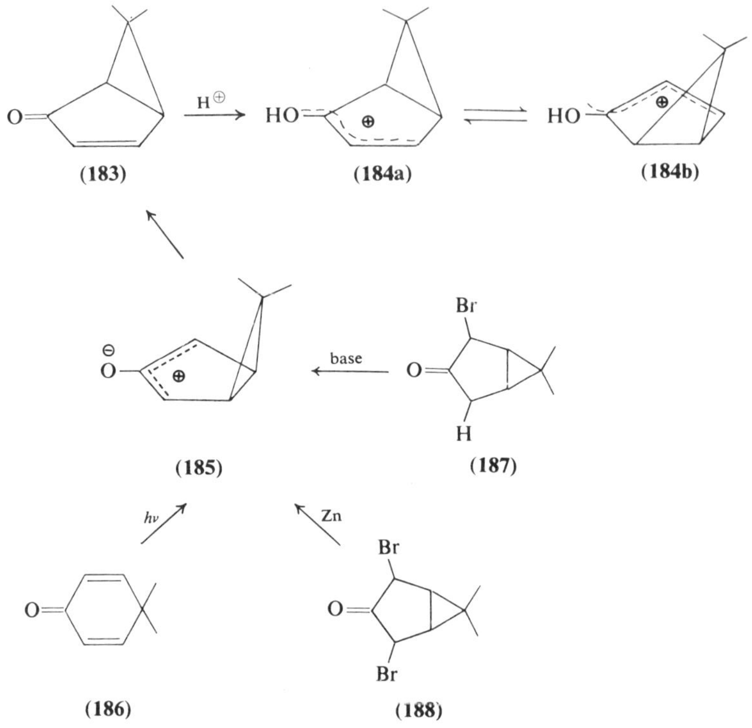

1970a,b,c; Lai and Miller, 1970; Schmid *et al.*, 1971; Borgulya *et al.*, 1973; Widmer *et al.*, 1973) as well as those characteristic of cationic π systems, e.g., [1,2], [1,4], [3,4] (Hansen and Schmid, 1968; Lai and Miller, 1970; Miller, 1970a,b,c; Borgulya *et al.*, 1973; Widmer *et al.*, 1973).

Suprafacial [1,4]-sigmatropic migrations of hydride ion are symmetry forbidden (Woodward and Hoffmann, 1970) but have been postulated to occur in a concerted fashion in several instances. The stereochemistry was unambiguously established in the rearrangement **189 → 190** (Noyori *et al.*, 1975). In the cationic polymerization of spiro[4,4]nona-1,3-diene **191**, a

(189)   (190)

(191)   (192)   (193)

[1,4]-sigmatropic shift (**192 → 193**) has been postulated to rationalize some of the observed skeletal structure of the polymer (Ohara *et al.*, 1973). It has been suggested (Epiotis, 1973) on the basis of theoretical arguments that, if the migrating group carries the bulk of the positive charge, a concerted [1,4]-sigmatropic migration of hydride (or alkyl group with retention of configuration) can proceed in a suprafacial manner. This condition is met in the latter example (**192 → 193**) if, as the authors suggest, there is direct participation of the counterion in the migrating step. In either case, a large exothermic driving force is expected to aid the migration.

Curiously, a [1,4]-hydride migration to a vinylic cation (**195 → 196**) has been suggested (Bly and Koock, 1969) as a possible explanation for the overall rearrangement **194 → 197**. Two comments can be made. Since the migration is to a linear cation, steric constraints may not be sufficient to prevent it from proceeding in the symmetry allowed *antarafacial* mode. Alternatively, it has been demonstrated (Kispert *et al.*, 1971; Wilcox *et al.*, 1973) that little conjugative interaction (orbital symmetry control) is transmitted across a cyclopropyl ring and it may be inappropriate to discuss the "suprafacial" or "antarafacial" character of the migration. [1,4]-Hydride migrations across a

(194) (195)

(196) (197)

saturated hydrocarbon skeleton, as in the 2,5-dimethylhex-2-yl cation (**198**) have been observed by NMR (Saunders and Stofko, 1973). A relatively low activation energy, 12 to 13 kcal/mole, has been determined for the process (**198**) (Saunders and Stofko, 1973).

(**198a**) (**198b**)

(**199a**) (**199b**)

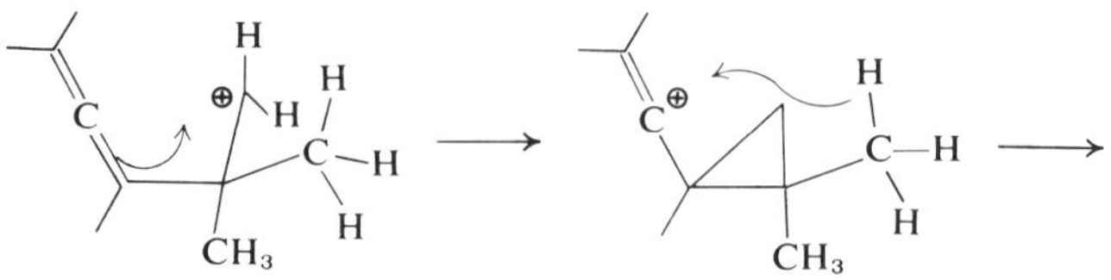

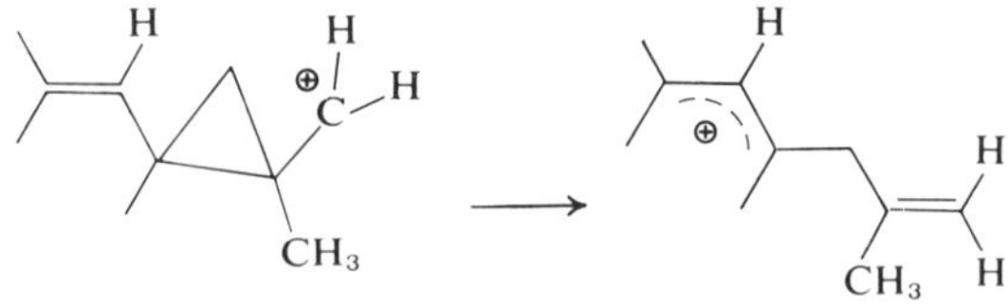

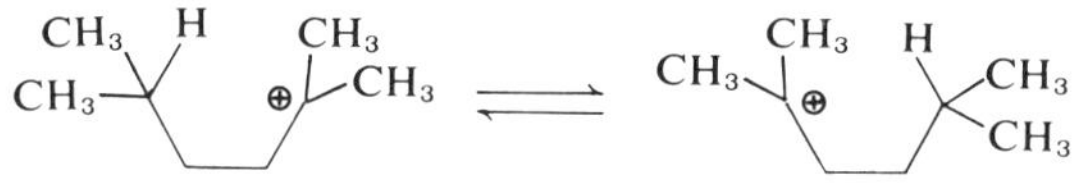

F. [1,5]-Sigmatropic Rearrangements

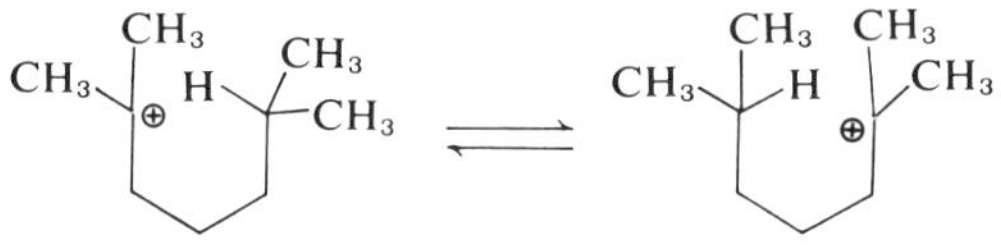

The [1,5]-hydride shift in 2,6-dimethylhept-2-yl cation (**199**) occurs more readily than the analogous shifts in the lower homologs (**171**, **198**) (Saunders and Stofko, 1973). The more favorable six-membered ring transition state for the [1,5] rearrangement may be more readily attainable than one might expect on statistical grounds due to a tendency of the long hydrocarbon chain cation to undergo internal solvation of the electron deficient site (Meyerson and Leitch, 1971). [1,5]-Sigmatropic shifts of alkyl groups in cationic systems are not observed.

1. CARBOCATION 63

G. [1,6]-SIGMATROPIC REARRANGEMENTS

The first report of a (symmetry allowed, suprafacial) [1,6]-hydride migration (**201** → **202**) was made in connection with the isomerization of 5,6-epoxycycloocta-1,3-diene (**200**) to cycloocta-2,4-dienone (**203**) (Heap *et al.*, 1969). A [1,6]-suprafacial shift of an alkyl group had been postulated to

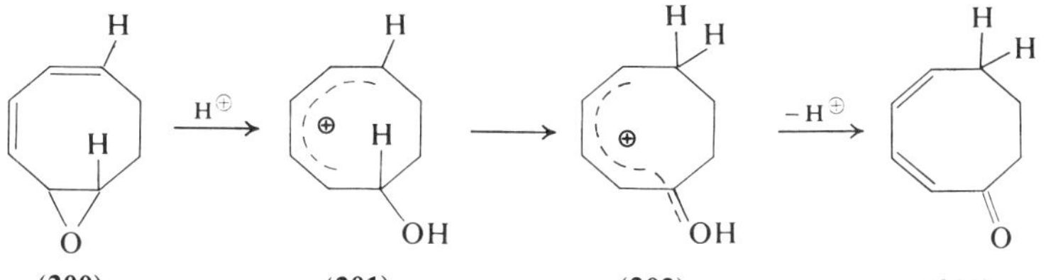

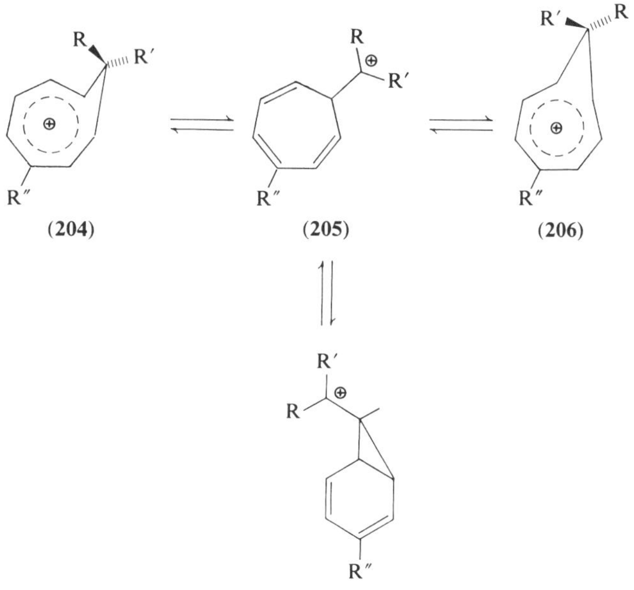

occur in homotropylium ions (**204** ⇌ **206**). However, recent theoretical computations (Hehre, 1972b, 1974) have led to an estimate of 43 kcal/mole for the degenerate rearrangement **204** ⇌ **206** (R = R' = R" = H) proceeding by the Woodward–Hoffmann allowed eclipsed structure **205** which interchanges exo and endo positions of the migrating center. An upper limit of 27 kcal/mole for this process has been determined by a deuterium labeling experiment (Berson and Jenkins, 1972) **204** ⇌ **206** (R = R' = H, R" = D). Another process, ring inversion through a planar cyclooctatrienyl cation, also interchanges exo and endo groups but requires less activation energy (22 kcal/mole; Winstein *et al.*, 1966). The relative stabilities of the ground state (**204**) and transition state (**205**) can be reversed if R, R' are cation stabilizing groups. Thus the 2,4,6-cycloheptatrienylcarbinyl cation (**205**) (R = NMe$_2$, R' = OEt, R" = H) has been shown (Betz and Daub, 1971) to be in dynamic equilibrium at room temperature with the slightly less stable bicyclo[4.1.0]hepta-2,4-dien-7-ylcarbinyl cation (**207**) (R = NMe$_2$, R' = OEt, R" = H).

H. [3,4]-Sigmatropic Rearrangements

Orbital symmetry considerations (Woodward and Hoffmann, 1970) lead to the prediction that the thermally allowed [3,4]-sigmatropic rearrangement (**208** → **209**) should occur suprafacially with respect to each π fragment. Although such rearrangements have been observed in alkyl-substituted dienones (Borgulya *et al.*, 1973; Widmer *et al.*, 1973) and related (Hansen and Schmid, 1968; Hansen *et al.*, 1968) systems, the stereochemistry has not been

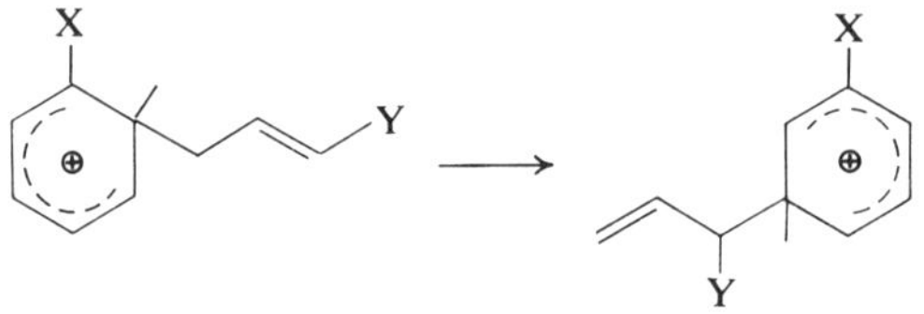

(**208**)   (**209**)

established. It has been speculated (Dewar, 1971) that [3,4]-sigmatropic rearrangements **210** → **211**, analogous to the Cope or Claisen rearrangements should proceed readily, but these have not yet been observed.

(**210**)   (**211**)

V. Pericyclic, Longicyclic, and Laticyclic Stabilization

Often, in cyclic systems, a formal carbonium ion is generated in close spatial proximity to one or more sources of loosely bound electrons with consequent stabilization. π Bonds (**212**) or strained σ bonds (**213**) are two such sources of electrons. In the terminology of Goldstein and Hoffmann, the interacting orbitals of **212** or **213** have a *pericyclic* topology which is stabilized

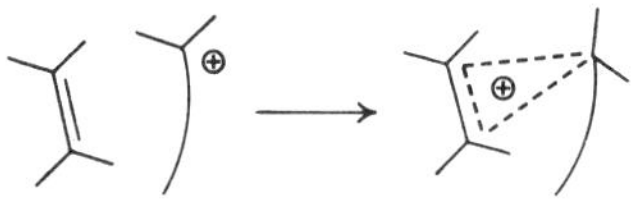

(212)

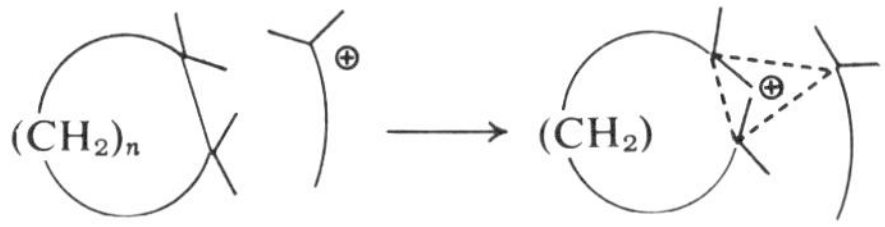

(213)
(*n* = 1, 2)

if, as in the present case, $4q+2$ electrons are involved (Goldstein and Hoffmann, 1971). The observations that, for example, **214** and **215** solvolyze more rapidly, by factors of 10^{11} (Tanida and Tsushima, 1970; Winstein and Shatavsky, 1956; Winstein *et al.*, 1963) and 10^{14} (Haywood-Farmer and Pincock, 1969; Battiste *et al.*, 1967; Tanida *et al.*, 1967), respectively, than

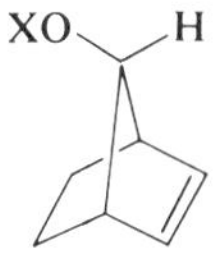

(214) 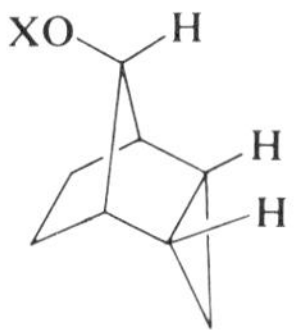   (215) 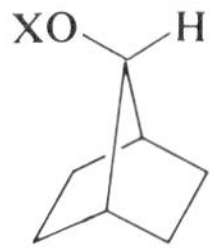   (216)

216 confirm the existence of transition states resembling **212** and **213** in which the charge has been extensively delocalized. Similar, though less dramatic, edge participation by cyclobutanes has also been observed (Sakai *et al.*, 1970). As discussed previously, such structures are normally not stable intermediates if the molecular geometry permits a collapse to more classical structures. The subject of the present section will be those cases in which the nuclear framework favors the stabilization of intermediates such as **212** and **213** or allows for rapid degenerate rearrangements with transition states such as **212** or **213**. Often, additional stabilization by more unusual orbital topologies such as laticyclic and longicyclic topologies (Goldstein and Hoffman, 1971) may be possible.

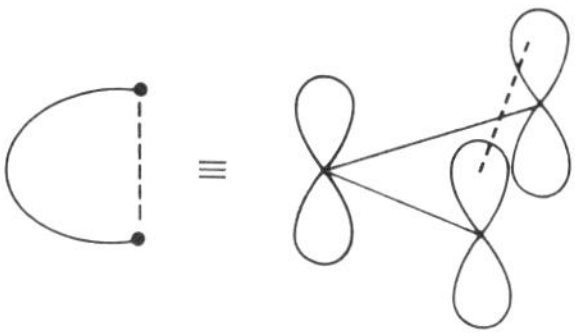

The extent of pericyclic stabilization of a single ribbon or of any number of ribbons depends on the extent of overlap between the ends of the interrupted cyclic array. This is governed by the separation and orientation of the three (in the cationic two electron case) orbitals of the π ribbon. As discussed previously, there is little evidence of a 1,3-interaction in acyclic allyl systems or in cyclic systems where the termini are separated by a saturated carbon chain of two or more atoms. Where only one carbon atom interrupts the cyclic conjugation, one has the special case of homoaromatic stabilization in the homocyclopropenyl cation system. NMR evidence suggests that the structure of the parent **167** is nonplanar (Olah *et al.*, 1974b) indicating substantial

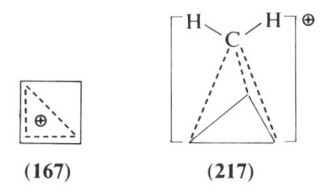

(167)   (217)

1,3-interaction, although little else is known about the structure from experimental work. The symmetrically bridged, pentacoordinated structure **217** has neither experimental (Olah *et al.*, 1974b) nor theoretical (Devaquet and Hehre, 1974) support.

A pericyclic array of two electrons interrupted at two points requires that one of the "ribbons" be a single *p* orbital. Numerous examples, such as **214**,

are known. Solvolysis studies of the bicyclo[*n*.2.1]alkenyl series **218** ($n = 1, 2, 3$) indicate that the extent of bishomoaromatic participation is a sensitive

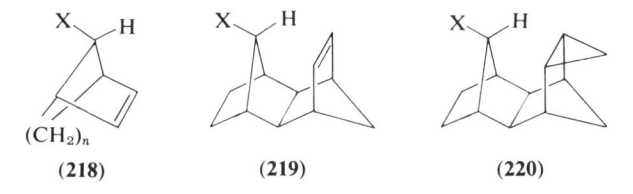

(218)   (219)   (220)

function of the geometry of the five-membered ring, the addition of each successive methylene group *retarding* the rate by three powers of ten (Hess, 1969; Masamune *et al.*, 1969). The special case of **218**, where $n = 0$, is discussed below. Considerable stabilization, as evidenced by enhanced solvolysis rates, can occur in systems such as **218** even if the intervening saturated links involve more than one carbon atom, if the molecular skeleton forces the π bond and (developing) cationic center into close spatial proximity, as for example in **219** (Battiste et al., 1970).

A pericyclic array of two electrons interrupted at three points requires that each of the "ribbons" be a single p orbital and yields the system **213** whose simplest manifestations are the trishomocyclopropenyl cationic derivatives (**221**). A common entry into the system **221** involves the initial generation of

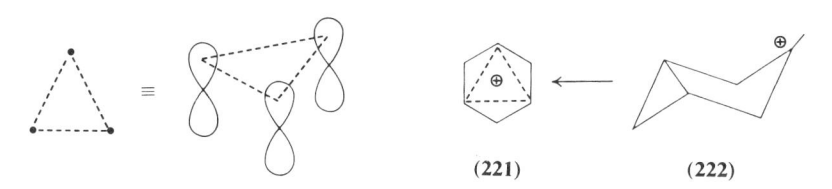

(221)   (222)

cations of the type **222** (Winstein et al., 1959; Coates and Kirkpatrick, 1970; Gassman and Pike, 1975; Masamune et al., 1974a,b; Haywood-Farmer and Pincock, 1969; Battiste et al., 1967; Tanida et al., 1967). Comparative solvolysis data such as was cited above for **214** and **215** suggest that the strained σ bond of cyclopropyl may be more effective at charge dispersal than would be a π bond in a similar location. The converse has been found in the case of the polycyclic structures such as **220** which solvolyze somewhat more slowly than **219** (Battiste et al., 1970). The reversal may well be due to minor structural variations between **219** and **220**.

The next higher aromatic cationic homolog is the well-known tropylium ion (**223**). Some of its homoaromatic systems, namely homotropylium cation (**224**) (Rosenberg et al., 1962; Winstein et al., 1965, 1966; Berson and Jenkins, 1972; Gasteiger and Huisgen, 1972; Huisgen and Gasteiger, 1972) and 1,4-bishomotropylium cation (**225**) (R = H) (Ahlberg et al., 1970b) and **225** (R = CH$_3$) (Ahlberg et al., 1970a) have also been prepared.

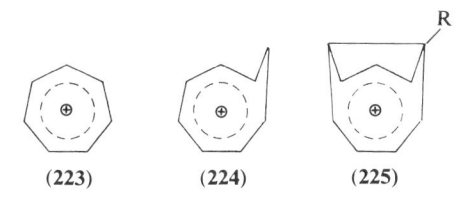

(223)   (224)   (225)

The possibility of *bicycloaromatic* (Goldstein, 1967) or longicyclic (Goldstein and Hoffmann, 1971) stabilization arises if a second π bond is added to **214**. The resulting symmetric structure of the cation **226** must occur at the expense of the single pericyclic interaction of the type discussed above which would be favored by the unsymmetric structure **227**. In fact, the structure of the 7-norbornadienyl cation is unsymmetrical (227, R = H). The observation of a considerable barrier, possibly greater than 19.6 kcal/mole (Brookhart

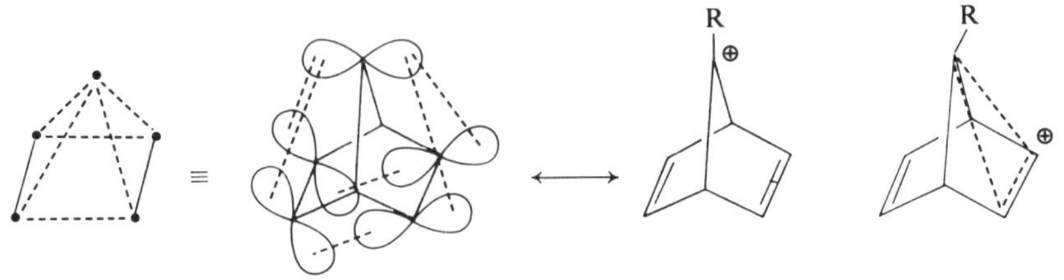

(226)   (227)

et al., 1967) in the parent molecule, has led to the consideration that substantial *laticyclic* (Goldstein and Hoffmann, 1971) stabilization may be operative, e.g., 228. That this is not the case for 7-norbornadienyl cation or other cations with potential laticyclic orbital topologies (such as 229 and 230)

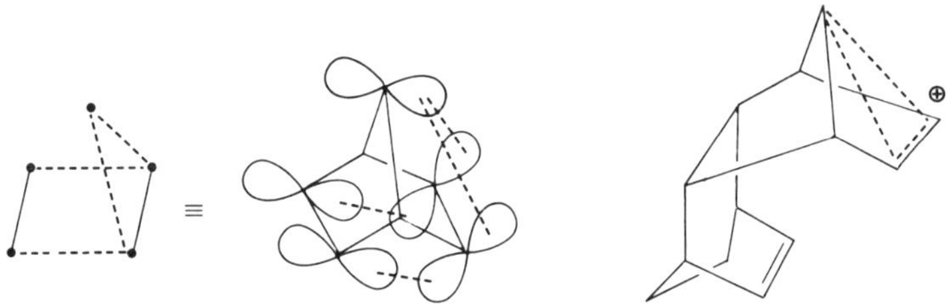

(228)   (229)

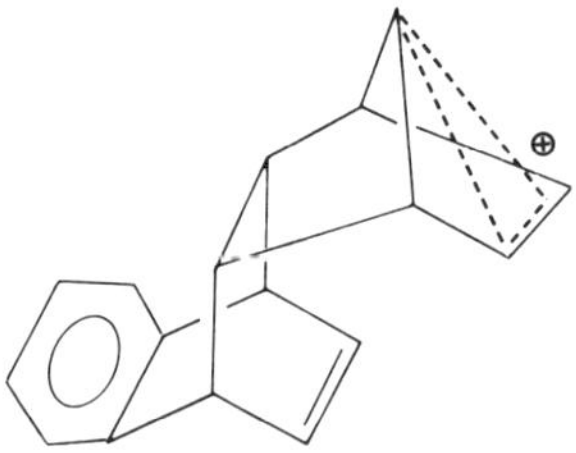

(230)

has been persuasively argued by Paquette and Dunkin (1975). Apparently, the electronic requirements of the cationic center are satisfied by the pericyclic interaction alone to such an extent that additional delocalization of charge by involvement of other more remote π linkages does not lead to appreciable extra stabilization. In a study of the solvolysis of *syn-endo*-tricyclo[3.2.1.0^{2,4}]oct-6-en-8-yl *p*-nitrobenzoate, it was demonstrated that pericyclic edge participation by cyclopropyl (231) is favored over participation

(231)   (232)   (233)

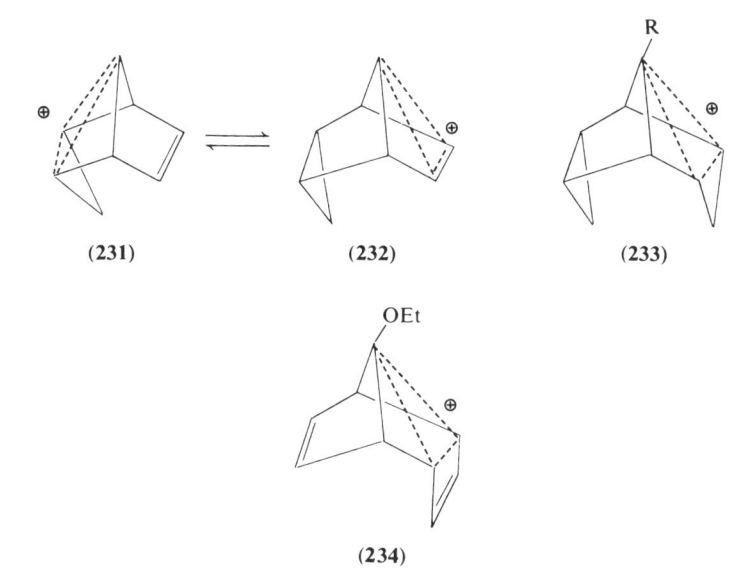

(234)

of the π bond **232** (Jovanovich and Lambert, 1971). As in the 7-norbornadienyl cation case, results were consistent with the equilibrium **231** ⇌ **232** being slow relative to the solvolysis reaction time scale. The barrier to bridge flipping in **233** (R = CH_3), the dicyclopropyl analog of **226**, has recently been shown to be 13.0 kcal/mole (Coates and Fretz, 1975) and is very similar to that reported for the methyl derivative of the 7-norbornadienyl cation (**226**) (R = CH_3), 12.6 kcal/mole (Brookhart *et al.*, 1967). In **234**, where pericyclic stabilization by the strained σ bond of a cyclobutene ring competes with the π bond, the former appears to be more effective (Schipper *et al.*, 1974).

Longicyclic *destabilization* can be invoked to explain the inability to form **235** from logical precursors (for a discussion, see Goldstein and Hoffmann,

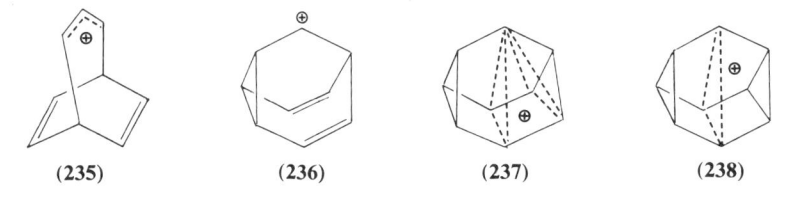

(235)   (236)   (237)   (238)

1971), its attempted formation leading apparently exclusively to the 9-barbaralyl cation (**236**). The latter itself can partake of longicyclic stabilization via **237** or of pericyclic stabilization via **238**. Experimental (Barborak and Schleyer, 1970; Barborak *et al.*, 1969) and theoretical (R. Hoffmann *et al.*, 1972) results are consistent with the symmetrical structure **238** for the 9-barbaralyl cation.

The prediction, from several semiempirical calculations (Kohlmer et al., 1973; Stohrer and Hoffmann, 1972), that the stable form of cations whose parent may be formulated $(CH)_5^+$ could have symmetrical structures of C_{4v}

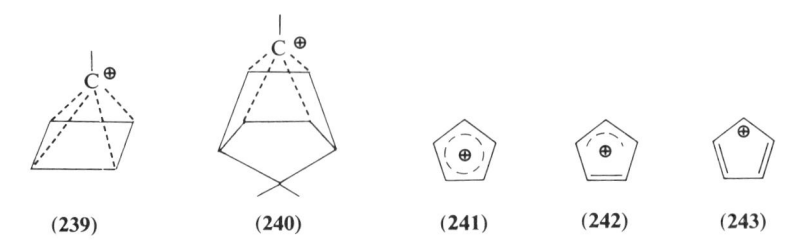

(239)   (240)   (241)   (242)   (243)

symmetry 239 has prompted many experimental attempts to synthesize the species (Masamune et al., 1972b) or its analogs (Masamune et al., 1972a,b,c; Hart and Kasuya, 1972, 1974, 1975a,b). Although the symmetrical structure for bishomo-$(CH)_5^+$-type cations (240) has proponents (Hart and Kasuya, 1972, 1974, 1975a,b), it appears premature to assign a nonclassical structure such as 239 to $(CH)_5^+$-type cations in general (Masamune et al., 1972b,c), especially since MINDO/3 (Dewar and Haddon, 1973) and ab initio MO (Hehre and Schleyer, 1973) calculations yield the result that the structure 239 is very unstable relative to more classical forms such as 241, 242, or 243, and presumably also 244. In the same vein, both theoretical (Jonkman and Nieuwpoort,

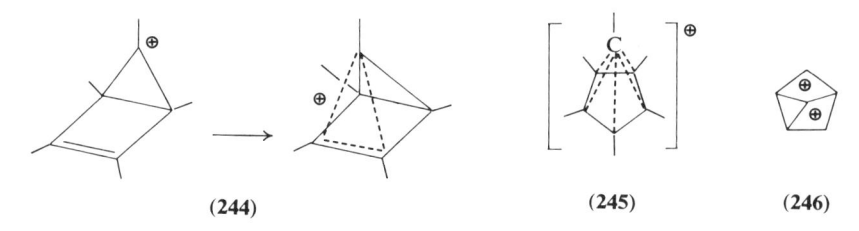

(244)   (245)   (246)

1973) and experimental (Hogeveen and Kwant, 1973, 1974) evidence lends support to the unusual pentagonal pyramidal structure (245) for $(CH)_6^{2+}$-type dications, which involve a carbon with a formal coordination number of six. However, an alternative structure 246 which can undergo rapid degenerate rearrangement by means of 1,2 shifts cannot be excluded.

The symmetrical species 239 and 245 are especially appealing in the light of the simple bonding schemes which can be evoked by considering interaction of an R–C$^+$ with the appropriate 4π electron cyclic system, cyclobutadiene (Stohrer and Hoffmann, 1972) and cyclopentadienyl cation, respectively.*

* Note that a similar interaction of H–C$^+$ with the 4π electron three-membered ring, cyclopropenyl anion, leads to a bonding description for tetrahedrane.

References

Ahlberg, P., Harris, D. L., and Winstein, S. (1970a). *J. Am. Chem. Soc.* **92**, 2146.
Ahlberg, P., Harris, D. L., and Winstein, S. (1970b). *J. Am. Chem. Soc.* **92**, 4454.
Akhtar, M., Chadwick, J. C., and Fray, G. I. (1972). *J. Chem. Soc., Perkin Trans.* 1 p. 1778.
Ausloos, P., Rebbert, R. E., Sieck, L. W., and Tiernan, T. O. (1972), *J. Am. Chem. Soc.* **94**, 8939.
Baird, M. S., and Reese, C. B. (1969). *J. Chem. Soc. C* p. 1803.
Baird, M. S., and Reese, C. B. (1970). *Chem. Commun.* p. 1644.
Baird, M. S., Lindsay, D. G., and Reese, C. B. (1969). *J. Chem. Soc. C* p. 1173.
Baird, N. C. (1972). *Tetrahedron* **28**, 2355.
Barborak, J. C., and Schleyer, P. v. R. (1970). *J. Am. Chem. Soc.* **92**, 3184.
Barborak, J. C., Daub, J., Follweiler, D. M., and Schleyer, P. v. R. (1969). *J. Am. Chem. Soc.* **91**, 7760.
Battiste, M. A., and Nebzydoski, J. W. (1970). *J. Am. Chem. Soc.* **92**, 4450.
Battiste, M., Deyrup, C. L., Pincock, R. E., and Haywood-Farmer, J. (1967). *J. Am. Chem. Soc.* **89**, 1954.
Battiste, M. A., Haywood-Farmer, J., Malkus, H., Seidl, P., and Winstein, S. (1970). *J. Am. Chem. Soc.* **92**, 2144.
Berson, J. A. (1972). *Acc. Chem. Res.* **5**, 406.
Berson, J. A., and Hasty, N. M., Jr. (1971). *J. Am. Chem. Soc.* **93**, 1549.
Berson, J. A., and Jenkins, J. A. (1972). *J. Am. Chem. Soc.* **94**, 8907.
Berson, J. A., and Salem, L. (1972). *J. Am. Chem. Soc.* **94**, 8917.
Berson, J. A., Bergman, R. G., Clark, G. M., and Wege, D. (1969). *J. Am. Chem. Soc.* **91**, 5601.
Betz, W., and Daub, J. (1971). *Angew. Chem., Int. Ed. Engl.* **10**, 269.
Biefield, C. G., Eick, H. A., and Hart, H. (1973). *Tetrahedron Lett.* p. 4507.
Bischof, P. K., and Dewar, M. J. S. (1975). *J. Am. Chem. Soc.* **97**, 2278.
Bladek, R., and Sorensen, T. S. (1972). *Can. J. Chem.* **50**, 2806.
Bly, R. S., and Koock, S. U. (1969). *J. Am. Chem. Soc.* **91**, 3299.
Boche, G., Hechtl, W., Huber, H., and Huisgen, R. (1967). *J. Am. Chem. Soc.* **89**, 3344.
Bollinger, J. M., Brinich, J. M., and Olah, G. A. (1970). *J. Am. Chem. Soc.* **92**, 4025.
Borgulya, J., Madija, R., Fahrni, P., Hansen, H. J., Schmid, H., and Barner, R. (1973). *Helv. Chim. Acta* **56**, 15.
Brennan, T. M., and Hill, R. K. (1968). *J. Am. Chem. Soc.* **90**, 5614.
Brookhart, M., Lustgarten, R. K., and Winstein, S. (1967). *J. Am. Chem. Soc.* **89**, 6352.
Brown, H. C., and Liu, K.-T. (1975). *J. Am. Chem. Soc.* **97**, 600.
Brown, R. S., and Traylor, T. G. (1973). *J. Am. Chem. Soc.* **95**, 8025.
Buss, V., Schleyer, P. v. R., and Allen, L. C. (1973). *Top. Stereochem.* **7**, 253.
Caine, D., Brake, P. F., DeBardelen, J. F., Jr., and Dawson, J. B. (1973). *J. Org. Chem.* **38**, 967.
Campbell, P. H., Chiu, N. W. K., Deugau, K., Miller, I. J., and Sorensen, T. S. (1969). *J. Am. Chem. Soc.* **91**, 6404.
Cantacuzène, D., and Tordeux, M. (1971). *Tetrahedron Lett.* p. 4807.
Carpenter, B. K. (1974). *J. Chem. Soc., Perkin Trans.* 2 p. 1.
Casanova, J., and Waegell, B. (1972). *Bull. Soc. Chim. Fr.* p. 2669.
Childs, R. F., and Parrington, B. (1970). *Chem. Commun.* p. 1540.
Childs, R. F., and Taguchi, V. (1970). *Chem. Commun.* p. 695.
Childs, R. F., and Winstein, S. (1968). *J. Am. Chem. Soc.* **90**, 7146.

Childs, R. F., Sakai, M., and Winstein, S. (1968). *J. Am. Chem. Soc.* **90**, 7144.
Chiu, N. W. K., and Sorensen, T. S. (1973). *Can. J. Chem.* **51**, 2776.
Chong, S.-L., and Franklin, J. L. (1972). *J. Am. Chem. Soc.* **94**, 6347.
Christensen, P. A. (1971). Ph. D. Thesis, University of Calgary, Calgary, Alberta, Canada (unpublished results).
Christensen, P. A., Huang, Y. Y., Meesters, A., and Sorensen, T. S. (1974). *Can. J. Chem.* **52**, 3424.
Clark, D. T., and Armstrong, D. R. (1969). *Theor. Chim. Acta* **13**, 365.
Clark, D. T., and Smale, G. (1969). *Tetrahedron* **25**, 13.
Clark, G. M., Hancock, K. G., and Zweifel, G. (1971). *J. Am. Chem. Soc.* **93**, 1308.
Coates, R. M., and Fretz, E. R. (1975). *J. Am. Chem. Soc.* **97**, 2538.
Coates, R. M., and Kirkpatrick, J. L. (1970). *J. Am. Chem. Soc.* **92**, 4883.
Coffey, P., and Jug, K. (1974). *Theor. Chim. Acta* **34**, 213.
Criegee, R., and Moschel, A. (1959). *Chem. Ber.* **92**, 2181.
Cristol, S. J., Sequeira, R. M., and De Puy, C. H. (1965). *J. Am. Chem. Soc.* **87**, 4007.
Cueille, G., and Jullien, R. (1972). *Bull. Soc. Chim. Fr.* p. 318.
Dauben, W. G., and Aoyagi, E. I. (1970). *Tetrahedron* **26**, 1249.
Dehmlow, E. V., and Schönefeld, J. (1971). *Justus Liebigs Ann. Chem.* **744**, 42.
Deno, N. C., and Pittman, C. U., Jr., (1964). *J. Am. Chem. Soc.* **86**, 1871.
Deno, N. C., Pittman, C. U., Jr., and Turner, J. O. (1965). *J. Am. Chem. Soc.* **87**, 2153.
De Puy, C. H. (1968). *Acc. Chem. Res.* **1**, 33.
De Puy, C. H., Schnack, L. G., and Hausser, J. W. (1966). *J. Am. Chem. Soc.* **88**, 3343.
Devaquet, A. J. P., and Hehre, W. J. (1974). *J. Am. Chem. Soc.* **96**, 3644.
Dewar, M. J. S. (1969). "The Molecular Orbital Theory of Organic Chemistry." McGraw-Hill, New York.
Dewar, M. J. S. (1971). *Angew. Chem., Int. Ed. Engl.* **10**, 761.
Dewar, M. J. S., and Haddon, R. C. (1973). *J. Am. Chem. Soc.* **95**, 5836.
Dewar, M. J. S., and Kirschner, S. (1971). *J. Am. Chem. Soc.* **93**, 4290, 4291, 4292.
Diaz, A. F., Harris, D. L., Sakai, M., and Winstein, S. (1971). *Tetrahedron Lett.* p. 303.
Dixon, D. A., and Lipscomb, W. N. (1973). *J. Am. Chem. Soc.* **95**, 2853.
Duffin, D., and Sutherland, J. K. (1970). *Chem. Commun.* p. 626.
Edelson, S. S., and Turro, N. J. (1970). *J. Am. Chem. Soc.* **92**, 2770.
Epiotis, N. D (1973). *J. Am. Chem. Soc.* **95**, 1206.
Erman, W. F., Treptow, R. S., Bakuzis, P., and Wenkert, E. (1971). *J. Am. Chem. Soc.* **93**, 657.
Farnum, D. G., and Wolf, A. D. (1974). *J. Am. Chem. Soc.* **96**, 5166.
Farnum, D. G., Mostashari, A., and Hagedorn, A. A. (1971). *J. Org. Chem.* **36**, 698.
Farnum, D. G., Mader, R. A., and Mehta, G. (1973). *J. Am. Chem. Soc.* **95**, 8692.
Fellenberger, K., Schöllkopf, U., Bahn, C. A., and Schleyer, P. v. R. (1972). *Tetrahedron Lett.* p. 359.
Filipescu, N., and Pavlik, J. W. (1970). *J. Am. Chem. Soc.* **92**, 6062.
Fong, F. K. (1974). *J. Am. Chem. Soc.* **96**, 7638.
Gajewski, J J., and Oberdier, J. P. (1972). *J. Am. Chem. Soc.* **94**, 6053.
Gascoigne, R. M. (1958). *J. Chem. Soc.* p. 876.
Gašić, M., Whalen, D., Johnson, B., and Winstein, S. (1967). *J. Am. Chem. Soc.* **89**, 6382.
Gassman, P. G., and Pike, W. C. (1975). *J. Am. Chem. Soc.* **97**, 1250.
Gassman, P. G., Dygos, D. K., and Trent, J. E. (1970). *J. Am. Chem. Soc.* **92**, 2084.
Gassman, P. G., Nishiguchi, I., and Yamamoto, H.-A. (1975). *J. Am. Chem. Soc.* **97**, 1600.

Gasteiger, J., and Huisgen, R. (1972). *Tetrahedron Lett.* p. 3665.
Ghosez, L., and de Perez, C. (1971). *Angew. Chem.* **83**, 171.
Goering, H. L., Chang, C.-S., and Clevenger, J. V. (1974). *J. Am. Chem. Soc.* **96**, 7602.
Goetz, D. W., and Allen, L. C. (1971). *Proc. Int. Congr. Pure Appl. Chem.* **1**, 51.
Goldstein, M. J. (1967). *J. Am. Chem. Soc.* **89**, 6357.
Goldstein, M. J., and Hoffmann, R. (1971). *J. Am. Chem. Soc.* **93**, 6193.
Graefe, J., and Mühlstädt, M. (1970). *Tetrahedron* **26**, 795.
Greenwood, G., and Hoffmann, H. M. R. (1972). *J. Org. Chem.* **37**, 611.
Griesbaum, K. (1966). *Angew. Chem.* **78**, 953.
Griesbaum, K., and Chu, Y. (1974). *Chem. Ber.* **107**, 838.
Griesbaum, K., Naegele, W., and Wanless, G. G. (1965). *J. Am. Chem. Soc.* **87**, 3151.
Griffin, R. H., and Jewett, J. G. (1970). *J. Am. Chem. Soc.* **92**, 1104.
Günther, H. (1970). *Tetrahedron Lett.* p. 5173.
Hanack, M., and Schneider, H. J. (1967). *Angew. Chem., Int. Ed. Engl.* **6**, 666.
Hansen, H. J., and Schmid, H. (1968). *Chem. Br.* **5**, 111.
Hansen, H. J., Sutter, B., and Schmid, H. (1968). *Helv. Chim. Acta* **51**, 828.
Hariharan, P. C., Lathan, W. A., and Pople, J. A. (1972). *Chem. Phys. Lett.* **14**, 385.
Hariharan, P. C., Radom, L., Pople, J. A., and Schleyer, P. v. R. (1974). *J. Am. Chem. Soc.* **96**, 599.
Hart, H., and Kasuya, M. (1972). *J. Am. Chem. Soc.* **94**, 8958.
Hart, H., and Kasuya, M. (1974). *J. Am. Chem. Soc.* **96**, 6436.
Hart, H., and Kasuya, M. (1975a). *J. Am. Chem. Soc.* **97**, 2450.
Hart, H., and Kasuya, M. (1975b). *J. Am. Chem. Soc.* **97**, 2459.
Hart, H., Collins, P. M., and Waring, A. J. (1966). *J. Am. Chem. Soc.* **88**, 1005.
Hart, H., Rodgers, T. R., and Griffiths, J. (1969). *J. Am. Chem. Soc.* **91**, 754.
Hausser, J. W., and Grubber, M. J. (1972). *J. Org. Chem.* **37**, 2648.
Hausser, J. W., and Uchic, J. T. (1972). *J. Org. Chem.* **37**, 4087.
Haywood-Farmer, J., and Pincock, R. E. (1969). *J. Am. Chem. Soc.* **91**, 3020.
Heap, N., Green, G. E., and Whitham, G. H. (1969). *J. Chem. Soc. C* p. 160.
Hehre, W. J. (1972a). *J. Am. Chem. Soc.* **94**, 5919.
Hehre, W. J. (1972b). *J. Am. Chem. Soc.* **94**, 8908.
Hehre, W. J. (1974). *J. Am. Chem. Soc.* **96**, 5207.
Hehre, W. J., and Hiberty, P. C. (1972). *J. Am. Chem. Soc.* **94**, 5917.
Hehre, W. J., and Schleyer, P. v. R. (1973). *J. Am. Chem. Soc.* **95**, 5837.
Hess, B. A. (1969). *J. Am. Chem. Soc.* **91**, 5657.
Hill, A. E., Greenwood, G., and Hoffmann, H. M. R. (1973). *J. Am. Chem. Soc.* **95**, 1338.
Hine, K. E., and Childs, R. F. (1971). *J. Am. Chem. Soc.* **93**, 2323.
Hoffmann, H. M. R. (1973). *Angew. Chem., Int. Ed. Engl.* **12**, 819.
Hoffmann, H. M. R., Joy, D. R., and Suter, A. K. (1968). *J. Chem. Soc. B* p. 57.
Hoffmann, H. M. R., Kernaghan, G. F. P., and Greenwood, G. (1971). *J. Chem. Soc. B* p. 2257.
Hoffmann, H. M. R., Clemens, K. E., Schmidt, E. A., and Smithers, R. H. (1972a). *J. Am. Chem. Soc.* **94**, 3201.
Hoffmann, H. M. R., Clemens, K. E., and Smithers, R. H. (1972b). *J. Am. Chem. Soc.* **94**, 3940.
Hoffmann, R. (1968). *J. Am. Chem. Soc.* **90**, 1475.
Hoffmann, R. (1970). *Tetrahedron Lett.* p. 2709.
Hoffmann, R. (1971). *Acc. Chem. Res.* **4**, 1.
Hoffmann, R., and Woodward, R. B. (1965). *J. Amer. Chem. Soc.* **87**, 2046.

Hoffmann, R., Stohrer, W.-D., and Goldstein, M. J. (1972). *Bull. Chem. Soc. Jpn.* **45**, 2513.
Hogeveen, H., and Gaasbeck, C. J. (1970). *Recl. Trav. Chim. Pays-Bas* **89**, 1079.
Hogeveen, H., and Kwant, P. W. (1973). *Tetrahedron Lett.* p. 1665.
Hogeveen, H., and Kwant, P. W. (1974). *J. Am. Chem. Soc.* **96**, 2208.
Huisgen, R., and Gasteiger, J. (1972). *Tetrahedron Lett.* p. 3661.
Isaev, I. S., Mamatyuk, V. I., Egorova, T. G., Kuzubova, L. I., and Koptyug, V. A. (1969). *Izv. Akad. Nauk SSSR, Ser. Khim*, p. 2089.
Itô, S., and Itoh, I. (1971). *Tetrahedron Lett.* p. 2969.
Itô, S. Mori, A., Saito, I., Sakan, K., Ishiyama, H., and Sasaki, K. (1973). *Tetrahedron Lett.* p. 2731.
Johnstone, R. A. W., and Ward, S. D. (1968). *J. Chem. Soc. C* p. 1805.
Jongejan, E., van Tilborg, W. J. M., Dusseau, C. H. V., Steinberg, H., and de Boer, T. J. (1972). *Tetrahedron Lett.* p. 2359.
Jonkman, H. T., and Nieuwpoort, W. C. (1973). *Tetrahedron Lett.* p. 1671.
Jovanovich, A. P., and Lambert, J. B. (1971). *J. Chem. Soc. B* p. 1129.
Kari, R. E., and Csizmadia, I. G. (1969). *J. Chem. Phys.* **59**, 1443.
Kirmse, W., and Gruber, W. (1973). *Chem. Ber.* **106**, 1365.
Kirmse, W., and Scheidt, F. (1970). *Chem. Ber.* **103**, 3711.
Kirmse, W., Gruber, W., and Krist, J. (1973). *Chem. Ber.* **106**, 1376.
Kispert, L. D., Engelman, C., Dyas, C., and Pittman, C. U., Jr. (1971). *J. Am. Chem. Soc.* **93**, 6948 (and references therein).
Klopman, G. (1969). *J. Am. Chem. Soc.* **91**, 89.
Kohlmer, H., Smith, H. O., and Schleyer, P. v. R. (1973). *J. Am. Chem. Soc.* **95**, 5834.
Kollmar, H., and Smith, H. O. (1972). *Angew. Chem., Int. Ed. Engl.* **11**, 635.
Koptyug, V. A., Kuzubova, L. I., Isaev, I. S., and Mamatyuk, V. I. (1970). *Zh. Org. Khim.* **6**, 2258.
Koptyug, V. A., Shleider, I. A., and Isaev, I. S. (1971). *Zh. Org. Khim.* **7**, 852.
Kurland, D. (1967). Ph.D. Dissertation, Harvard University, Cambridge, Massachusetts.
Kusuda, K., West, R., and Mallikarjuna Rao, V. N. (1971). *J. Am. Chem. Soc.* **93**, 3627.
Kutzelnigg, W. (1967). *Tetrahedron Lett.* p. 4965.
Lai, K.-H., and Miller, B. (1970). *Chem. Commun.* p. 1072.
Lambert, J. B., Koeng, F. R., and Jovanovich, A. P. (1972). *J. Org. Chem.* **37**, 374.
Lambert, J. B., Jovanovich, A. P., Hamersma, J. W., Koeng, F. R., and Oliver, S. S. (1973). *J. Am. Chem. Soc.* **95**, 1570.
Lathan, W. A., Hehre, W. J., and Pople, J. A. (1971). *J. Am. Chem. Soc.* **93**, 808.
Ledlie, D. B. (1972). *J. Org. Chem.* **37**, 1439.
Lehr, R. (1968). Ph.D. Dissertation, Harvard University, Cambridge, Massachusetts.
Liberles, A., Greenberg, A., and Lesk, A. (1972). *J. Am. Chem. Soc.* **94**, 8684.
Lillien, I., and Handloser, L. (1969a). *J. Org. Chem.* **34**, 3058.
Lillien, I., and Handloser, L. (1969b). *Tetrahedron Lett.* p. 1035.
Lodder, A. E., Buck, H. M., and Oosterhoff, L. J. (1970). *Recl. Trav. Chim. Pays-Bas* **89**, 1229.
Lossing, F. P., and Semeliuk, G. P. (1970). *Can. J. Chem.* **48**, 955.
Lustgarten, R. K., Brookhart, M., and Winstein, S. (1968). *J. Am. Chem. Soc.* **90**, 7364.
McAdoo, D. J., McLafferty, F. W., and Bente, P. F., III. (1972). *J. Am. Chem. Soc.* **94**, 2027.
McDonald, R. N., and Steppel, R. N. (1969). *J. Am. Chem. Soc.* **91**, 782.

McDonald, R. N., and Tabor, T. E. (1967). *J. Am. Chem. Soc.* **89**, 6573.
McLafferty, F. W., and Winkler, J. (1974). *J. Am. Chem. Soc.* **96**, 5182.
Majerski, Z., and Schleyer, P. v. R. (1971). *J. Am. Chem. Soc.* **93**, 665 (and references therein).
Masamune, S., Takada, S., Nakatsuka, N., Vukov, R., and Cain, E. N. (1969). *J. Am. Chem. Soc.* **91**, 4322.
Masamune, S., Vukov, R., Bennett, M. J., and Purdham, J. (1972a). *J. Am. Chem. Soc.* **94**, 8239.
Masamune, S., Sakai, M., and Ona, H. (1972b). *J. Am. Chem. Soc.* **94**, 8955.
Masamune, S., Sakai, M., Ona, H., and Jones, A. J. (1972c). *J. Am. Chem. Soc.* **94**, 8956.
Masamune, S., Sakai, M., Kemp-Jones, A. V., and Nakashima, T. (1974a). *Can. J. Chem.* **52**, 855.
Masamune, S., Sakai, M., and Kemp-Jones, A. V. (1974b). *Can. J. Chem.* **52**, 858.
Merlet, P., Peyerimhoff, S. D., Buenker, R. J., and Shih, S. (1974). *J. Am. Chem. Soc.* **96**, 959.
Meyerson, S., and Leitch, L. C. (1971). *J. Am. Chem. Soc.* **93**, 2244.
Miller, B. (1969). *J. Am. Chem. Soc.* **91**, 2170.
Miller, B. (1970a). *J. Am. Chem. Soc.* **92**, 432.
Miller, B. (1970b). *J. Am. Chem. Soc.* **92**, 6246.
Miller, B. (1970c). *J. Am. Chem. Soc.* **92**, 6252.
Monti, H., and Bertrand, M. (1973). *Tetrahedron* **29**, 1565.
Nickon, A., and Weglein, R. C. (1975). *J. Am. Chem. Soc.* **97**, 1271.
Noyori, R., Makino, S., and Takaya, H. (1971). *J. Am. Chem. Soc.* **93**, 1272.
Noyori, R., Ohnishi, Y., and Katô, M. (1972). *J. Am. Chem. Soc.* **94**, 5105.
Noyori, R., Yokoyama, K., and Hayakawa, Y. (1973). *J. Am. Chem. Soc.* **95**, 2722.
Noyori, R., Ohnishi, Y., and Katô, M. (1975). *J. Am. Chem. Soc.* **97**, 928.
Ogura, K., and Matsura, T. (1970). *Bull. Chem. Soc. Jpn.* **43**, 2891.
Ohara, O., Aso, C., and Kunitake, T. (1973). *Polym. J.* **5**, 49.
Ohloff, G., Schulte-Elte, K. H., and Demole, E. (1971). *Helv. Chim. Acta* **54**, 2813.
Okamoto, T., and Shudo, K. (1973). *Tetrahedron Lett.* p. 4533.
Olah, G. A. (1972). *J. Am. Chem. Soc.* **94**, 808.
Olah, G. A. (1975). "Halonium Ions in Organic Chemistry." Wiley (Interscience), New York.
Olah, G. A., and Liang, G. (1975). *J. Am. Chem. Soc.* **97**, 2236.
Olah, G. A., and Porter, R. D. (1970). *J. Am. Chem. Soc.* **92**, 7627.
Olah, G. A., and Porter, R. D. (1971). *J. Am. Chem. Soc.* **93**, 6877.
Olah, G. A., Commeyras, A., and Lui, C. Y. (1968). *J. Am. Chem. Soc.* **90**, 3882.
Olah, G. A., White, A. M., DeMember, J. R., Commeyras, A., and Lui, C. Y. (1970). *J. Am. Chem. Soc.* **92**, 4627.
Olah, G. A., DeMember, J. R., Lui, C. Y., and Porter, R. D. (1971a). *J. Am. Chem. Soc.* **93**, 1442.
Olah, G. A., Clifford, P. R., Halpern, Y., and Johanson, R. G. (1971b). *J. Am. Chem. Soc.* **93**, 4219.
Olah, G. A., Mateescu, G. D., and Riemenschneider, J. L. (1972). *J. Am. Chem. Soc.* **94**, 2529.
Olah, G. A., Liang, G., Mateescu, G. D., and Riemenschneider, J. L. (1973). *J. Am. Chem. Soc.* **95**, 8698.
Olah, G. A., Schilling, P., Westerman, P. W., and Lin, H. C. (1974a). *J. Am. Chem. Soc.* **96**, 3581.
Olah, G. A., Staral, J. S., and Liang, G. (1974b). *J. Am. Chem. Soc.* **96**, 6233.

Olah, G. A., Liang, G., and Staral, J. S. (1974c). *J. Am. Chem. Soc.* **96**, 8112.
Olsen, J. F., Kang, S., and Burnelle, L. (1971). *J. Mol. Struct.* **9**, 305.
Ostlund, N. S., and Bowen, M. F. (1975). *Theoret. Chem. Acta* **40**, 1880.
Östman, B., and Sjöberg, S. (1970). *Tetrahedron Lett.* p. 3137.
Paquette, L. A., and Broadhurst, M. J. (1973). *J. Org. Chem.* **38**, 1893.
Paquette, L. A., and Dunkin, I. R. (1975). *J. Am. Chem. Soc.* **97**, 2243.
Paquette, L. A., and Henzel, K. A. (1973). *J. Am. Chem. Soc.* **95**, 2726.
Parham, W. E. (1968). *Rec. Chem. Prog.* **29**, 3.
Parham, W. E., and Yong, K. S. (1970). *J. Org. Chem.* **35**, 683.
Parham, W. E., Johnson, D. R., Hughes, C. T., Meilahn, M. K., and Rinehart, J. K. (1970). *J. Org. Chem.* **35**, 1048.
Parrington, B., and Childs, R. F. (1970). *Chem. Commun.* p. 1581.
Pasto, D. J., Chen, A. F.-T., Ciurdaru, G., and Paquette, L. A. (1973). *J. Org. Chem.* **33**, 1015.
Pearson, R. G. (1971). *Acc. Chem. Res.* **4**, 152.
Peter-Katalinic, J., Zsindely, J., and Schmid, H. (1973). *Helv. Chim. Acta* **56**, 2796.
Peyerimhoff, S. D., and Buenker, R. J. (1969). *J. Chem. Phys.* **51**, 2528.
Pfeiffer, G. V., and Jewett, J. G. (1970). *J. Am. Chem. Soc.* **92**, 2143.
Pittman, C. U., Jr., and Miller, W. G. (1973). *J. Am. Chem. Soc.* **95**, 2947.
Potter, C. D., Friedrich, E. C., and Winstein, S. (1970). *J. Am. Chem. Soc.* **92**, 4274.
Poulter, C. D., and Winstein, S. (1972). *J. Am. Chem. Soc.* **94**, 2297.
Radom, L., Pople, J. A., Buss, V., and Schleyer, P. v. R. (1972a). *J. Am. Chem. Soc.* **94**, 311.
Radom, L., Pople, J. A., and Schleyer, P. v. R. (1972b). *J. Am. Chem. Soc.* **94**, 5935.
Radom, L., Hariharan, P. C., Pople, J. A., and Schleyer, P. v. R. (1973a). *J. Am. Chem. Soc.* **95**, 6531.
Radom, L., Pople, J. A., and Schleyer, P. v. R. (1973b). *J. Am. Chem. Soc.* **95**, 8193.
Rajeswari, K. (1971). Ph.D. Thesis, University of Calgary, Calgary, Alberta, Canada (unpublished results).
Rajeswari, K., and Sorensen, T. S. (1971). *J. Am. Chem. Soc.* **93**, 4222.
Rajeswari, K., and Sorensen, T. S. (1973). *J. Am. Chem. Soc.* **95**, 1239.
Reese, C. B., and Shaw, A. (1970a). *Chem. Commun.* p. 1365.
Reese, C. B., and Shaw, A. (1970b). *J. Am. Chem. Soc.* **92**, 2566.
Reese, C. B., and Strebles, M. R. D. (1972). *Tetrahedron Lett.* p. 4427.
Richey, H. G., Jr. (1972). *Carbonium Ions* **3**, 1201–1294.
Richey, H. G., Jr., and Buckley, N. C. (1963). *J. Am. Chem. Soc.* **85**, 3057.
Rodewald, L. B., and Lewis, M. C. (1971). *Tetrahedron* **27**, 5273.
Rosenberg, J. L., Mahler, J. E., and Pettit, R. (1962), *J. Am. Chem. Soc.* **84**, 2842.
Sakai, M., Diaz, A., and Winstein, S. (1970). *J. Am. Chem. Soc.* **92**, 4452.
Salem, L. (1970). *J. Am. Chem. Soc.* **90**, 543.
Santelli, M., and Bertrand, M. (1974). *Tetrahedron* **30**, 257.
Saunders, M., and Stofko, J. J., Jr. (1973). *J. Amer. Chem. Soc.* **95**, 252.
Saunders, M., Vogel, P., Hagen, E. L., and Rosenfeld, J. (1973). *Acc. Chem. Res.* **6**, 53.
Schipper, P., Driessen, P. B. J., de Haan, J. W., and Buck, H. M. (1974). *J. Am. Chem. Soc.* **96**, 4706.
Schleyer, P. v. R., Van Dine, G. W., Schöllkopf, U., and Paust, J. (1966). *J. Am. Chem. Soc.* **88**, 2868.
Schleyer, P. v. R. (1967). *Abst., 20th Nat. Org. Symp. Am. Chem. Soc., 1967* p. 8.
Schleyer, P. v. R., Su, T. M., Saunders, M., and Rosenfeld, J. C. (1969). *J. Am. Chem. Soc.* **91**, 5174.
Schmid, M., Hansen, H. J., and Schmid, H. (1971). *Helv. Chim. Acta* **54**, 937.

Schmidt, R. R. (1973). *Angew. Chem., Int. Ed. Engl.* **12**, 212.
Schöllkopf, U. (1968). *Angew. Chem., Int. Ed. Engl.* **7**, 588.
Schöllkopf, U., Fellenberger, K., Patsch, M., Schleyer, P. v. R., Su, T., and Van Dine, G. W. (1967). *Tetrahedron Lett.* p. 3639.
Schöllkopf, U., Ruban, E., Tonne, P., and Riedel, K. (1970). *Tetrahedron Lett.* p. 5037.
Schuster, D. I., and Liu, K.-C. (1971). *J. Am. Chem. Soc.* **93**, 6711.
Seyferth, D., Jula, T. F., Mueller, D. C., Mazerolles, P., Manuel, G., and Thoumas, F. (1970). *J. Am. Chem. Soc.* **92**, 657.
Shanshal, M. (1972). *Z. Naturforsch., Teil A* **27**, 1665.
Shono, T., Fujita, K., and Kumai, S. (1973). *Tetrahedron Lett.* p. 3125.
Shoppee, C. W., and Cooke, B. J. A. (1972). *J. Chem. Soc., Perkin Trans. 1* p. 2271.
Shoppee, C. W., and Lack, R. E. (1969). *J. Chem. Soc. C* p. 1346.
Smirnov-Zamkov, I. V. (1952). *Dokl. Akad. Nauk SSSR* **83**, 869.
Sorensen, T. S. (1964). *Can. J. Chem.* **42**, 2781.
Sorensen, T. S. (1965a). *Can. J. Chem.* **43**, 2746.
Sorensen, T. S. (1965b). *J. Am. Chem. Soc.* **87**, 5075.
Stohrer, W.-D., and Hoffmann, R. (1972). *J. Am. Chem. Soc.* **94**, 1661.
Story, P. R., and Fahrenholtz, S. R. (1964). *J. Am. Chem. Soc.* **86**, 527.
Story, P. R., and Fahrenholtz, S. R. (1966). *J. Am. Chem. Soc.* **88**, 374.
Sustmann, R., Williams, J. E., Dewar, M. J. S., Allen, L. C., and Schleyer, P. v. R. (1969). *J. Am. Chem. Soc.* **91**, 5350.
Tanida, H., and Tsushima, T. (1970). *J. Am. Chem. Soc.* **92**, 3397.
Tanida, H., Tsuji, T., and Irie, T. (1967). *J. Am. Chem. Soc.* **89**, 1953.
Tee, O. S., and Yates, K. (1972). *J. Am. Chem. Soc.* **94**, 3074.
Traylor, T. G., Hanstein, W., Berwin, H. J., Clinton, N. A., and Brown, R. S. (1971). *J. Am. Chem. Soc.* **93**, 5715.
Trindle, C., and Sinanoğlu, O. (1969). *J. Am. Chem. Soc.* **91**, 4054.
Trost, B. M., and Keeley, D. E. (1974). *J. Am. Chem. Soc.* **96**, 1252 (and references therein).
Tufariello, J. J., Bayer, A. C., and Spadaro, J. J., Jr. (1972). *Tetrahedron Lett.* p. 363.
Turbitt, T. D., and Watts, W. E. (1974). *J. Chem. Soc., Perkin Trans. 2* p. 195.
Turro, N. J. (1969). *Acc. Chem. Res.* **2**, 25.
Turro, N. J., Edelson, S. S., Williams, J. R., Darling, T. R., and Hammond, W. B. (1969). *J. Am. Chem. Soc.* **91**, 2283.
van der Hout-Lodder, A. E., Buck, H. M., and de Haan, J. W. (1972). *Recl. Trav. Chim. Pays-Bas* **91**, 164.
van der Hout-Lodder, A. E., Buck, H. M., de Haan, J. W., and van de Ven, L. J. M. (1973). *Recl. Trav. Chim. Pays-Bas* **92**, 1040.
van Tamelen, E. E., and Cole, T. M., Jr. (1971). *J. Am. Chem. Soc.* **93**, 6158.
van Tamelen, E. E., Cole, T. M., Jr., Greeley, R. H., and Schumacher, H. (1968). *J. Am. Chem. Soc.* **90**, 1372.
van Tamelen, E. E., Greeley, R. H., and Schumacher, H. (1971). *J. Am. Chem. Soc.* **93**, 6151.
Vogel, P., Saunders, M., Hasty, N. M., Jr., and Berson, J. A. (1971). *J. Am. Chem. Soc.* **93**, 1557.
Wagner, H.-U., and Gompper, R. (1971a). *Tetrahedron Lett.* p. 4061.
Wagner, H.-U., and Gompper, R. (1971b). *Tetrahedron Lett.* p. 4065.
Weiss, R. G. (1971). *Tetrahedron* **27**, 271.
Whalen, D., Găsić, M., Johnson, B., Jones, H., and Winstein, S. (1967). *J. Am. Chem. Soc.* **89**, 6384.
Whitham, G. H., and Wright, M. (1967). *Chem. Commun.* p. 294.

Wiberg, K. B., and Fenoglio, R. A. (1963). *Tetrahedron Lett.* p. 1273.
Wiberg, K. B., and Nakahira, T. (1971). *J. Am. Chem. Soc.* **93**, 5193.
Wiberg, K. B., and Pfeiffer, J. G. (1970). *J. Am. Chem. Soc.* **92**, 553.
Wiberg, K. B., and Szeimies, G. (1970). *J. Am. Chem. Soc.* **92**, 571.
Wiberg, K. B., Williams, V. Z., Jr., and Friedrich, L. E. (1970a). *J. Am. Chem. Soc.* **92**, 564.
Wiberg, K. B., Fenoglio, R. A., Williams, V. Z., Jr., and Ubersax, R. W. (1970b). *J. Am. Chem. Soc.* **92**, 568.
Widmer, U., Zsindely, J., Hansen, H. J., and Schmid, H. (1973). *Helv. Chim. Acta* **56**, 75.
Wilcox, C. F., Loew, L. M., and Hoffmann, R. (1973). *J. Am. Chem. Soc.* **95**, 8192.
Williams, J. E., Jr., Buss, V., Allen, L. C., Schleyer, P. v. R., Lathan, W. A., Hehre, W. J., and Pople, J. A. (1970). *J. Am. Chem. Soc.* **92**, 2141.
Williams, J. E., Jr., Buss. V., and Allen, L. C. (1971). *J. Am. Chem. Soc.* **93**, 6867.
Wilt, J. W., and Malloy, T. P. (1970). *J. Am. Chem. Soc.* **92**, 4747.
Winstein, S. (1969). *Qt. Rev., Chem. Soc.* **23**, 141.
Winstein, S., and Shatavsky, M. (1956). *J. Am. Chem. Soc.* **78**, 592.
Winstein, S., Sonnenberg, J., and de Vries, L. (1959). *J. Am. Chem. Soc.* **81**, 6523.
Winstein, S., Lewin, A. L., and Pande, K. C. (1963). *J. Am. Chem. Soc.* **85**, 2324.
Winstein, S., Kaesz, H. D., Kreiter, C. G., and Friedrich, E. C. (1965). *J. Am. Chem. Soc.* **87**, 3267.
Winstein, S., Kreiter, C. G., and Brauman, J. F. (1966). *J. Amer. Chem. Soc.* **88**, 2047.
Woodward, R. B., and Hoffmann, R. (1965). *J. Am. Chem. Soc.* **87**, 395.
Woodward, R. B., and Hoffmann, R. (1969). *Angew. Chem., Int. Ed. Engl.* **8**, 781.
Woodward, R. B., and Hoffmann, R. (1970). "The Conservation of Orbital Symmetry." Academic Press, New York.
Yamaguchi, K., and Fueno, T. (1973). *Chem. Phys. Lett.* **22**, 471.
Zimmerman, H. E., and Crumrine, D. S. (1968). *J. Am. Chem. Soc.* **90**, 5612.
Zurawski, B., Ahlrichs, R., and Kutzelnigg, W. (1973). *Chem. Phys. Lett.* **21**, 309.

2

Pericyclic Reactions of Cumulenes*

LÉON GHOSEZ and MARTIN J. O'DONNELL

I. Introduction 79
II. Electronic Structure of Cumulenes 80
III. [2 + 2] Cycloadditions of Ketenes to Alkenes 85
   A. Theoretical Analysis 85
   B. Periselectivity and Regioselectivity of the Cycloadditions . . 89
   C. Stereochemistry 93
   D. Influence of Substituents, Solvents, and Steric Factors on the Ketenophilic Reactivity 98
   E. Secondary Deuterium Isotope Effects 102
   F. Cycloreversions 103
IV. Ketene Dimerization 104
V. Ketene–Allene Cycloadditions 104
VI. [2 + 2] Cycloadditions of Allenes 109
   A. General 109
   B. Allene Dimerizations 109
   C. Cycloaddition of Allenes to Olefins 113
   D. Degenerate Rearrangements 115
VII. [2 + 2] Cycloadditions of Ketenimines 117
VIII. [2 + 2] Cycloadditions of Keteniminium Salts 118
   A. General 118
   B. Reactions with Olefins and Acetylenes 120
   C. Reactions with Dienes 123
IX. [2 + 4] Cycloadditions of Cumulenes 126
   A. Structure and Dienophilic Reactivity 126
   B. Intermolecular [2 + 4] Cycloadditions 127
   C. Intramolecular [2 + 4] Cycloadditions 130
X. Cycloadditions of Vinylcumulenes 130
   References 135

I. Introduction

More than fifty years have elapsed since H. Staudinger (1912) discovered the ability of ketenes to react with olefins to yield cyclobutanones. One of his most intriguing observations was the formation of a four-membered ring to the exclusion of a six-membered Diels–Alder adduct when diphenylketene

* This chapter is dedicated to Professor R. B. Woodward on the occasion of his 60th birthday.

was reacted with cyclopentadiene. Yet, it was not until quite recently that this tendency of ketenes to form four-membered rings by cycloaddition was rationalized by Woodward and Hoffmann (1969) in an analysis based on the Principle of the Conservation of Orbital Symmetry. This new light shed on old facts stimulated further experimental and theoretical studies on [2 + 2] cycloadditions of ketenes and other cumulenes which now make possible a better understanding of their mechanisms as well as their use as a predictable synthetic tool.

The participation of a cumulene in a cycloaddition is obviously not limited to [2 + 2] cycloadditions. Several reactions are known involving the π bonds of cumulenes as dipolarophilic or dienophilic partners in [4 + 2] cycloadditions. Moreover, if the cumulene is conjugated with a double bond, it forms a reactive diene which can undergo a Diels–Alder cycloaddition. Cycloadditions of cumulenes often provide simple routes to cyclic compounds otherwise difficult to prepare and are, therefore, of considerable synthetic interest.

A review of the pericyclic reactions of all cumulenes is beyond the scope of an article of this length. Our aim is to offer a critical discussion of selected cumulenes as cycloaddition partners with respect to their electronic structure. The topics to be described have been selected according to personal taste. Rather than describing only well-established mechanistic pictures we have preferred to raise controversial points which, we hope, will stimulate new studies. The review is by no means comprehensive even within this limited scope. Thus, we shall discuss in detail the ketene cycloadditions which have been subjected to extensive studies and offer a broad spectrum of mechanistic possibilities available in cycloadditions. We hope to show how variations within the cumulene structure, namely, in going from ketene to allene, ketenimine and keteniminium, causes changes in their behavior in pericyclic reactions. The chemistry of these cumulenes has already been reviewed in some detail (Hanford and Sauer, 1946; Lacey, 1960; Roberts and Sharts, 1962; Wilson and Goldhamer, 1963; Fisher, 1964; Huisgen et al., 1964; Lacey, 1964; Petrov and Fedorova, 1964; Griesbaum, 1966; Mavrov and Kucherov, 1967; Taylor, 1967; Ulrich, 1967; Bertrand and Le Gras, 1968; Borrmann, 1968; Gompper, 1969; Luknitskii and Vovsi, 1969; Baldwin and Fleming, 1970; Brady, 1971; Krow, 1971; Gilchrist and Storr, 1972; Okamoto, 1972; Ghosez and Marchand-Brynaert, 1976).

II. Electronic Structure of Cumulenes

The structural characteristic of cumulenes is the presence of π-electron systems adjacent to each other. Allenes are the simplest representatives with an all-carbon skeleton. They contain two π-electron systems whose

TABLE I. A Classification of Cumulenes according to the
Number of Electrons in the Orthogonal π Systems

$_n 2 \perp _n 2$	$_n 2 \perp _n 4$	$_n 4 \perp _n 4$
$>\overset{\oplus}{N}=C=\overset{\oplus}{N}<$	$>\overset{\oplus}{N}=C=O$	$O=C=O$
$>C=C=\overset{\oplus}{N}<$	$>\overset{\oplus}{N}=C=N-$	$-N=C=O$
$>C=C=C<$	$>C=C=O$	$-N=C=N-$
	$>C=C=N-$	

nodal planes are perpendicular. Replacement of one of the C=C< bonds by a C=$\overset{\oplus}{N}$< leads to a charged heterocumulene, the keteniminium system, which is isoelectronic with allene. These cumulenes form a first group in which each π-electron system contains two electrons (Table I).

The dication formed by replacing both C=C< bonds in allene by C=$\overset{\oplus}{N}$< bonds is still an unknown representative of this first group of cumulenes. A somewhat different situation is encountered when a C=C< bond in allene is replaced by a C=N— or C=O group. The π-electron systems of the C=N— or C=O bonds still contain two electrons, but the orthogonal π electrons of the C=C bond interact with a lone pair of the heteroatom so that the resulting π systems contain four electrons and are isoelectronic to an enamine and an enolate anion, respectively. The cations derived by replacement of C=C< by C=$\overset{\oplus}{N}$< in ketenes or ketenimines also belong to this second class of heterocumulenes (Table I). Finally, carbon dioxide, isocyanates, and carbodiimides are well-known representatives of a third class of cumulenes characterized by two orthogonal π systems containing four electrons each.

The differences between each group are apparent when considering the structures from the point of view of valence bond theory. Allene is essentially described by a single structure whereas the keteniminium ion contains contributions from two resonance structures of which the iminium form dominates.

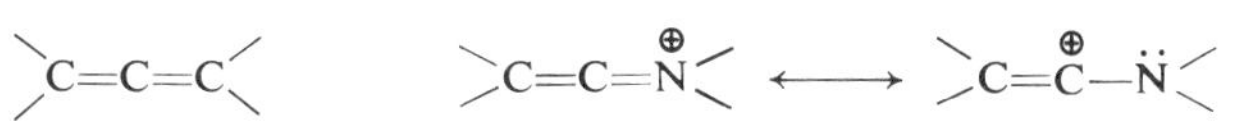

$$>C=C=C< \qquad >C=C=\overset{\oplus}{N}< \longleftrightarrow >C=\overset{\oplus}{C}-\ddot{N}<$$

On the other hand, cumulenes of the second group contain significant contributions of three resonance structures as shown, for example, in the case of ketene.

$$\text{\textbackslash}C{=}C{=}\overset{..}{O} \longleftrightarrow \text{\textbackslash}C{=}\overset{\oplus}{C}{-}\overline{\underline{O}}|^{\ominus} \longleftrightarrow \text{\textbackslash}\overline{C}^{\ominus}{-}C{\equiv}\overset{\oplus}{O}|$$

The structures of allenes and ketenes have been subjected to several theoretical studies (Pullmann and Berthier, 1949; Orville-Thomas and Jones, 1960; Pople and Gordon, 1967; André et al., 1969; Buenker, 1969; Del Bene and Jaffé, 1969; Del Bene, 1969; Hehre et al., 1969; Letcher et al., 1969; Hopkinson, 1973; van Dongen et al., 1973; Weimann and Christoffersen, 1973; Hopkinson and Csizmadia, 1974; Houk et al., 1974). Recently, the electronic structures of ketenimine and keteniminium ion have been examined theoretically (*ab initio* calculations, STO-3G basis set) and compared with ketene and allene (E. Sonveaux, L. Ghosez, J. M. André, and A. Delhalle, unpublished results, 1975). The main results of a complete geometry search (assuming a colinear C(2)–C(1)–X arrangement and hydrogens on C(2) and X in orthogonal planes) are listed in Fig. 1.

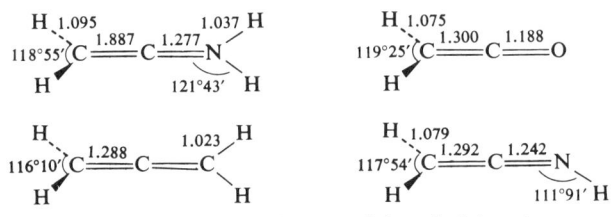

Fig. 1. Theoretical structures of ketenimine and keteniminium ions compared with ketene and allene.

These theoretical results compare favorably with the experimental values for ketene (Johnson and Strandberg, 1952; Arendale and Fletcher, 1953; Cox et al., 1959) and allene (Stoicheff, 1955; Almenningen et al., 1959; Eaton and Thompson, 1959).

The relative contribution of the different valence bond structures, discussed above, is reflected in the net p orbital populations shown in Fig. 2. In allene the π bonds are only weakly polarized in the direction $C(2)^{\delta-}{-}C(1)^{\delta+}$. As expected the $\pi_{C=X}$ bonds in keteniminium, ketene and ketenimine are strongly polarized toward the heteroatoms. In keteniminium this polarization causes a high electron deficiency in the C_{1p} orbital which induces a displacement of charge in the orthogonal $\pi_{C=C}$ system in the direction $C(2)^{\delta+}{-}C(1)^{\delta-}$ opposite that observed for isoelectronic allene. The electronic population of the C(2)–C(1) bond in ketene and ketenimine clearly shows that the heteroatom behaves as a π donor and, thus, that resonance structures where the C(2) atom is negatively charged, contribute significantly to the ground state structure of these heterocumulenes. Interestingly, the oxygen atom behaves as a better donor than does the NH group.

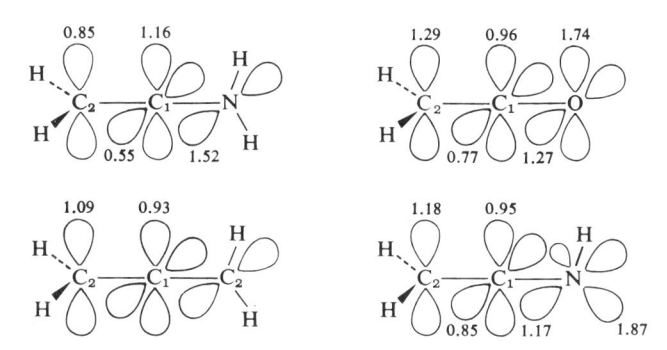

Fig. 2. Orbital diagrams for ketenimine, ketene, keteniminium ions, and allene.

The results of these computations agree very well with several experimentally measured properties of these cumulenes such as dipole moments (Lawrence and Strandberg, 1951; Johnson and Strandberg, 1952), ^{1}H and ^{13}C magnetic resonance spectra (Allred et al., 1965; Weiss and Flygare, 1966; Steur et al., 1971; Crandall and Sajka, 1972; Reilly and Krow, 1972; Firl and Runge, 1973; Olah and Westerman, 1973; van Dongen et al., 1973; Grishin et al., 1974; Firl and Runge, 1974; A. M. Hesbain-Frisque and L. Ghosez, unpublished results, 1974), and infrared spectra (Pitzer and Strickler, 1964; Johns et al., 1972; Sabin and Kim, 1972).

The energies and coefficients of the frontier molecular orbitals, which are expected to play an important role in the chemical reactivity of cumulenes, have been determined (Houk et al., 1974; E. Sonveaux, L. Ghosez, J. M. André, and A. Delhalle, unpublished results, 1975) and are shown in Fig. 3.

These studies of the ground state structures of cumulenes provide insight into their reactivity. Thus, the double bonds of allene are expected to behave in a somewhat similar fashion as isolated olefinic double bonds. Experimentally it is observed that allene is only slightly more reactive than simple olefins in cycloadditions, though it does show a greater tendency to dimerize.

The keteniminium ion behaves as a very strongly electrophilic molecule as a result of the very low energy of the lowest unoccupied molecular orbital (LUMO). This enhanced electrophilicity even compared with simpler iminium ions has been experimentally demonstrated by the quantitative capture of keteniminium ions by either chloride or iodide ion to form the corresponding α-haloenamines (Ghosez et al., 1969; Ghosez and Marchand-Brynaert, 1976).

(X = Cl, I)

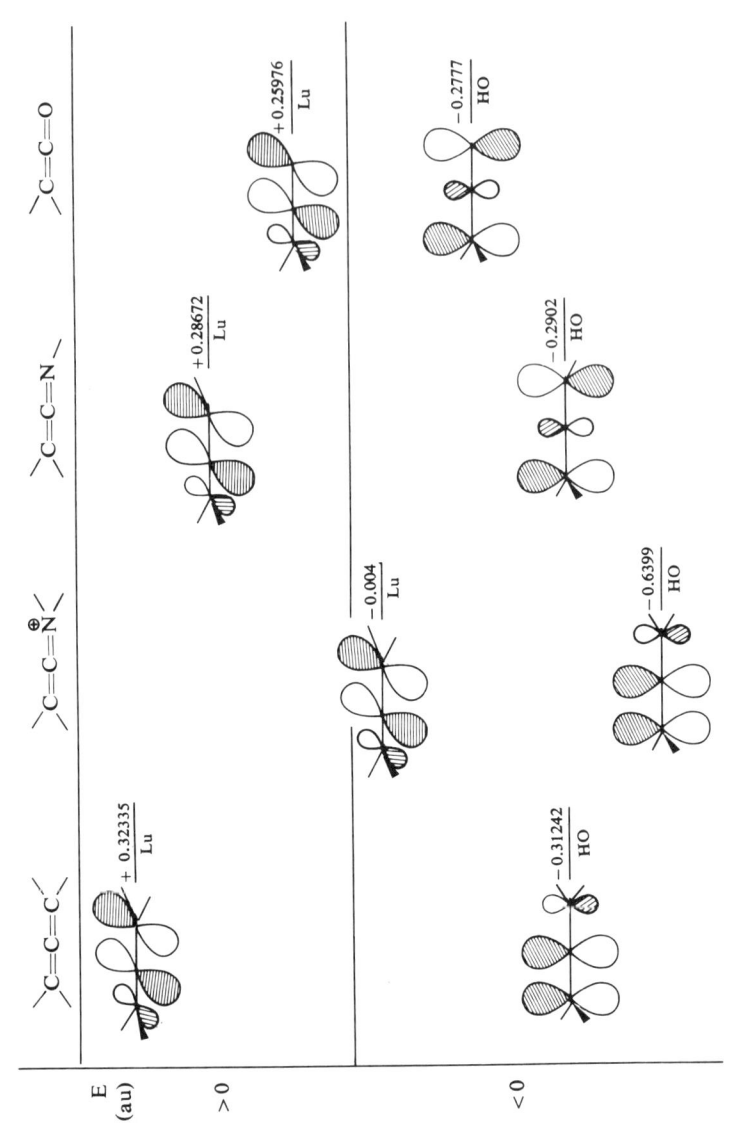

Fig. 3. Energies and coefficients of the frontier MO's.

Cumulenes of the second group such as ketenes or ketenimines can be regarded as bifunctional reagents containing orthogonal electrophilic and nucleophilic systems. The high electrophilicity of ketene results from the

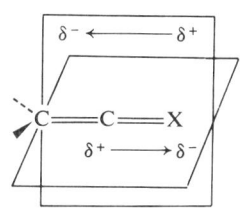

low-lying LUMO ($\pi^*_{C=O}$) which will readily accept electrons from a nucleophilic reagent. Moreover, the relatively high-lying highest occupied molecular orbital (HOMO) is expected to react readily with an appropriate ketenophile LUMO. As we shall discuss in the following sections, it is precisely this combination of electrophilic character at C(1) and nucleophilic character at C(2) which is responsible for the high tendency of ketene to dimerize as well as for its ability to readily undergo [2 + 2] cycloadditions to a variety of unsaturated substrates. This 1,2-dipolar character is less pronounced in ketenimines as shown by the higher energy of the LUMO and the lower energy of the HOMO as compared with the corresponding ketenes (E. Sonveaux, L. Ghosez, J. M. André, and A. Delhalle, unpublished results, 1975).

III. [2 + 2] Cycloadditions of Ketenes to Alkenes

A. Theoretical Analysis

According to the selection rules for pericyclic reactions (Woodward and Hoffmann, 1969), [2 + 2] cycloadditions are symmetry allowed when the interaction of the two electronic systems takes place in such a manner that one component is used suprafacially and the other one antarafacially. Thus, the [$_\pi 2_s + _\pi 2_a$] dimerization of two ethylenes is an allowed process whereas a [$_\pi 2_s + _\pi 2_s$] interaction is forbidden (Fig. 4). A [$_\pi 2_s + _\pi 2_a$] interaction

[$_\pi 2_s + _\pi 2_s$]   [$_\pi 2_s + _\pi 2_a$]
"Forbidden"   "Allowed"

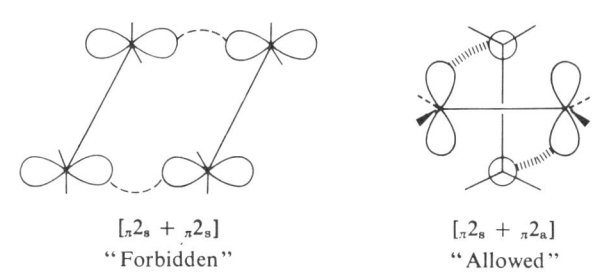

Fig. 4. "Forbidden" and "allowed" interactions.

requires an orthogonal approach of the reaction partners and leads to a transition state which should not be easily accessible since orbital overlap is poor and important steric interactions are present. Special structural features might, however, facilitate such pericyclic [2 + 2] combinations of two reactants. One such feature is twisting of the π bonds of the reactants which was suggested (Woodward and Hoffmann, 1969) to be responsible for the facile dimerization of bicyclo[4.2.2]deca-*trans*-3-*cis*-7,9-triene, a photoadduct of butadiene plus benzene (Kraft and Koltzenburgh, 1967a,b).

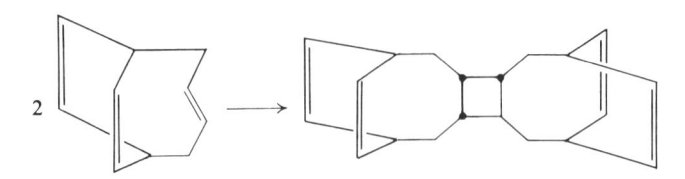

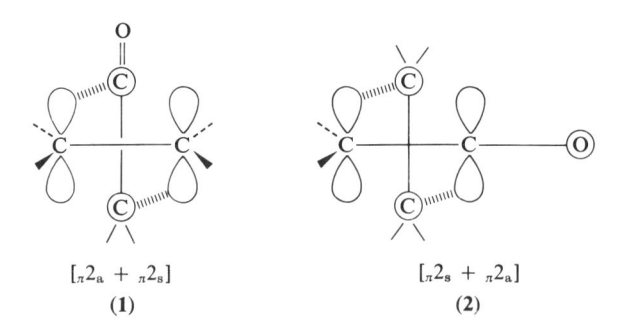

The possible transition states for a $[_\pi 2_s + {}_\pi 2_a]$ combination of a ketene with an olefin are depicted in Fig. 5. In transition state **1**, the ketene reacts in a suprafacial manner whereas, in transition state **2**, it reacts antarafacially. It can be easily seen that both transition states **1** and **2** are substantially more favorable in respect to steric hindrance than is the corresponding transition state for the cycloaddition of two olefins since the C(1) atom of the ketene is bonded to only two ligands. On the other hand, Woodward and Hoffmann (1969) suggested that in transition state **2**, the π system of the olefin is properly oriented to give a favorable secondary interaction with the low-lying $\pi^*_{C=O}$ of the ketene. This can be formulated in another way by considering the reaction as a six-electron cycloaddition, e.g., $[_\pi 2_s + {}_\pi 2_s + {}_\pi 2_s]$ (Fig. 6, **3**). However, caution should be used in applying the general rule for pericyclic reactions in this case, since the cycloaddition involves two orthogonal orbitals on the same carbon atom. In this case all the analyses are not equivalent (Fig. 6, **3**

$[_\pi 2_a + {}_\pi 2_s]$
(1)

$[_\pi 2_s + {}_\pi 2_a]$
(2)

Fig. 5. Possible transition states for a $[_\pi 2_s + {}_\pi 2_a]$ combination.

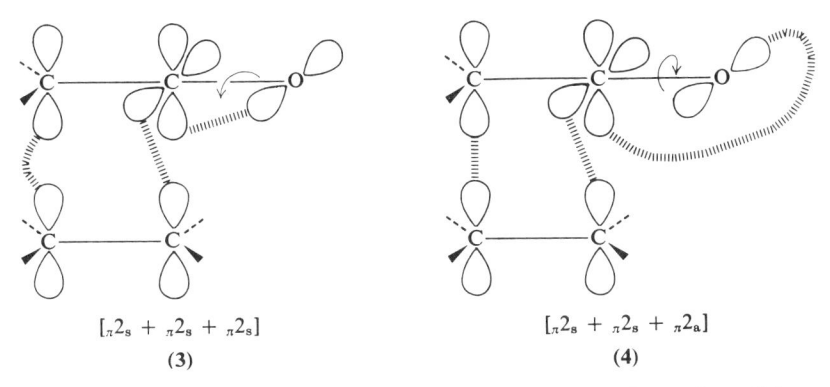

Fig. 6. Possible transition states for $[_\pi 2_s + {}_\pi 2_s + {}_\pi 2_s]$ and $[_\pi 2_s + {}_\pi 2_s + {}_\pi 2_a]$.

and **4**) (Nguyen Trong Anh, 1970), and a simple application of the rule results in ambiguities. The carbene–olefin reaction is another case where such restrictions may be applied.

The Woodward–Hoffmann rules can also be obtained in a simple way by requiring the transition state of an allowed reaction to be aromatic (Evans and Warhurst, 1938; Dewar, 1966, 1971; Zimmerman, 1966, 1969, 1971; Mulder and Oosterhoff, 1970a,b). This approach has been applied to cycloadditions of ketenes to olefins. The reaction is explained in terms of a transition state in which the *p* atomic orbital of the olefin overlaps with two orthogonal *p* atomic orbitals at C(1) and C(2), respectively, the coupling of the two systems being effected by a *p* atomic orbital of the oxygen. The signs of the atomic orbitals can be chosen so that they overlap in phase. The transition state is thus of the Hückel type (Fig. 7) and, since it involves six electrons, it is aromatic and the reaction is allowed.

All these descriptions emphasize the crucial role of the carbonyl group of the ketenes in the cycloaddition, although bonding occurs normally at the

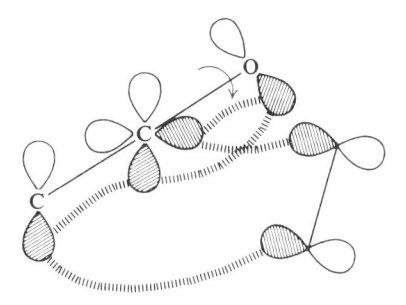

Fig. 7. Hückel type transition state.

C(2) atom rather that at the oxygen atom. A semiquantitative perturbational treatment (Sustmann *et al.*, 1972) of parallel and orthogonal complexes gives a more quantitative insight of the early stages of the cycloaddition, showing that an orthogonal interaction gains considerably from the interaction with the carbonyl π bond of the ketene. In the early stages of the reaction the dominant interaction occurs between the HOMO of the olefin and the LUMO of the ketene. The energy gap between this HOMO–LUMO pair is smaller than is the corresponding energy separation between the LUMO of the olefin and the HOMO of the ketene (Fig. 3).

This interaction should result in a polarization of the orthogonal π system and favor electron transfer from the ketene HOMO to the ketenophile LUMO so that bonding can occur at C(2) where the largest coefficient in the HOMO of the ketene is found. The electronic motions depicted in Fig. 8 stress very clearly the driving forces for the cycloaddition which result from the strong "pull–push" character of the ketene molecule (E. Sonveaux and L. Ghosez, unpublished results, 1974). It is interesting to observe that, in this respect, ketene resembles singlet carbenes which are also composed of orthogonal "pull" and "push" electronic systems and also display a high reactivity in $[2_s + 2_a]$ pericyclic cycloadditions.

Such models utilizing perturbation or configurational interaction calculations (Salem, 1968; Dewar, 1971; Fukui, 1971; Epiotis, 1972a,b, 1973a,b,c, 1974; Herndon, 1972; Herndon *et al.*, 1974; Houk, 1972, 1973a,b; Sustmann, 1971; Sustmann *et al.*, 1972; Inagaki *et al.*, 1974) have helped our understanding of regiospecificity, substituent effects and perispecificity of the reactions, but they do not indicate whether the orthogonal complex—ketene + olefin—will yield the adduct in a concerted fashion or prefer a multi-step pathway. Thus, it is expected that electron-releasing substituents on the ketenophile and electron-attracting groups on the ketene will increase the rate of formation of the bond at C(1) of the ketene as a consequence of the reduced separation between the olefin HOMO (which is raised) and the ketene LUMO (which is lowered). However, another consequence of the introduction of such substituents is an increase of the separation of the olefin LUMO

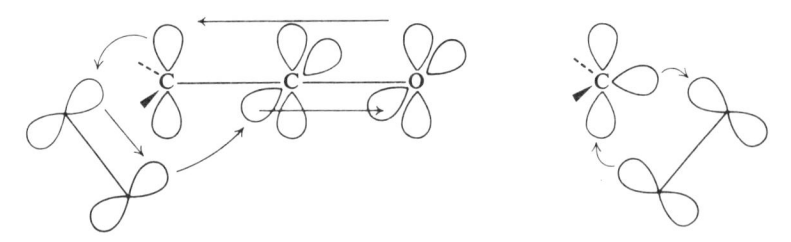

Fig. 8. Ketenes and carbenes as "push–pull" molecules.

and the ketene HOMO, and, therefore, a reduced interaction between these frontier orbitals. If this effect is sufficiently pronounced, bonding at C(2) might not occur concertedly with that at C(1), and 1,4-dipolar species (5) could be formed along the reaction path.

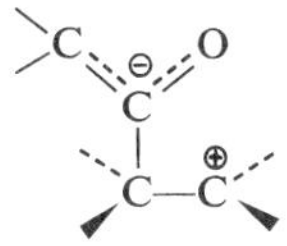

$$\begin{array}{c} \diagup C \cdots \overset{\ominus}{C} \cdots O \\ \vdots \quad | \quad \overset{\oplus}{\cdot} \\ \diagup C - C \diagdown \end{array}$$

(5)

In the absence of calculations of the total potential energy surface, one should be careful not to overuse such theoretical models to predict the details of a mechanistic pathway. The perturbational model discussed above focuses attention only on the early stages of the energy profile.

Steric effects are also expected to play an important role in the choice of a reaction path, and, in general, it is expected that steric hindrance will increase the energy of the highly organized transition state of the concerted process more than that of a stepwise reaction.

Experiments sometimes allow a distinction to be made between concerted or stepwise pathways when an intermediate can be trapped. In many cases, however, the concertedness of a process is mainly inferred from "negative" evidence, the absence of a detectable intermediate, or from indirect arguments resulting from stereochemical or kinetic studies or the analysis of kinetic isotope effects.

B. PERISELECTIVITY AND REGIOSELECTIVITY OF THE CYCLOADDITIONS

The pioneering studies of H. Staudinger (1912) have been confirmed by a vast body of experimental data showing the preference of ketenes to react with olefinic substrates across the C=C rather than the C=O double bond. This unusual capacity of ketenes to yield cyclobutanones is illustrated by the perispecific addition of butylethylketene to butadiene (Martin et al., 1965).

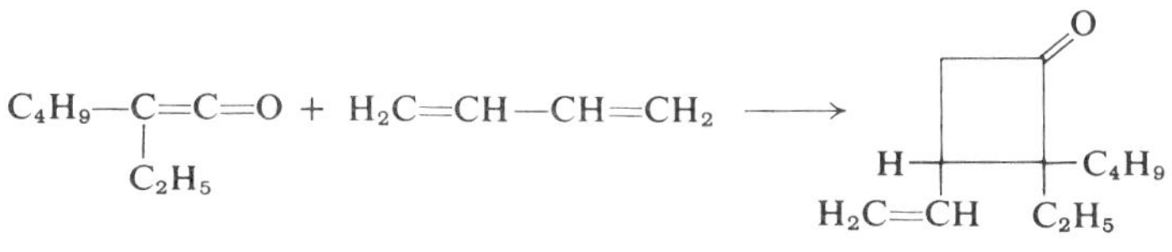

$$C_4H_9\!-\!\underset{\underset{C_2H_5}{|}}{C}\!=\!C\!=\!O \;+\; H_2C\!=\!CH\!-\!CH\!=\!CH_2 \longrightarrow$$

(cyclobutanone product with H, C$_4$H$_9$, H$_2$C=CH, C$_2$H$_5$ substituents)

The perispecificity is also observed when the diene is constrained in a cisoid conformation which is expected to favor the formation of [4 + 2]-adducts. Thus, cyclopentadiene reacts smoothly with ketenes to yield

exclusively [2 + 2]-cycloadducts. Several of these cycloadditions are of special synthetic interest.
   Stevens et al. (1965), Ghosez et al. (1966, 1971).

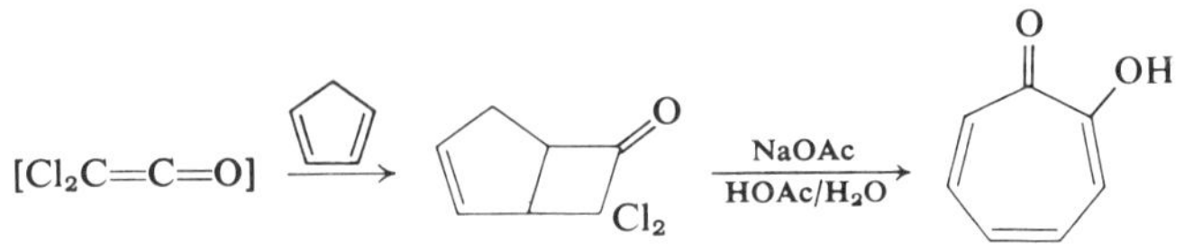

Cossement et al. (1974).

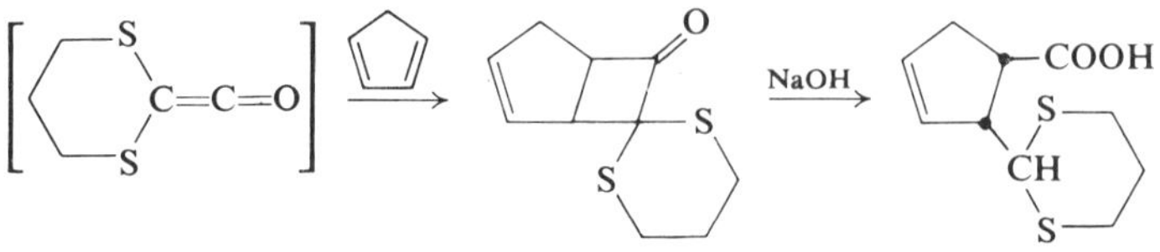

The second sequence—cycloaddition + ring opening—allows for the formal addition of two carbon chains with different functionalities to an olefinic bond, a new method of carbon chain lengthening termed "vicinal carbofunctionalization" (Ghosez et al., 1974).

Usually, the reactions of ketenes with simple alkyl-substituted olefins proceed rather slowly, and, often, oligomerization or polymerization of the ketene occurs more rapidly than the cycloaddition. Diphenyl ketene and butyl ethyl ketene both display a relatively low tendency to dimerize; accordingly, they react successfully with simple alkenes (Martin et al., 1965). Ethylene itself was found to cycloadd to diphenylketene (Huisgen and Feiler, 1969). Substitution of aryl, alkoxy, or amine groups at the olefinic

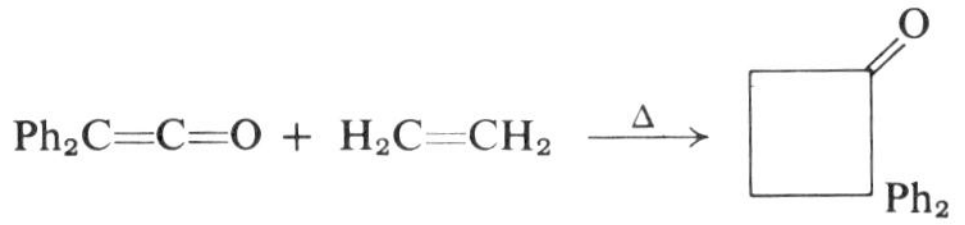

$$Ph_2C=C=O + H_2C=CH_2 \xrightarrow{\Delta}$$

double bond increases considerably the reactivity toward ketenes. In most cases, the cycloadditions were shown to be perispecific and regiospecific. The direction of addition is in agreement with predictions made on the basis of the perturbation theory: bonding always occurs between C(1) of the ketene and the C(β) of the olefin $\overline{X} - C^\alpha = C^\beta$ where the largest coefficients of the interacting frontier orbitals (LUMO of ketene and HOMO of olefin) are found. Thus, for example, with propene, styrene, or ethyl vinyl ether, as well as with a simple acetylene such as phenylacetylene, the adducts **6, 7, 8**, and **9**, have been observed (Hasek et al., 1964; Huisgen and Feiler, 1969;

Gheorgiu et al., 1973). Interestingly, the addition of dichloroketene to 2-cholestene and 4-*tert*-butylcyclohexene exhibits an unusual degree of

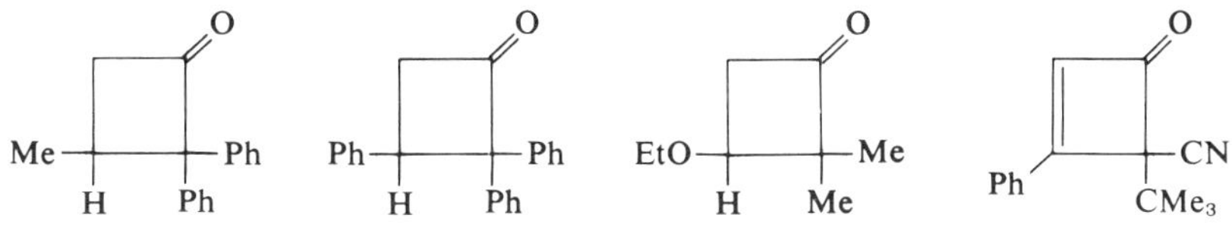

(6)   (7)   (8)   (9)

regioselectivity; namely, the newly formed bond of the carbonyl carbon prefers to be axial with respect to the six-membered ring in the chair conformation (Fletcher and Hassner, 1970; Hassner et al., 1971). The preferred formation of **10** has been easily accounted for by considering orbital overlap involved during the symmetry-allowed [$_\pi2_s + {_\pi}2_a$] process.

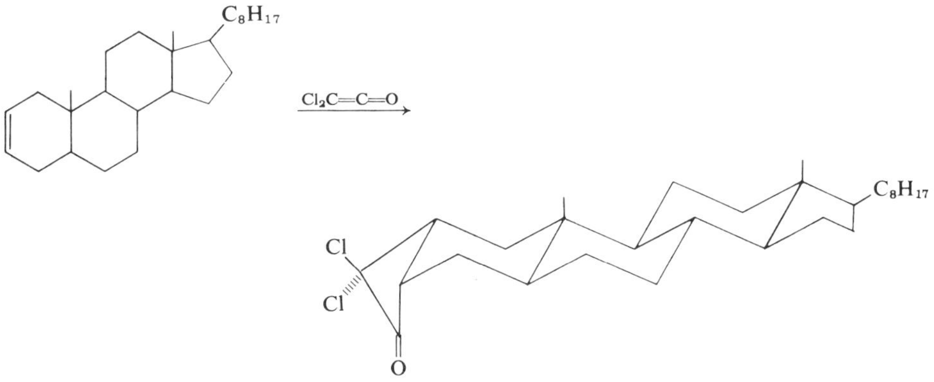

(10) (>80%)

It is of mechanistic significance that regioisomeric cyclobutanones (**11**) and (**12**) are isolated from the reactions of 2,2-dimethyl-1-methylenecyclopropane (**13**) with either dimethyl or diphenyl ketene (Maurin and Bertrand, 1970; Dumont, 1972). This result can hardly be reconciled with a mechanism involving the formation of a dipolar intermediate which would be expected to rearrange and lead to five-membered rings, as shown, for instance, by the

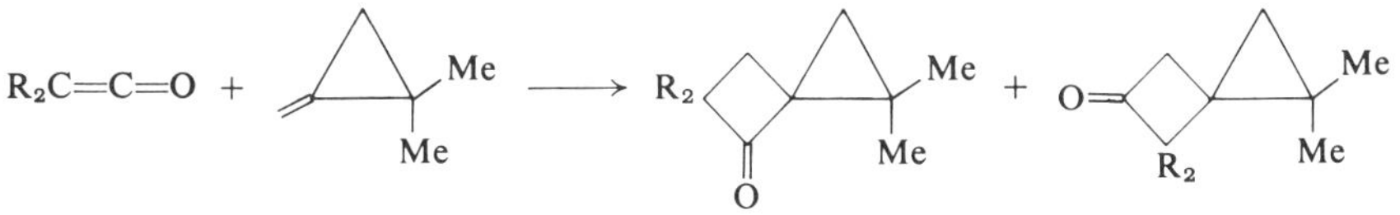

R$_2$C=C=O + (13) → (11) + (12)

R = Me, Ph

addition of **13** to chlorosulfonylisocyanate (Dumont, 1972). On the other hand, the formation of two regioisomeric cyclobutanones can be readily rationalized in terms of counterbalancing electronic and steric factors in a [$_\pi 2_s + _\pi 2_a$] process.

In some cases, however, the mechanistic picture of the ketene–olefin reaction appears to be much more complicated. Thus, the highly nucleophilic enamines react very readily with ketenes to yield 1:1 and 1:2 adducts. For example, dimethyl ketene combines with *N*-isobutenylpyrrolidine to give a cyclobutanone (**14**) and a 2:1 adduct which proved to have the structure of a γ-methylene-δ-lactone **15** (Hasek *et al.*, 1964). The formation of **15** is best interpreted as resulting from a 1,4-dipolar addition of an intermediate such as **16** to the electron-rich double bond of the ketene, especially because the cyclobutanone (**14**) is not converted to **15** by reaction with excess ketene. A beautiful kinetic analysis (Huisgen and Otto, 1969) has established that two mechanisms participate in the formation of cyclobutanone (**14**). In cyclohexane the reaction is approximately 92% concerted whereas, in acetonitrile, it is only 53% concerted. The cycloadditions of ketenes to ynamines apparently

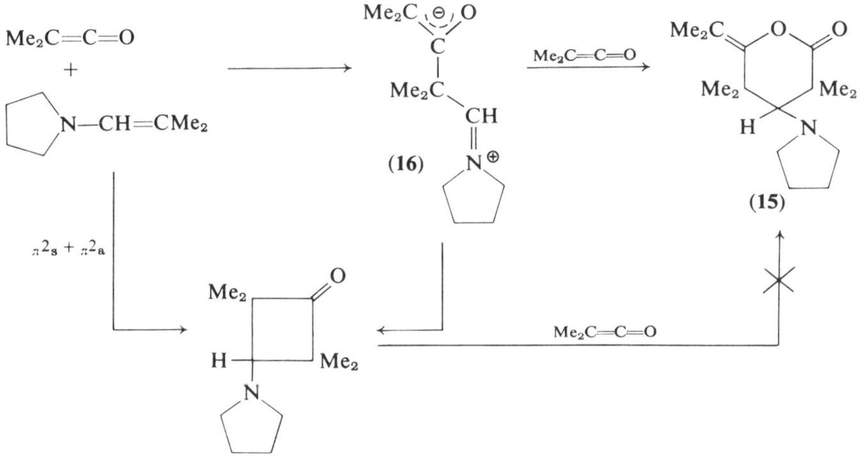

proceed through the formation of a zwitterionic intermediate which can cyclize in two ways (Kuehne and Sheeran, 1968; Delaunois and Ghosez, 1969). The concurrent formation of 1:1 and 2:1 adducts from diphenyl ketene and norbornene in acetonitrile has also been explained in terms of parallel mechanistic pathways (Feiler *et al.*, 1973).

A similar trend from a concerted to a stepwise mechanism is observed when the anion-stabilizing ability of the ketene substituents is increased, as shown by the brilliant studies of England and Krespan (1970a–d) on the

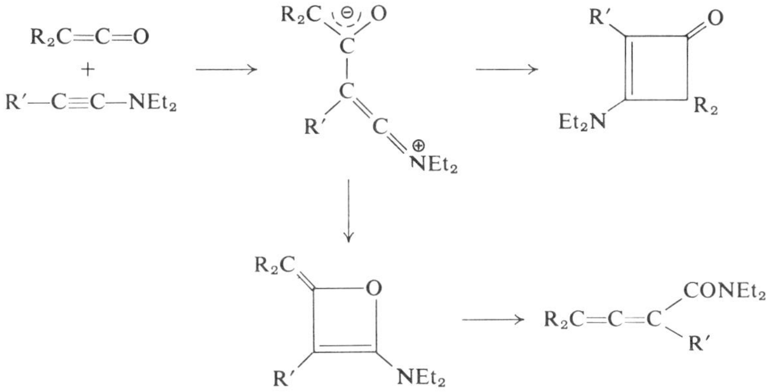

$R_2C=C=O$
+
$R'—C\equiv C—NEt_2$

$\longrightarrow$

[Scheme showing zwitterionic intermediate with R_2C, C, R', NEt_2 groups]

$\longrightarrow$

[Cyclobutenone with R', Et₂N, R₂, O]

$\downarrow$

[Oxete with R₂C, R', NEt₂, O]

$\longrightarrow$ $R_2C=C=C\begin{smallmatrix}CONEt_2\\R'\end{smallmatrix}$

cycloadditions of bis(trifluoromethyl) ketene. Thus, whereas ketenes and enol ethers yield cyclobutanones perispecifically, bis(trifluoromethyl) ketene was found to react with ethyl vinyl ether in hexane at 0° to form an oxetane (**17**), which, on warming, yields the corresponding cyclobutanone (**18**) and a linear ketone (**19**). These observations suggest that the presence of two CF_3 groups

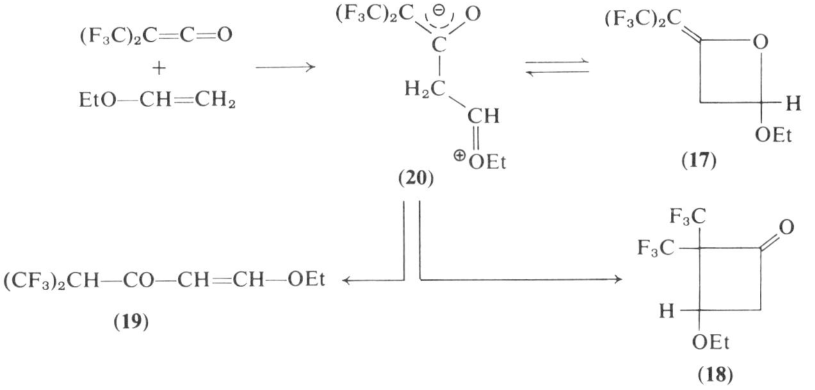

$(F_3C)_2C=C=O$
+
$EtO—CH=CH_2$

$\longrightarrow$

[Zwitterion structure (20) with $(F_3C)_2C$, H_2C, CH, $^\oplus OEt$]

(**20**)

$\rightleftharpoons$

[Oxetane (17) with $(F_3C)_2C$, O, H, OEt]

(**17**)

$(CF_3)_2CH—CO—CH=CH—OEt$ $\longleftarrow$

(**19**)

[Cyclobutanone (18) with F_3C, F_3C, H, OEt, O]

(**18**)

on the ketene may be sufficient to favor the stepwise mechanism due to the ability of the CF_3 groups to stabilize the negative charge on an intermediate zwitterion (**20**).

C. Stereochemistry

The [2 + 2] cycloaddition of a ketene to an olefinic substrate offers two stereochemical variables of interest: (a) first, the relative configuration of the carbon atoms in the adduct which formerly were parts of the olefinic bond, and (b) the configuration at the carbon atom of the adduct, C_α, which bears the ketene substituents.

1. *Configuration at the Former Olefinic Carbon Atoms*

A large amount of experimental data show that, in general, the [2 + 2] cycloadditions of ketene to olefins and dienes occur with complete retention of configuration at the former olefinic carbon atoms. Thus, *cis*- and *trans*-enol ethers were shown (Huisgen *et al.*, 1964, 1969; Martin *et al.*, 1965) to add in a cis-stereospecific fashion to diphenyl and dimethyl ketenes. The cycloadditions

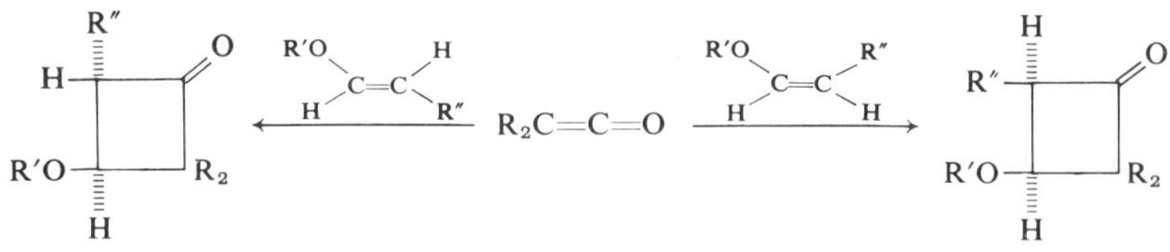

of dichloroketene and diphenyl ketene to the geometric isomers of cyclooctene were also shown (Montaigne and Ghosez, 1968; Dumont, 1972) to be cis-stereospecific in solvents of low or high polarities. Retention of configuration

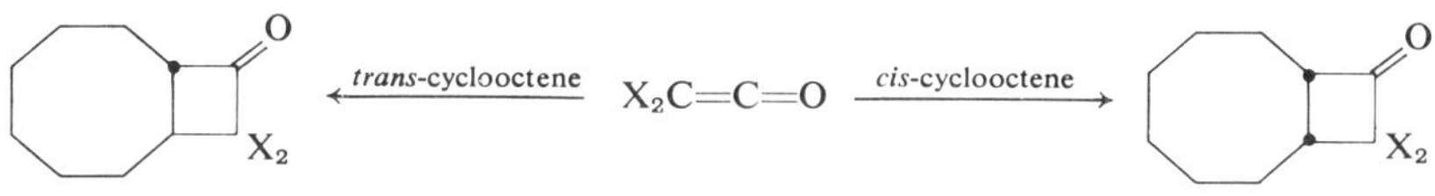

X = Cl, Ph

for the double-bond substituents has also been observed for the following cycloadditions: diphenyl ketene + *cis*- and *trans*-β-deuteriostyrene (Baldwin and Kapecki, 1970b), diphenyl ketene and ethoxyketene + *cis*- and *trans*-butene (Huisgen and Feiler, 1969; Do Minh and Strausz, 1970), *tert*-butyl-cyanoketene + *trans*-cyclooctene (Weyler *et al.*, 1972). Such cis-stereospecificity is entirely consistent with a concerted mechanism in which the π bond of the ketenophile is used suprafacially. Partial loss of configuration was, however, reported in the energetically difficult reaction of *trans*-but-2-ene with dimethyl ketene (Frey and Isaacs, 1970).

Does stereospecificity exclude the formation of short-lived dipolar intermediates between reactants and products along the reaction coordinate? To answer this question, it is necessary to examine the possible evolution of the intermediates in some detail (Fig. 9). It can be safely assumed that the first bond will result from the interaction of the most nucleophilic carbon of the olefin on the carbonyl group of the ketene. It will be seen immediately that the resulting dipolar intermediate (**21**) directly benefits from the delocalization of its anionic portion. From a conformation such as **21** which should be favored by coulombic attraction and through bond coupling (Hoffmann, 1971; Gleiter, 1974) as well as for

Fig. 9. Reactions of **21** to **25**.

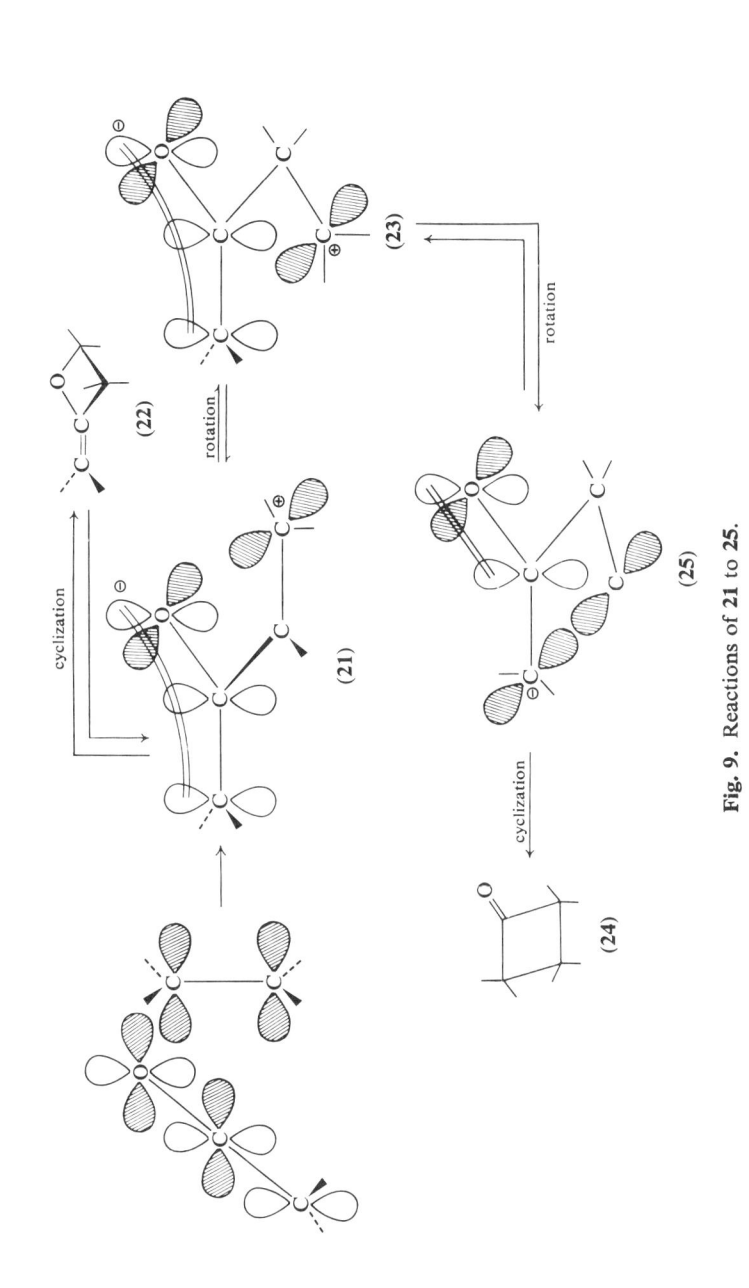

steric reasons, the zwitterion can close, possibly in a reversible reaction, to an oxetane (**22**). Neither **21** nor the less stable **23** can lead to the cyclobutanone **24**. This cyclization requires a 90° rotation around the former C–C bond of the ketene which disrupts the delocalization of the electron pair to yield a higher energy dipole (**25**). Thus, in the intermediate dipolar ion, several rotations around single bonds must occur before it reaches the appropriate conformation for cyclization to a cyclobutanone. Under the circumstances, one can hardly expect that the configuration along the former olefinic C=C bond should be retained. Therefore, we believe that both the absence of oxetane and the retention of configuration, especially in polar solvents, implies the absence of a dipolar intermediate in the ketene–olefin cycloaddition. Supporting this, one finds the results obtained by England and Krespan (1970c) concerning the behavior of bis(trifluoromethyl) ketene toward enol ethers. With *cis*-propenyl propyl ether in hexane at 25°, they obtained stereospecifically an oxetane (**26**), a cyclobutanone (**27**), and a small amount of a linear ketone (**28**). Furthermore, the stereospecificity is no

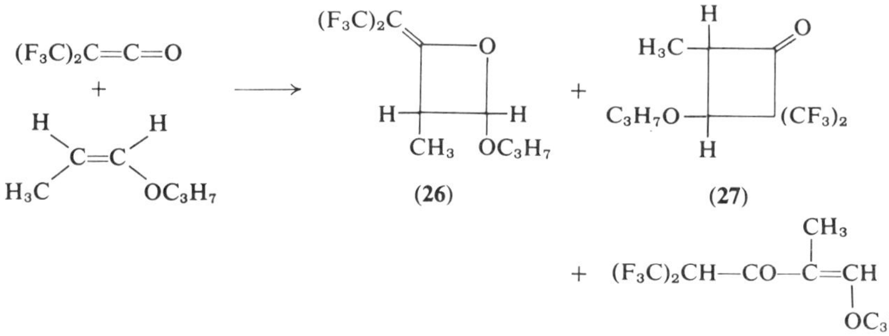

$(F_3C)_2C=C=O$

$+$

$\underset{H_3C}{\overset{H}{\diagdown}}C=C\underset{OC_3H_7}{\overset{H}{\diagup}}$

$\longrightarrow$

[oxetane structure with $(F_3C)_2C$, O, H, H, CH$_3$, OC$_3$H$_7$]
(**26**)

$+$

[cyclobutanone structure with H, H$_3$C, O, C$_3$H$_7$O, (CF$_3$)$_2$, H]
(**27**)

$+ \ (F_3C)_2CH-CO-\underset{OC_3H_7}{\overset{CH_3}{\underset{|}{\overset{|}{C}}}}=CH$

(**28**)

longer observed when one changes to solvents of higher polarity. These results are further indications that the reaction of bis(trifluoromethyl) ketene, in contrast to those of other ketenes, involves dipolar intermediates with finite lifetimes.

It might have been expected that formation of stabilized dipolar or diradical intermediates would be particularly favored in the case of a reaction between a ketene and a diene and that loss of configuration would be possible in this case (Montgomery *et al.*, 1964; Bartlett, 1971). Still, diphenylketene was found to add stereospecifically to the cis double bond of *cis,trans*-hexadiene at room temperature (Dumont, 1972).

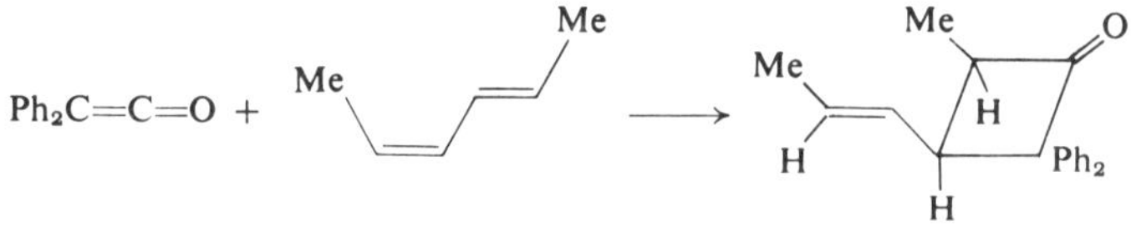

$Ph_2C=C=O \ +$ [cis,trans-hexadiene structure with Me groups] $\longrightarrow$ [cyclobutanone product structure with Me, H, Ph$_2$, O]

2. Configuration at Carbon Atom α to the Carbonyl Group

The cycloaddition of aldoketenes to cyclopentadiene proceeds with a high stereoselectivity to yield preferentially the isomer with the substituent of the ketene in the endo configuration (Brady et al., 1969, 1970; Brady and Hoff, 1970; Brady and Roe, 1970; Brook et al., 1970a,b; Rey et al., 1970; P. Vannes and L. Ghosez, unpublished results, 1970). When both substituents are

X = alkyl, aryl, F, Cl, N$_3$,

different from hydrogen, the cycloaddition yields predominantly the adduct with the larger substituent in the endo position. This stereoselectivity increases with the relative size of X and Y (Fig. 10). These results have been interpreted to indicate the participation of a ketene as the antarafacial partner in a [$_\pi2_s + _\pi2_a$] cycloaddition. The theoretical analysis of Dewar and Zimmerman also offers an attractive rationale for the preferred configuration of the adduct. In such a concerted process, the ketene approaches cyclopentadiene in an orthogonal or "skew" fashion with the larger of the two substituents oriented preferentially away from the ring (Fig. 11). However, the pertinence of the arguments is perhaps somewhat weakened by the observation that, when the endo- and exo-bicyclo[3.2.0]hept-2-en-6-ones could be equilibrated (Rey et al.,

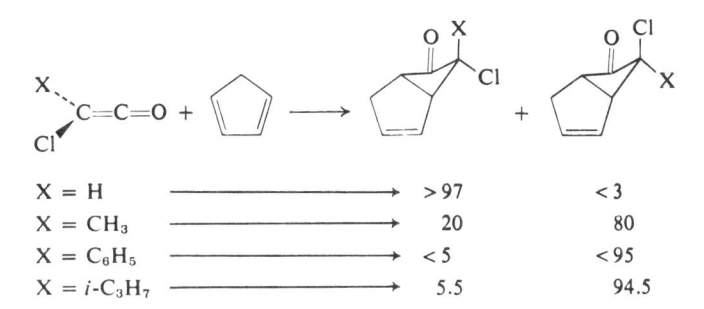

X			
X = H	→	> 97	< 3
X = CH$_3$	→	20	80
X = C$_6$H$_5$	→	< 5	< 95
X = i-C$_3$H$_7$	→	5.5	94.5

Fig. 10. Stereoselectivity increase with the relative size of X and Y.

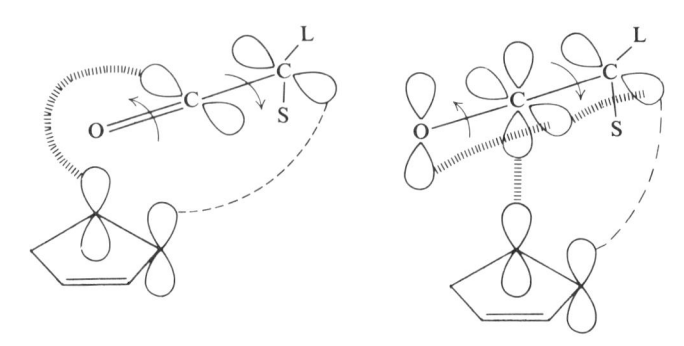

Fig. 11. Preferential orientation away from the ring as the ketene approaches cyclopentadiene.

1970; Roussel, 1970; Brook et al., 1974), *a strong thermodynamic bias was found for the endo adduct. Thus, the cycloaddition often produces the most stable product*. This is, however, not the case when one of the two ketene substituents is isopropyl or *tert*-butyl since the stereochemical composition of the adducts differs considerably from the thermodynamic composition.

Both stereochemical variables have been followed in the reaction of the torsionally strained *trans*-cyclooctene with unsymmetrically substituted ketenes (Roussel, 1970; Weyler et al., 1972; Cossement, 1973). The results are compatible with the theoretical models discussed above and further support the participation of the olefin as the suprafacial partner and of the ketene as the antarafacial partner. Indeed, the trans configuration of cyclooctene is maintained in the bicyclic adduct and the larger ketene substituent is found preferentially or exclusively in a cis position with respect to the adjacent CH_2, as shown, for instance, for the case of chloroketene.

The addition of ethoxyketene to *cis*-butene follows a similar stereochemical pattern (Do Minh and Strausz, 1970).

D. Influence of Substituents, Solvents, and Steric Factors on the Ketenophilic Reactivity

The dependence of the reaction rate upon the structure of the ketenophile or the nature of the reaction medium, as well as the values of the activation parameters, often provide important information concerning the mechanism of a reaction. Direct kinetic measurements have been possible for the cycloadditions involving the stable diphenyl ketene. In most other cases, the high degree of instability of the ketene (which often is formed *in situ*) requires the use of competition experiments to determine the relative rates.

The favorable effect of electron-releasing substituents at the ketenophilic double bond, which has already been established qualitatively (see Section III, B), is illustrated by comparing the second-order rate constants for the reaction of diphenyl ketene with unsaturated substrates (Table II) (Huisgen *et al.*, 1968). Thus, in going from cyclopentene to 2,3-dihydrofuran, there is a rate increase of about 1 to 85,000 and the replacement of an ethoxy- by a pyrrolidine grouping in 1-ethoxyisobutene results in a further increase of about 10^6 in reaction rate. The rate constants, measured by Huisgen *et al.* (1968), extend over a range of 10^7. At the bottom of the scale, one finds simple olefins. Electron-poor double bonds do not react with ketenes under these conditions. A similar trend has also been observed for the reactions of dimethyl ketene with olefins at 100° (Isaacs and Stanbury, 1973). The cycloadditions of diphenyl ketene with 4-substituted styrenes and with 1,1-diarylethylenes in bromobenzene at 20° gave linear Hammett plots with ρ equal to -0.78 and -0.73, respectively (Baldwin and Kapecki, 1970a), whereas the corresponding reactions of dimethyl ketene with substituted styrenes at 100° were reported to give $\rho = -1.4$ (Isaacs and Stanbury, 1973). Both the relative rates of Table II and the signs of ρ indicate that, in the transition state, positive charge develops in the olefinic portion of the molecule. The magnitudes of ρ values are quite different from those observed for the ionic process and are therefore not compatible with the formation of a 1,4-dipolar intermediate. On the other hand, they fit quite well with the hypothesis of a concerted reaction or possibly, with a diradical mechanism with some dipolar character. However, this latter mechanism can hardly sustain such observations as (a) the slower rate of addition of 1,1-diphenyl

TABLE II. Rates of Cycloaddition of Diphenyl Ketene in Benzonitrile at 40.3°

Ketenophile	$10^4 k_2$ (liter/mole sec)	Ketenophile	$10^4 k_2$ (liter/mole sec)
1-Pyrrolidinoisobutene	240,000	α-Methoxystyrene	12
2,3-Dihydrofuran	1040	*trans*-Propenyl propyl ether	0.6
Cyclopentadiene	372.5	1-Ethoxyisobutene	0.29
2-Ethoxypropene	227	Styrene	0.23
1-Morpholinoisobutene	169	α-Methylstyrene	0.038
cis-Propenyl propyl ether	110	Cyclopentene	0.0122
Ethyl vinyl ether	45	Acrylonitrile	No reaction

ethylene as compared to styrene, (b) the inertness of electrophilic olefins and (c) the total stereospecificity of the reaction of diphenyl ketene with cis,trans-2,4-hexadiene.

The rather moderate solvent dependence for the reaction of diphenyl ketene to butyl vinyl ether: k_2 acetonitrile/k_2 hexane = 160 at 31.4° (Huisgen et al., 1968), and the absence of solvent effect for the reaction of the same ketene with styrene at 120° (Baldwin and Kapecki, 1970a) are further evidence for a concerted reaction in which the rates of formation of σ bonds become more unequal as the electron-releasing ability of the olefin substituent is increased.

As expected for such concerted cycloadditions, the activation enthalpies are rather low, but the activation entropies are highly negative and are indicative of being highly sensitive to steric factors. Observation of such effects may help in gaining insight into the geometry of the transition state. Thus, the higher reactivity of cis-alkenes compared to the related trans isomers is an interesting phenomenon since the reverse order of reactivity is usually observed for the Diels–Alder or 1,3-dipolar cycloadditions (Huisgen et al., 1964, and references cited therein). We have already mentioned the preferred addition of diphenyl ketene to the cis double bond of cis,trans-2,4-hexadiene (Dumont, 1972). The relative reactivity of cis:trans-propenyl propyl ether is 184:1 (Table II), whereas ethoxyketene, dimethyl ketene, or carbonyl dithiolane react faster with cis-2-butene than with its trans isomer (Do Minh and Strausz, 1970; Isaacs and Stanbury, 1973; J. P. Dejonghe and L. Ghosez, unpublished results, 1974). A detailed kinetic analysis of the cycloadditions of diphenyl and alkyl phenyl ketenes to cis- and trans-enol ethers further confirmed this singular preference of ketenes for cis-1,2-disubstituted olefins but also allowed deduction of the preferred structures for the orientation complexes (Fig. 12) which were considered to be satisfactory models for the respective transition states (Huisgen and Mayr, 1975a,b). Thus, methyl phenyl ketene prefers to react with cis- and trans-propenyl ether complexes A (88%) and B (100%) whereas phenyl tert-butyl ketene exclusively uses the pathways via A' and B'. These observations can be convincingly accounted for on the basis of a transition state resulting from an orthogonal (skew) approach of the two reactants. No differences of steric origin should be

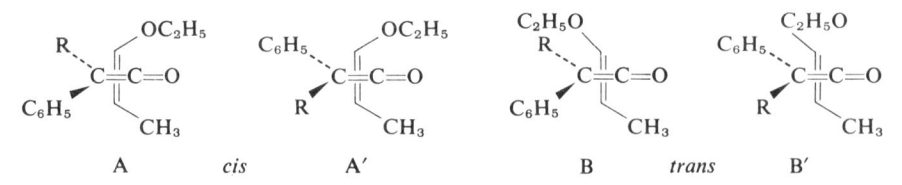

Fig. 12. Preferred structures for the orientation complexes.

observed between *cis*- and *trans*-olefin–ketene complexes resulting from a parallel approach. Furthermore, the cis preference is apparently not observed in [2 + 2] cycloadditions involving zwitterionic intermediates. Thus, the cycloaddition rates of tetracyanoethylene with *cis*- and *trans*-1-alkenyl alkyl ethers are almost equal (Huisgen and Steiner, 1973). The reactivity sequence is, however, reversed for the geometric isomers of cyclooctene; even at 20° the *trans* isomer combines instantaneously and quantitatively with diphenyl ketene, whereas the same reaction with the cis isomer requires 24 hr at 100° (L. Ghosez and W. Dumont, unpublished results, 1972). The partial relief of strain of *trans*-cylooctene is perhaps not the sole factor responsible for its anomalously high ketenophilicity. The torsion imposed at the olefinic bond in an eight-membered ring is indeed expected to facilitate overlap between the interacting orbitals in the orthogonal (skew) complex of the reactants. It is also of mechanistic significance that highly strained olefins such as hexamethyl(Dewar benzene) and 1,2,2-trimethylcyclopropene were recovered unchanged after having been exposed to diphenyl ketene at 20° for six months (L. Ghosez and W. Dumont, unpublished results, 1972). Here, the steric crowding resulting from the presence of the methyl groups on the rings prevents the reactants from attaining the geometry of the transition state for the concerted process. It is worth noting that both olefins react rapidly with chlorosulfonylisocyanate by a stepwise pathway which, as expected, is less sensitive to steric hindrance (Paquette and Krow, 1969; Barton *et al.*, 1970).

The low ketenophilic activity of the strained double bond of the bicyclo-[2.2.1]hept-2-ene system is also of interest especially since this type of olefin exhibits very high dienophilic and dipolarophilic character (Huisgen *et al.*, 1964). This low ketenophilicity has been demonstrated, e.g., by allowing the double bonds of dicyclopentadiene to compete for ketenes. The results obtained with this diolefin are shown in Table III and are compared with data obtained for other cycloadducts. The preference of diphenylketene and dichloroketene for reaction at the five-membered ring double bond is best explained by steric factors. The orthogonal approach of the ketene toward the double bond of the bicyclic system is indeed severely hindered by the hydrogen of the methylene bridge. Remarkably, in another cycloaddition involving an antarafacial interaction (i.e., the addition of dichlorocarbene to olefins), the same reactivity pattern is observed (Ghosez *et al.*, 1964)! On the other hand, bis(trifluoromethyl) ketene or chlorosulfonylisocyanate react preferentially at the strained double bond. This observation further confirms the initial formation of an intermediate as the rate-determining step. Relief of strain is then the dominant factor affecting the rate. This is also the case for [$_\pi 4_s + {}_\pi 2_s$]-concerted cycloaddition involving the formation of five- or six-membered rings as shown, for instance, by the exclusive reaction of phenylazide at the strained double bond of dicyclopentadiene.

TABLE III. Relative Reactivities of the Double Bonds of Dicyclopentadiene in Cycloadditions

Reagents	$T(°)$	% Addition (relative) on the strained double bond	References
Diphenyl ketene	20–100	12.5–23	L. Ghosez and W. Dumont (unpublished results, 1972)
Dichloroketene	50–60	0	Ghosez et al. (1971)
Bis(trifluoromethyl) ketene	85–90	85	L. Ghosez and W. Dumont (unpublished results, 1972)
Dichlorocarbene	0	0	Ghosez et al. (1964)
Chlorosulfonylisocyanate	0	100	Moriconi and Crawford (1968)
Phenylazide	25	100	Alder and Stein (1931) Diels and König (1938) Huisgen et al. (1964)

E. Secondary Deuterium Isotope Effects

A concerted cycloaddition is expected to show inverse secondary deuterium isotope effects for both carbons of the olefinic substrate whereas in a stepwise reaction the carbon which forms the initial bond should show the inverse isotope effect while the other carbon (a radical center) would give a k_H/k_D of one. In addition, in a concerted cycloaddition, the difference in the two inverse secondary deuterium isotope effects might be used as a measure of the asymmetry in the transition state.

Katz and Dessau (1963) found a secondary-deuterium isotope effect of 0.885 ± 0.03 on the product ratio in the cycloaddition of 1-deuteriocyclohexene and diphenyl ketene. Since the effect at each olefinic carbon was not determined, it can only be stated that at the product determining transition state, bond formation has not occurred to the same extent at both olefinic centers.

Intermolecular competition studies were used by Baldwin and Kapecki (1969, 1970b) to measure the isotope effects on both the α- and β-carbons of styrene in its cycloaddition with diphenyl ketene. The value of 0.91 obtained for the β-carbon of styrene is entirely consistent with the expected hybridization change $sp^2 \rightarrow sp^3$. However, a normal secondary deuterium isotope effect (k_H/k_D:1.23) was found for the α-carbon of styrene—a value which is not consistent with the theory developed for either a concerted or a stepwise pathway. On the other hand, Isaacs and Hatcher (1974) have measured the secondary deuterium isotope effect for the cycloaddition of dimethyl ketene to α-deuteriostyrene by a competitive reaction technique. The effect was found to give $k_H/k_D = 0.8$, indicating considerable bond formation between the α-carbon of styrene and the ketene in the transition state. The radical differrence between the two observed α-isotope effects (1.23 vs. 0.8) has not yet received a satisfactory explanation.

F. Cycloreversions

Information concerning the mechanism of cumulene cycloadditions can also be obtained by consideration of the microscopic reverse of the cycloaddition reaction. The gas-phase thermal cycloreversion of cyclobutanone derivatives to ketenes and olefins has been studied in an effort to define the nature of the transition state of this process. Two different elimination

$$\text{cyclobutanone} \xrightarrow{\Delta} CH_2{=}CH_2 + CH_2{=}C{=}O$$

pathways are, in principle, available to an unsymmetrically substituted cyclobutanone. The particular route followed depends, to a large extent, on the nature of the substituents present in the starting cyclic ketone. Many of the same mechanistic tools which have been discussed in the previous sections have been used in the study of the mechanism of the cycloreversion.

Pyrolysis at 325° of either *cis-* or *trans-*2,3-dimethylcyclobutanone occurred with at least 98% retention of configuration to give ketene plus *cis-* or *trans-*2-butene, respectively (Carless and Lee, 1970; Metcalf *et al.*, 1972). The observed stereospecificity excludes the possible intermediacy of a biradical with an appreciable lifetime.

Kinetic studies have provided activation parameters for the cycloreversion reactions of cyclobutanones. The nature and position of substituents can markedly alter the rate of the cyclobutanone decomposition. Thus, alkyl groups in the 3-position have a small stabilizing effect on the transition state (i.e., they lower the activation energy for cycloreversion by approximately 3 kcal mole^{-1}) whereas 3-vinyl or 3-alkoxy substituents stabilize the transition state to a much greater extent (11 and 12 kcal mole^{-1}, respectively). On the other hand, alkyl groups α to the carbonyl do not seem to exhibit any stabilizing effect (Egger, 1973). In fact, 2,2,4,4-tetramethylcyclobutanone decomposes considerably more slowly than does cyclobutanone itself (Frey and Hoph, 1973). These results have been incorporated into a mechanistic model (Egger, 1973) which pictures the cycloreversion as a concerted, nonsynchronous process involving considerable charge separation in an asymmetrical transition state. Localized charge density in this "quasi-zwitterionic" transition state is believed to account for the effect of various substituents on the rate of decomposition.

IV. Ketene Dimerization

As a consequence of their strong pull–push character (see Sections II and III, A), ketenes show a high tendency to oligomerize or to polymerize. Substituted ketenes dimerize to give predominantly or exclusively cyclobutane-1,3-diones as head-to-tail dimers while ketene itself gives γ-methylene-β-propiolactone (Farnum *et al.*, 1965, and references cited therein). The mechanism of dimerization of dimethyl ketenes has been subjected to a kinetic investigation (Huisgen and Otto, 1968) to determine the activation parameters and the effect of solvent polarity on the rate. It was observed that a large change in solvent polarity (CCl_4 to CH_3CN) increased the rate of reaction only by factor of 30. If a zwitterion were formed as a reaction intermediate,

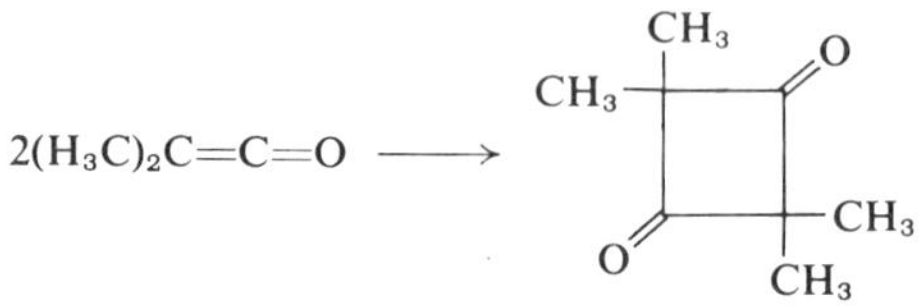

$$2(H_3C)_2C=C=O \longrightarrow$$

a much larger rate increase would have been expected in going from a nonpolar to a polar reaction medium. If, on the other hand, the reaction occurred via a symmetrical transition state, a small reverse solvent effect might have been expected since dimethyl ketene has a dipole moment whereas the product does not. The observed small normal influence of solvent polarity on the rate led the authors to conclude that the reaction follows a concerted course involving an unsymmetrical, charge-separated transition state. The activation parameters for the dimerization of dimethyl ketene in benzonitrile ($\Delta H^{\ddagger} = 10.8$ kcal mole^{-1}; $\Delta S^{\ddagger} = -42$ eu) provide further necessary but not sufficient evidence for a concerted pathway.

V. Ketene–Allene Cycloadditions

The product resulting from the cycloaddition of allenes with ketenes is, in most cases, the 2-methylenecyclobutanone adduct (Martin *et al.*, 1965; Bertrand *et al.*, 1968a, 1975a). The reaction is thus generally regio- and perispecific. An exception to this general rule was found in the reaction of 1,3-dimethylallene with bis(trifluoromethyl) ketene at 100° to give 65% of the

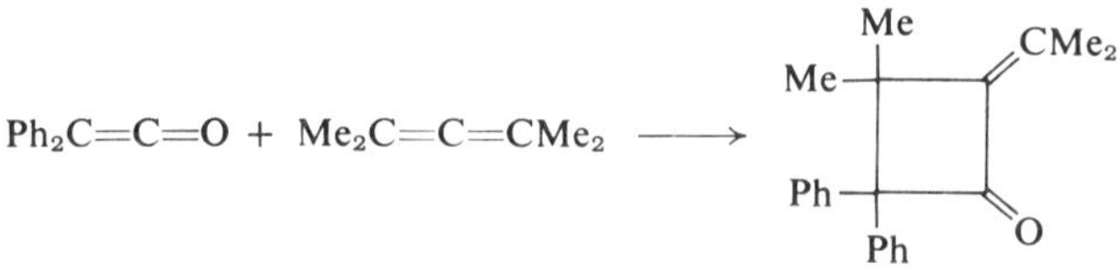

$$Ph_2C=C=O + Me_2C=C=CMe_2 \longrightarrow$$

expected product (**29**) as well as 17% of oxetane (**30**) (England and Krespan, 1970d). A ring-opened product (**31**) was the major product in a mixture of

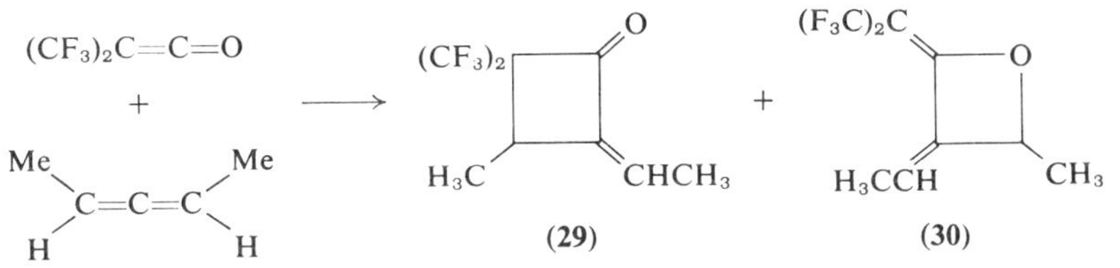

four adducts obtained from the reaction of *tert*-butylcyanoketene with 1-methyl-1-*tert*-butylallene (Bampfield and Brook, 1974b). Such a product could arise from either an intermolecular ene reaction or a prototropic shift from a zwitterionic reaction intermediate (**32**).

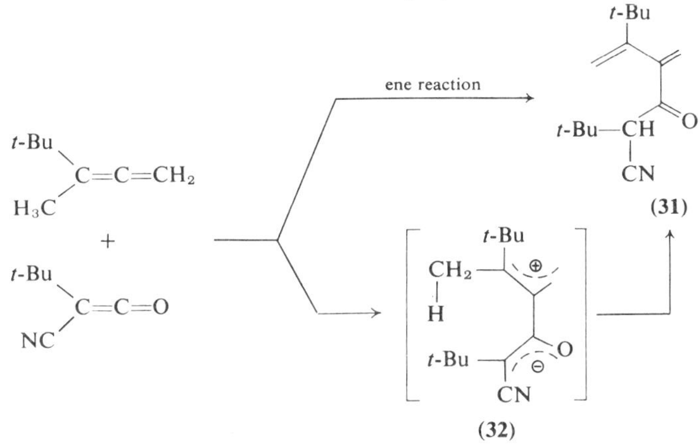

In past years, several research groups have reported elegant studies concerned with the determination of the stereochemistry of ketene–allene cycloadditions. Initial reports (Weyler *et al.*, 1972; Bertrand *et al.*, 1972a) made use of partially resolved 1,2-cyclononadiene. Because of the possibility of restricted rotation in this cyclic allene or its adducts, subsequent studies were made with partially resolved (*R*)-(−)-1,3-dimethylallene to give two products, a (*Z*)- and an (*E*)-tetramethyl-2-methylenecyclobutanone (**33**) and (**34**), respectively (Bertrand *et al.*, 1972b, 1975a,b). It was shown that the *Z* adduct

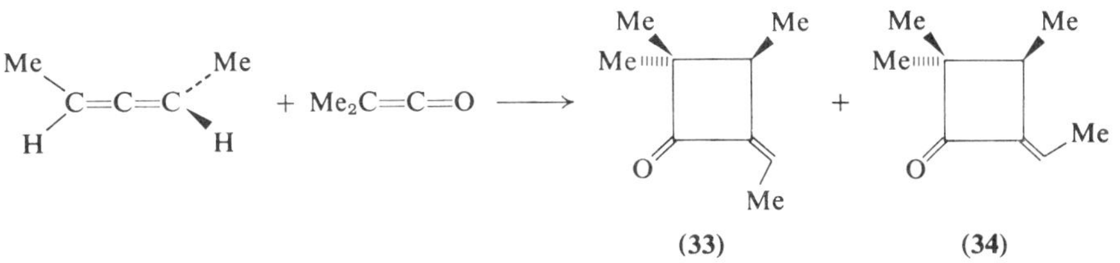

(33) is the kinetic product and that the two products are not interconverted under the reaction conditions. The absolute configuration at C(3) was shown to be *R* for both products. The measurement of the optical activity after each adduct had been converted to the same product (hydrogenation followed by base-catalyzed epimerization) revealed a net loss of activity for the *Z* isomer (33) compared with the *E* product (34). This implies that stereochemical integrity is maintained to a greater extent in the formation of 34. On the other hand, the reaction of partially resolved (*R*)-(−)-1,3-di-*tert*-butylallene gives about equal amounts of *Z* and *E* adducts which show nearly equal optical purities. In addition, it was found that the cycloaddition of dimethyl-ketenes to allenes was only weakly accelerated by an increased polarity of the solvent. Thus, a sevenfold increase in rate was observed for the cycloaddition of dimethyl ketene to 1,2-cyclononadiene in going from hexane to acetonitrile (M. Bertrand and J. Le Gras, unpublished results, 1975). Furthermore, the activation entropies of these cycloadditions are highly negative (M. Bertrand and J. Le Gras, unpublished results, 1975). These data indicate a tight transition state with little charge-separation in the rate-determining step. A model based on competition between various orbital symmetry allowed

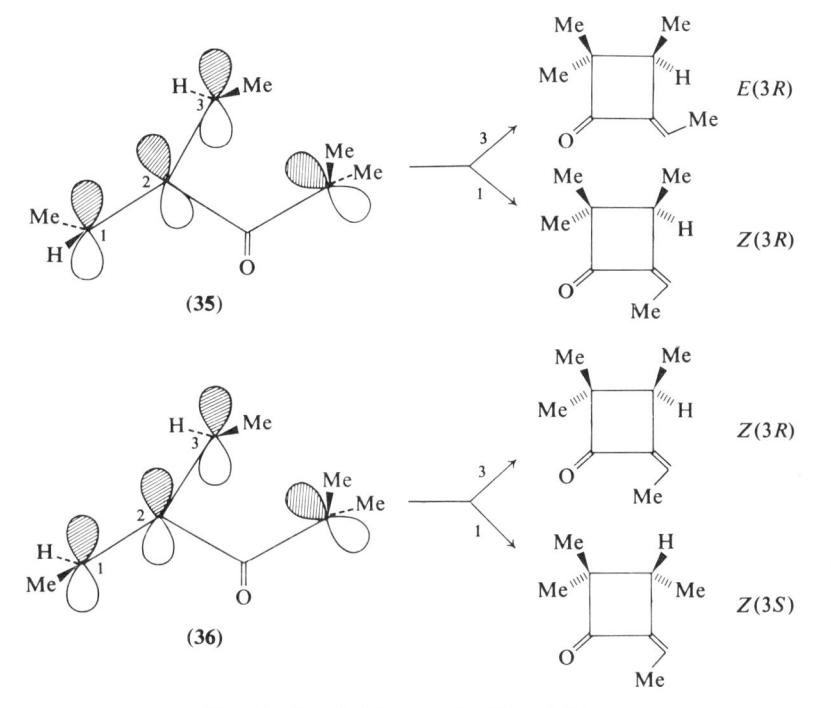

Fig. 13. *Z* and *E* isomers for 35 and 36.

processes including *inter alia* a ketene reacting antarafacially and an allene reacting suprafacially (Bertrand *et al.*, 1972a,b), appears unlikely on the basis of recent theoretical analyses based on perturbation theory or configuration interaction (see Section III,A). More recently M. Bertrand and J. Le Gras (unpublished results, 1975) have proposed the competitive formation of intermediates **35** and **36** resulting from the orthogonal approach of the ketene from either side of the chiral allene (Fig. 13). This model explains the formation of Z and E isomers from the same intermediates as well as the difference in optical purity of the Z and E ketones.

The [2 + 2] cycloaddition reactions of allenes with *tert*-butylcyanoketene have been studied in some detail (Duncan *et al.*, 1973; Bampfield and Brook, 1974a). The reaction of this ketene with optically enriched (S)-(+)-1,3-dimethylallene at room temperature gave four products (**37–40**) in the ratio of 3.8:21.3:41.3:33.7, respectively. The E adducts (**37**) and (**38**) showed circular dichroism curves while the Z adducts (**39**) and (**40**) did not. The lack of activity in the Z products implies that these adducts are derived from an

achiral transition state or intermediate while the active products come from a process in which chirality is maintained throughout the course of the reaction. The interaction of competing concerted processes which would give active products cannot be completely ruled out, but the possibility of the activity of both Z adducts being exactly canceled by competing concerted reactions does seem unlikely. As in previous cases, the optical purities of starting materials and products are not known so a quantitative estimate of the relative importance of the various processes involved is impossible. The results were rationalized by invoking a two-step mechanism with initial least hindered approach and bonding to give **41** (Fig. 14). Closure of this intermediate would give active **37**, or rotation *a* would lead to the orthogonal zwitterion **41** from which closure would give active E and Z products. If rotation *b* occurs faster than closure of **41**, the planar zwitterion **43** would be formed from which only racemic products would be obtained. Thus, steric factors have played a major role in determining the direction of rotation and the ease with which various product-forming steps occur. In a similar manner, approach of the ketene with the *tert*-butyl down would yield active **38** from a chiral intermediate and inactive **39** from a planar zwitterion.

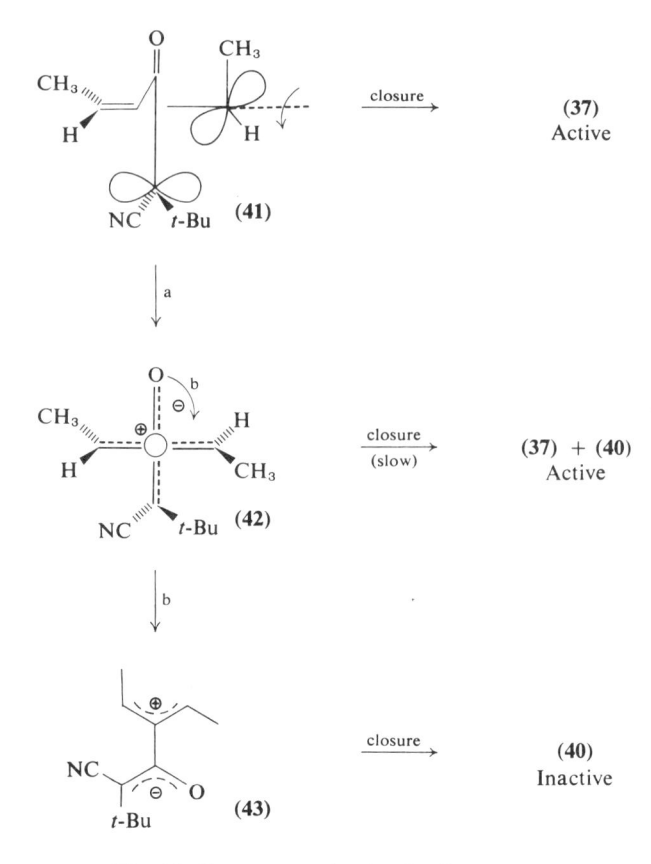

Fig. 14. Two-step mechanism leading to 41.

The sensitivity of the reaction course of ketene–allene cycloadditions to changes in substituents is demonstrated by the recent results of Bampfield et al. (1975). The reaction of *tert*-butylcyanoketene with 1,3-diphenylallene gave only the *E* isomers (corresponding to **37** and **38** in the previous discussion) of the substituted methylenecyclobutanones. Retention of at least some optical activity in the products derived from reaction with optically enriched 1,3-diphenylallene implies that the products are derived, at least in part, from a chiral intermediate or transition state.

As can be seen from the above discussion, the exact mechanism of ketene–allene cycloadditions has yet to be defined unambiguously. It is hoped that further experiments will shed new light on this problem and provide a mechanistic picture which can be used as a predictive tool by future researchers.

VI. [2 + 2] Cycloadditions of Allenes

A. General

The heats of hydrogenation of allene (71 kcal mole^{-1}) and propene (30 kcal mole^{-1}) measured by Kistiakowsky *et al.* (1936) indicate that the double bonds in the cumulene have a higher energy content than do isolated double bonds. Consequently, it is not surprising that allenes also show a higher tendency to dimerize or to undergo [2 + 2] cycloadditions than do simple olefins. However, the dimerization of allenes is always much slower than that of the related ketenes. This can be readily understood by comparing the energies of the frontier molecular orbitals of both cumulenes (Fig. 3). The strong "pull–push" character of ketenes (very low LUMO and very high HOMO) is no longer observed for simple allenes which have a high-energy LUMO and a low-lying HOMO. Consequently, the driving force for dimerization is much less pronounced in allene. Furthermore, simple allenes show no tendency to react with electron-rich olefins, but, on heating, form adducts with olefins which are reactive in radical reactions. Experimental facts on these cycloadditions and the corresponding cycloreversions are abundant and indicative of nonconcerted processes involving diradical intermediates. A recent and excellent discussion of these mechanistic studies is available (Baldwin and Fleming, 1970). Only the salient features will be reviewed here.

B. Allene Dimerizations

The secondary kinetic deuterium isotope effects in the dimerization of allene were used as a mechanistic probe to elucidate the reaction pathway (Dolbier and Dai, 1968, 1970a,b; Dai and Dolbier, 1972). The absence of an intermolecular secondary deuterium isotope effect (k_H/k_D intermolecular = 1.01) implies that the rate-determining step of the reaction does not involve the labeled centers. The observed intramolecular secondary deuterium isotope effect (k_H/k_D intramolecular = 1.14), on the other hand, points to isotopic discrimination at some point along the reaction pathway. The presence of normal secondary isotope effect in the second (intramolecular) case was reconfirmed by an independent investigation (Moore *et al.*, 1972). If the allene dimerization were a concerted process, the same isotope effect would have been expected for both of the above reactions, since a concerted reaction implies the existence of a single energy barrier along the reaction pathway. The fact that the isotope effects are different led to the conclusion that the reaction involves more than one energy barrier and is therefore nonconcerted.

$$\frac{k_H}{k_D} \text{ (inter)} = \frac{[d_0]}{\frac{1}{2}[d_4]} = \frac{[d_0]}{[d_8]} = 1.01$$

allyl H	4	2	0
vinyl H	0	2	4

$$\frac{k_H}{k_D} \text{ (intra)} = \frac{\text{allyl H}}{\text{vinyl H}} = 1.14$$

The stereoselectivity of the dimerization of racemic and optically active 1,2-cyclononadiene was studied by Moore and co-workers (1969). The dimerization was conducted neat at 125° to give the products shown below. The stereochemical consequence is the cis addition to one allene (used in a suprafacial manner) and the trans addition to the other (antarafacial). The high stereoselectivity can be explained either by a concerted pathway or by a

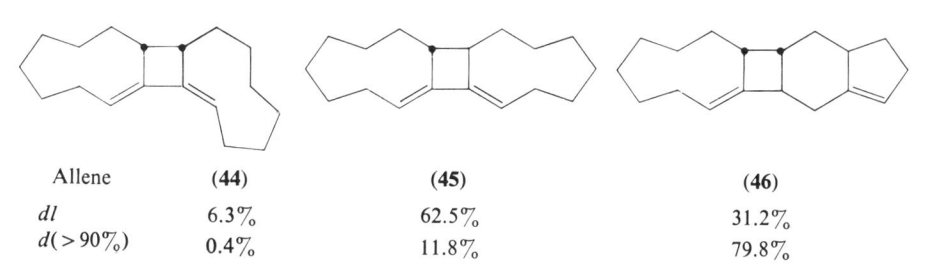

Allene	(44)	(45)	(46)
dl	6.3%	62.5%	31.2%
d(>90%)	0.4%	11.8%	79.8%

multistep process in which each step is stereospecific. Both models would predict the formation of **44** and **45** from racemic starting material and the exclusive formation of **46** from optically active allene. The presence of **46** in the racemic dimerization implies a difference in rate for the competing process: $(d + l)$ reacts about four times as fast as $(d + d)$. A multistep pathway would be possible by a least-hindered orthogonal approach of the two allenes with bonding at the central carbons followed by disrotatory motion, as shown below, to give an orthogonal tetramethyleneethane (TME) intermediate. Conrotatory closure of this species or rotation about the central bond to form a planar TME (**47**) followed by conrotatory closure gives the observed product. The other products are derived from completely analogous processes (Moore et al., 1969; Baldwin and Fleming, 1970). The rotations can be reversed (conrotation in the initial step followed by disrotatory ring closure) without affecting the results. The former process (dis then con) was favored in the light of results (Gajewski and Shih, 1969, 1972b)

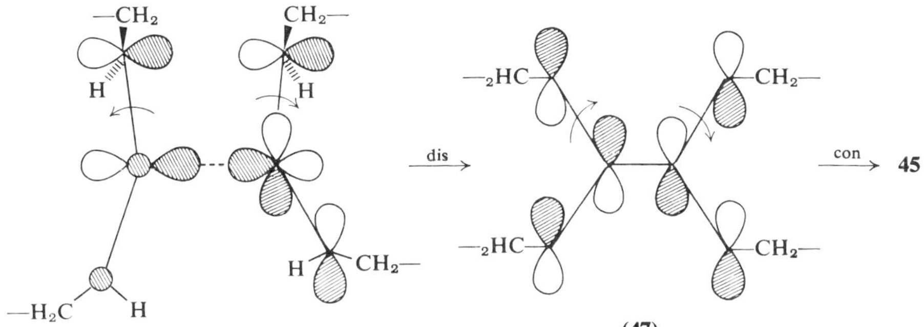

which established the preference for conrotatory ring opening (and closure) in the degenerate dimethylenecyclobutane rearrangement.

Much of the effort concerned with determining the mechanism of allene dimerization has been directed toward proof of the existence of the tetramethyleneethane diradical. Dowd (1970) observed a triplet in the ESR spectrum of perdeuteriotetramethyleneethane at 77°K, ($t_{1/2} = 20$ min). An organometallic complex, assigned structure **48**, was isolated when allene was reacted with triiron dodecacarbonyl (Nakamura, 1966). Other organometallic derivatives of TME have since been prepared (Ben-Shoshan and Pettit, 1968; Davis, 1968; Keim, 1968; Hewitt et al., 1969; Hughes and Powell, 1969; Racanelli et al., 1969; Hewitt and de Boer, 1971; Okamoto, 1971; Hughes and Powell, 1973). The attempted dimerization of 1,3-diadamantylallene

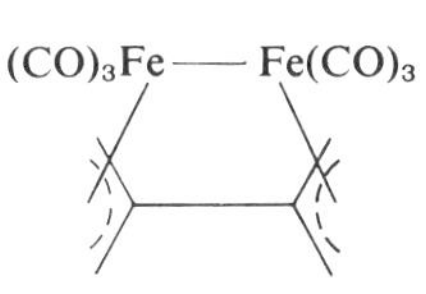

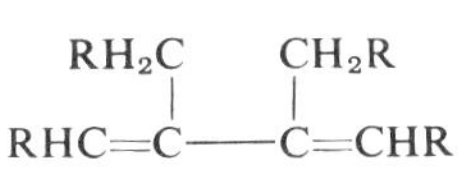

(Jacobs and Kammerer, 1972) led to the trapping of an intermediate rather than dimerization. Reaction at 250° gave only starting material whereas at 305° only tars were obtained. When the allene was heated at 250° in p-diisopropylbenzene (**49**) (R = 1-adamantyl), was isolated in 60% yield. This

product presumably results from a hydrogen transfer to a TME reaction intermediate which was incapable of ring closure to normal products because of the bulky adamantyl groups.

The labeled tetramethyleneethane precursor (50) was pyrolyzed or photolyzed to give a dimethylenecyclobutane with one *tert*-butyl group on an acyclic methylene and the other on the cyclobutane ring (Beetz and Kellogg, 1973). This result can be explained only by a "turning over" of a reaction intermediate which therefore implies the existence of an effectively orthogonal TME species at some point along the reaction coordinate.

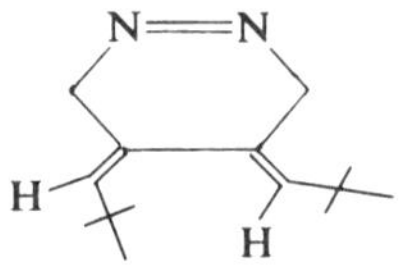

(50) 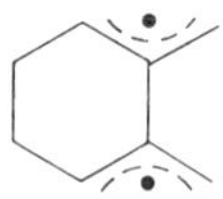   (51)

The intermolecular cycloaddition reactions of 2,3-dimethylenecyclohexadiene (51), prepared by solution pyrolysis at 60° of 2,3-dimethylenebicyclo[2.2.0]hexane, are in complete agreement with diradical reaction mechanisms (Chang and Bauld, 1972; Bauld and Chang, 1972). The gas phase and solution thermolyses of various compounds (52–55) which could lead to a tetramethyleneethane, were studied to determine the nature of a possible intermediate species (Roth and Erker, 1973a,b; Roth *et al.*, 1973). Gas phase pyrolysis (100°–180°) gave a similar distribution of intramolecular

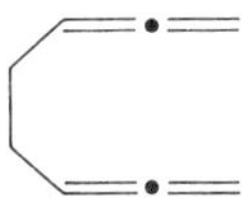

(52) 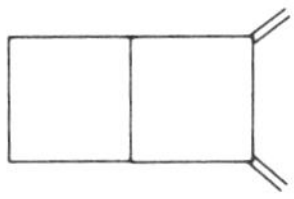   (53) 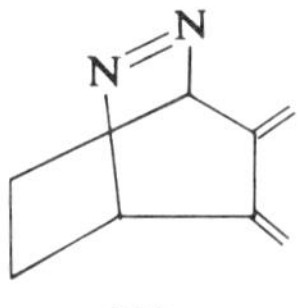   (54) 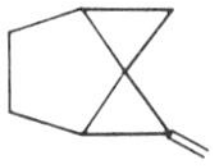   (55)

reaction products for all four reactants, implying, therefore, the intermediacy of a TME species. The presence of triplet intermediates in the solution thermolysis which gives mostly dimeric and polymeric products, was confirmed by ESR and CIDNP experiments. The product studies of Grimme and Rother (1973) were the basis of a discussion concerning the geometry "(planar" or "chair") of the intermediate diradical. It was concluded that a planar diradical, generated from 53, undergoes ring closure faster than it comes to conformational equilibrium with a chair diradical.

The relationship between the allene dimerization and the 1,2-dimethylenecyclobutane rearrangement has been an interesting problem since the implication of a perpendicular TME-type structure in the rearrangement of 56 (Gajewski and Shih, 1967, 1972b). Thermolysis at 278°, followed by suitable degradation experiments, showed that a mixture of the three products (**56**, **57**, and **58**) was formed in a ratio of 58:24:14, respectively. The *ca* 2:1 ratio of the last two isomers is exactly that predicted from a statistical closure of an orthogonal TME intermediate.

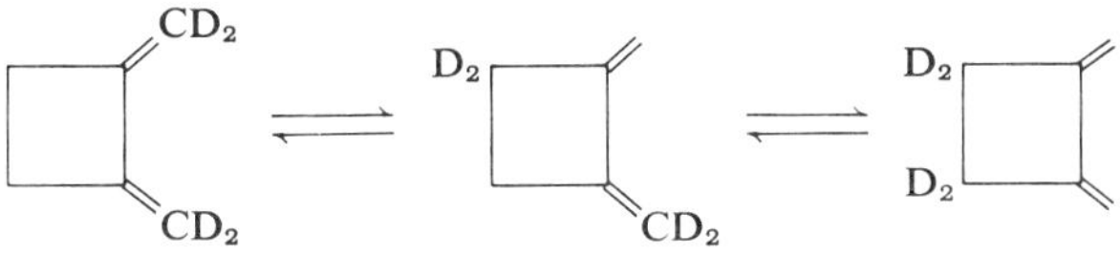

$$\text{(56)} \rightleftharpoons \text{(57)} \rightleftharpoons \text{(58)}$$

A necessary but not sufficient condition for the presence and reaction of the same intermediate for the dimerization and the degenerate rearrangement is the observation of identical products and the same product distribution for both reactions. The product distribution in the dimerization of methylallene (seven different nongeminal dimethyl-1,2-dimethylenecyclobutanes were obtained) was quite different from that observed in the rearrangement of either *cis*- or *trans*-1,2-dimethyl-1,2-dimethylenecyclobutane. It is especially interesting to note that the syn,syn isomer (**59**) is a major product in the

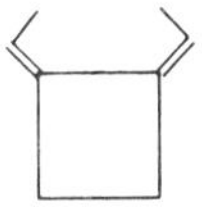

(**59**)

dimerization but it is not formed at all in the degenerate rearrangement (Gajewski and Shih, 1969, 1972a,b). No definitive conclusion can be drawn from this result, but it does certainly appear questionable that the two reactions involve the same reaction pathway.

C. Cycloaddition of Allenes to Olefins

The cycloaddition of activated olefins with allenes has been studied to determine the course of the reaction to form methylenecyclobutanes. The reaction of allene with various mono- or 1,1-disubstituted olefins (substituents: nitrile, carboxyl, carboalkoxyaldehyde, or aryl) gives 3-substituted methylenecyclobutanes, the products expected from the most stable diradical intermediate (Cripps *et al.*, 1959).

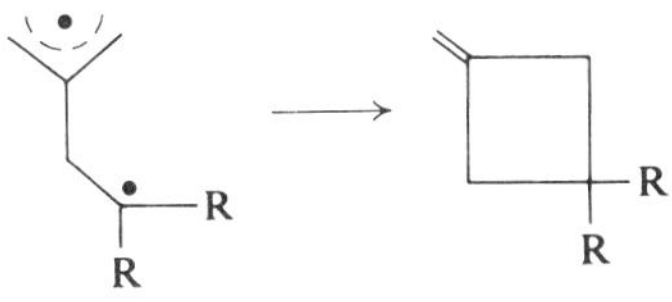

In the reaction of allene with polyhalo olefins, a minor product, which is not consistent with a normal diradical pathway, is formed in small amounts. With 1,1-dichlorodifluoroethylene, **60** is obtained in 5% yield (along with the normal product). This minor adduct has been suggested to arise from a

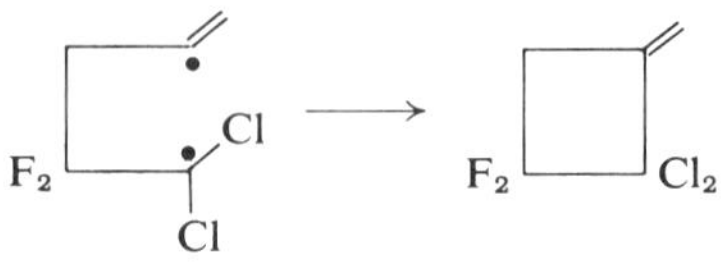

completely different type of diradical species, such as **61** (Taylor and Warburton, 1967; Taylor et al., 1972).

The olefinic component cycloadds in an essentially stereospecific manner to allene. This conclusion is supported by the observed retention of configuration in the products resulting from the addition of either dimethyl fumarate or dimethyl maleate to 1,1-dimethylallene (Kiefer and Okamura, 1968).

Isotope studies completely analogous to those discussed earlier for allene dimerization, gave an intramolecular kinetic secondary deuterium isotope effect of $k_H/k_D = 1.13$–1.21 and an intermolecular isotope effect of $k_H/k_D = 1.05$ for the [2 + 2] cycloaddition of allene and acrylonitrile (Dai and Dolbier, 1972). As before, these results imply a multistep pathway because the rate-determining isotope effect (intermolecular) is different from the product-determining effect (intramolecular).

A stepwise reaction following orbital symmetry control to the intermediate was proposed to account for the four products obtained when (R)-$(-)$-1,3-dimethylallene was reacted with acrylonitrile (Baldwin and Roy, 1969; Baldwin and Fleming, 1970). Thus, if all orbitals are used in a suprafacial

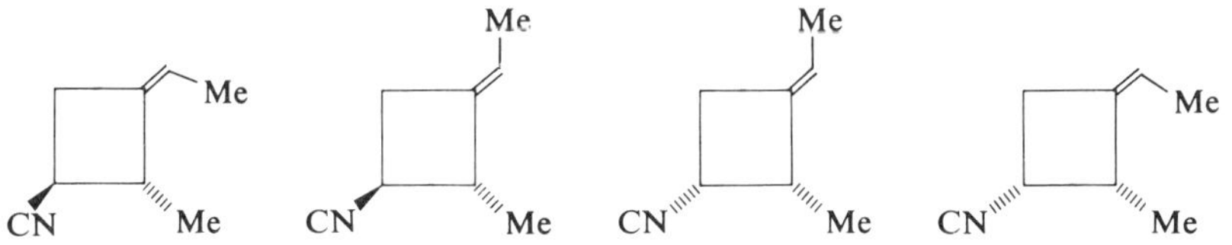

manner, least hindered approach to the olefin in either of two ways followed by bonding and rotation of a methylene to form the stabilized allylic unit would give intermediates **62** and **63**. Disrotatory closure of **62** would give the first two products shown above while closure of **63** would give the last two methylenecyclobutanes. As pointed out by the authors, this model accounts for the stereochemical results as well as the secondary deuterium isotope effects observed in the cyloaddition reaction of allenes and olefins.

(62)   (63)

D. Degenerate Rearrangements

The cycloadducts of allenes plus olefins or ketenes (methylenecyclobutanes and 2-methylenecyclobutanones, respectively) undergo degenerate thermal rearrangements on heating. As in the previous discussed case of the degenerate rearrangement of 1,2-dimethylenecyclobutanes, it is of interest to determine if there is any relationship between these degenerate rearrangements and the related cycloadditions.

In a very elegant set of experiments, Baldwin and Fleming (1972, 1973a,b,c) studied the degenerate rearrangement of methylenecyclobutanes to determine the mechanistic course of the reaction. Rates were determined for the kinetic thermal interconversion of **64–67** at 332° in the gas phase (Fig. 15). Starting

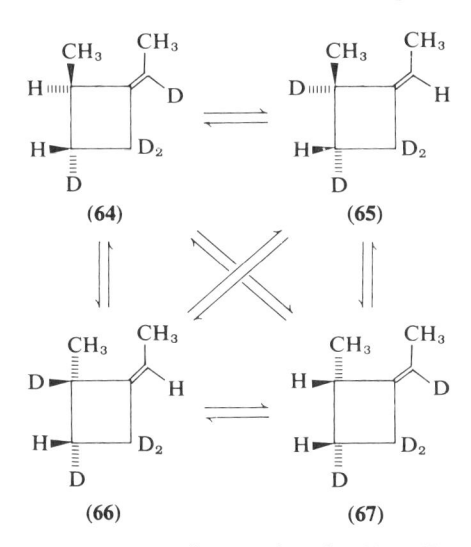

Fig. 15. Reversible reactions for **64** to **67**.

with **64**, three new products were observed: a product (**65**) derived from an orbital symmetry allowed 1,3-carbon migration, another (**66**) from an orbital symmetry forbidden process and a third (**67**) from the epimerization about C(2) of the starting compound. From the rate constants it was calculated that the 1,3-carbon migration occurs by an orbital symmetry allowed process 77% of the time. In addition, it was estimated that the 1,3 shift occurs with antarafacial utilization of the allylic component 65% of the time. These results imply that the methylenecyclobutane rearrangement, unlike the cycloaddition of allenes and olefins, occurs by a concerted pathway.

The two products observed in the thermolysis of methylenecyclobutanone (**68**) at 330° can be obtained in an orbital symmetry allowed fashion by using

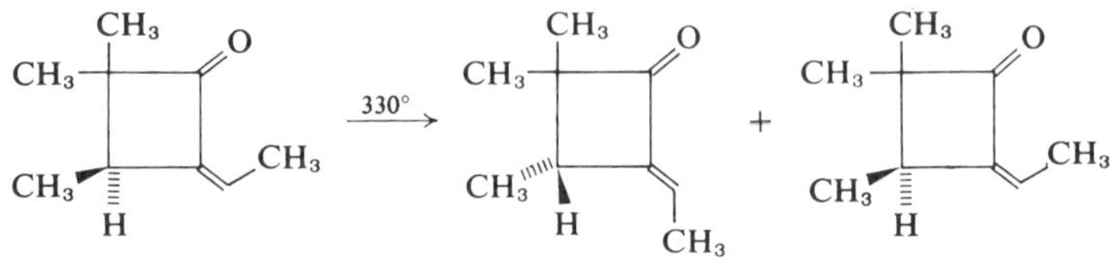

the allylic component in an antarafacial manner (Bertrand *et al.*, 1974). This result is in agreement with those described above for the degenerate methylenecyclobutane rearrangement.

Thermolysis of the partially deuterated adduct **69**, derived from cycloaddition of *tert*-butylcyanoketene with tetramethylallene was studied by

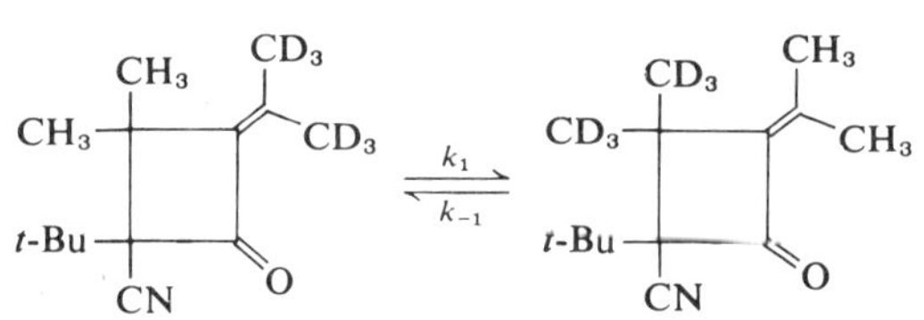

Bampfield and Brook (1974b). At 165° the following kinetic data were obtained: $k_1/k_{-1} = 1$; $E_a = 39.3$ kcal mole^{-1}; log A = 15.2. Prolonged heating of the nondeuterated adduct at 180° led to a ring-opened cross-conjugated dienone similar to **31**. The authors concluded that the degenerate rearrangement of **69** and the formation of a ring-opened product occur through a zwitterionic intermediate similar to **32**.

The last two examples indicate that the complete mechanistic picture of the methylenecyclobutanone rearrangement is still in question.

VII. [2 + 2] Cycloadditions of Ketenimines

The pull–push character of the ketene system has been considered (see Section III,A) to be the dominant factor responsible for the special ability of ketenes to undergo [2 + 2] cycloadditions across the C=C bond either by a concerted mechanism or via the formation of a 1,4 dipolar intermediate. Inspection of the relative energies of the frontier molecular orbitals of ketenimines and ketenes (Fig. 3) suggests that ketenimines should be much less reactive than ketenes in [2 + 2] cycloadditions to olefins or dienes. As a matter of fact, no reaction between a ketenimine and an electron-rich olefin has yet been described. However, ketenimines have been shown to react with the strongly nucleophilic triple bond of ynamines (Ghosez and de Perez, 1971). Diphenyl ketene N-methylimine and 1-diethylaminopropyne yielded a mixture of 1:1 and 2:1 adducts after seven days reaction in acetonitrile. The monoadduct resulted from an addition across the C=N bond of the

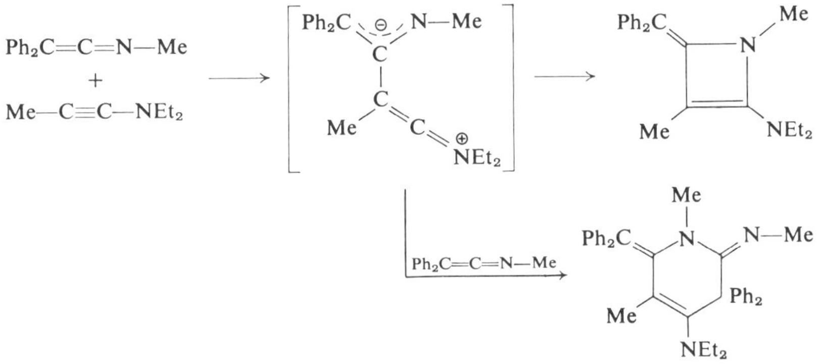

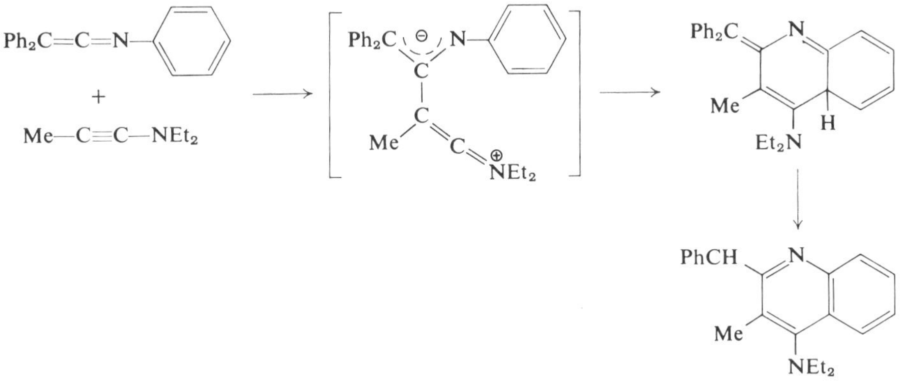

ketenimine. Diphenyl ketene N-phenylimine reacted differently and gave a quinoline derivative. These observations are best rationalized on the basis of a mechanism involving the initial formation of a stabilized 1,4 dipole which either cyclizes to a 6- or 4-membered ring or is trapped by a second molecule of ketenimine.

Interestingly, [2 + 2] cycloadditions involving ketenimines as nucleophilic components have also been encountered (Krow, 1971). Typical examples are the addition of diphenyl ketene *p*-tolylimine to bis(trifluoromethyl) ketone (Weidler-Kubanek and Litt, 1968) and to phenylisocyanate (Naser-Ud-Din et al., 1973).

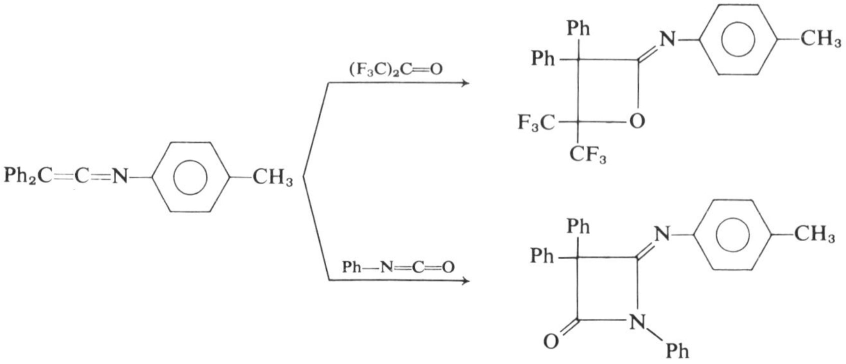

VIII. [2 + 2] Cycloadditions of Keteniminium Salts

A. GENERAL

The isolation and identification of stable, long-lived keteniminium salts is of fairly recent origin. These new heterocumulenes (Ghosez and Marchand-Brynaert, 1976) are highly electrophilic molecules which can be obtained only in the presence of weakly nucleophilic counterions (BF_4^-, $ZnCl_3^-$, PF_6^-) in inert solvents (Weingarten, 1970; Marchand-Brynaert and Ghosez, 1972; Sidani et al., 1974). In contrast with other iminium salts (Böhme and Haake, 1976; Jutz, 1976; Janousek and Viehe, 1976) they are not stable toward halide ions and give the corresponding α-haloenamines. However, this reaction is readily reversed in the presence of an appropriate Lewis acid and, in fact, the readily available α-haloenamines have been found to be the most practical source of keteniminium salts. Thus for example, tetramethyl keteniminium salts have been prepared from tetramethyl-α-chloro-, fluoro-, and iodo-enamines.

The exceptionally low-lying position of the LUMO (Fig. 3) accounts, in molecular orbital terms, for the high electrophilic character of keteniminium salts. On this basis one might be tempted to consider keteniminium salts simply as activated derivatives of ketenes and expect that they will be even more reactive in [2 + 2] cycloadditions to olefins or acetylenes. However,

2. PERICYCLIC REACTIONS OF CUMULENES 119

$$Me_2C=C\begin{smallmatrix}Cl\\NMe_2\end{smallmatrix} + MX \longrightarrow Me_2C=C=\overset{\oplus}{N}Me_2 \quad X^{\ominus}$$

with I_2 giving $Me_2C=C(I)NMe_2$ and with BF_3 giving $Me_2C=C(F)NMe_2$.

MX = AgBF$_4$, AgPF$_6$, ZnCl$_2$, AlCl$_3$, FeCl$_3$.

analysis of the behavior of keteniminium salts toward olefins on the basis of initial perturbations or polarities of the two orthogonal π systems should be approached with caution. Clearly, the most important perturbation at the early stages of the reaction results from a transfer of electrons from the olefin HOMO to the keteniminium LUMO. However, in contrast with ketenes, keteniminium ions have a strongly bonding HOMO (Fig. 3) and, therefore, the secondary interaction between the LUMO of the olefin and the HOMO of the keteniminium will be weak, at least at the beginning of the reaction. Indeed, whereas keteniminium salts show a higher reactivity than ketene in [2 + 2] cycloadditions to olefins and acetylenes, the difference between the electronic structures of the two heterocumulenes reveals itself in several aspects. Thus, whereas dimethylketene dimerizes extremely rapidly even at room temperature, tetramethyl keteniminium tetrafluoroborate has been found to be stable at room temperature (C. t'Kint and L. Ghosez, unpublished results, 1974) as a result of its lower 1,2-dipolar character. Furthermore, like the isoelectronic allenes, keteniminium salts give [4 + 2] adducts with cisoid dienes (Marchand-Brynaert and Ghosez, 1974).

An interesting mechanistic possibility (E. Sonveaux and L. Ghosez, unpublished results, 1974) can be seen from inspection of the π electronic distribution and the coefficients of the frontier molecular orbitals in keteniminium ions (Fig. 3). It involves the concerted formation of two σ bonds at the central carbon atom of the keteniminium to give a cationic intermediate (**70**) which can cyclize to a quaternary azaspiropentane (**71**) (Fig. 16). In this first step, the keteniminium ion reacts as a carbene (1,1 pull–push reagent). The next step would be the rearrangement of **70** to the observed product, a cyclobutaniminium salt. Such rearrangements of heterospiropentanes to four-membered rings are well documented (Trost, 1974; Crandall and Conover, 1974; Braun and Seebach, 1974). The theoretical studies of Wagner and Gompper (1970) led to a similar hypothesis for the cycloadditions of vinyl cations and ketenes. These mechanistic hypotheses still await experimental testing.

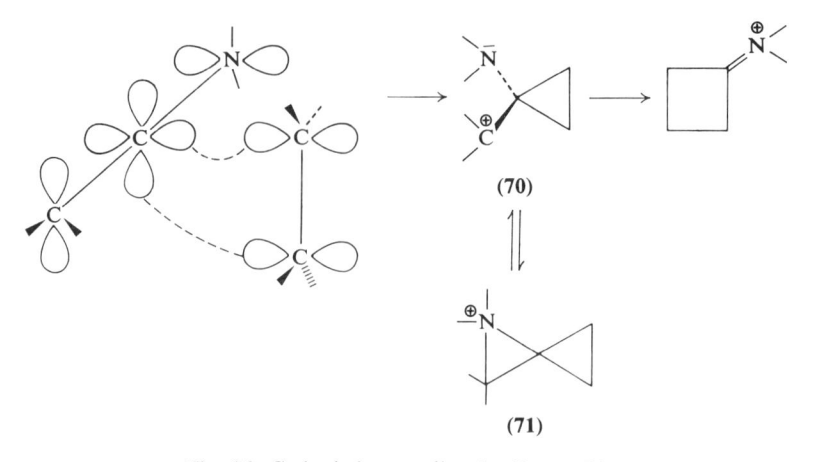

Fig. 16. Cationic intermediate leading to 71.

B. REACTIONS WITH OLEFINS AND ACETYLENES

The high reactivity of keteniminium salts in [2 + 2] cycloadditions is demonstrated by the facile reaction of tetramethyl keteniminium (TMK) tetrafluoroborate (72) with ethylene and acetylene at room temperature and atmospheric pressure to yield the corresponding adducts in 73 and 77% yields, respectively (Sidani et al., 1974; Hoornaert et al., 1975). High yields of cyclobutaniminium salts were also obtained from the reaction of TMK with several mono- and 1,2-disubstituted olefins (Marchand-Brynaert and Ghosez, 1972). Hydrolysis of these adducts with aqueous base gave the corresponding cyclobutanones. As shown by the regiospecific reaction of

R—CH=CH$_2$ + TMK ⟶

R = CH$_3$, C$_6$H$_5$

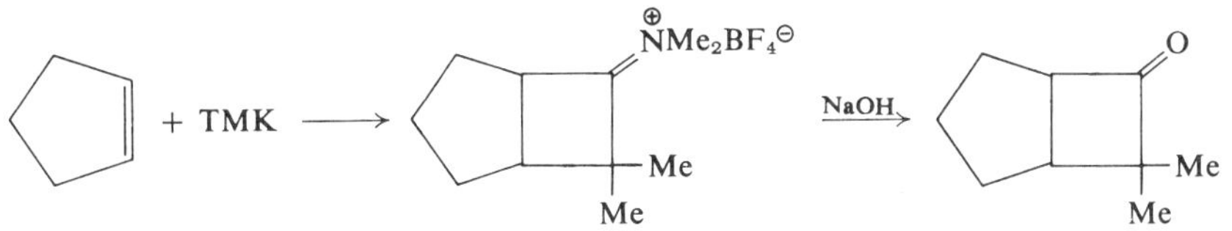

TMK with propene and styrene, the direction of addition is that expected from the union of the two atoms of highest frontier orbital density (HOMO of the olefin and LUMO of the keteniminium).

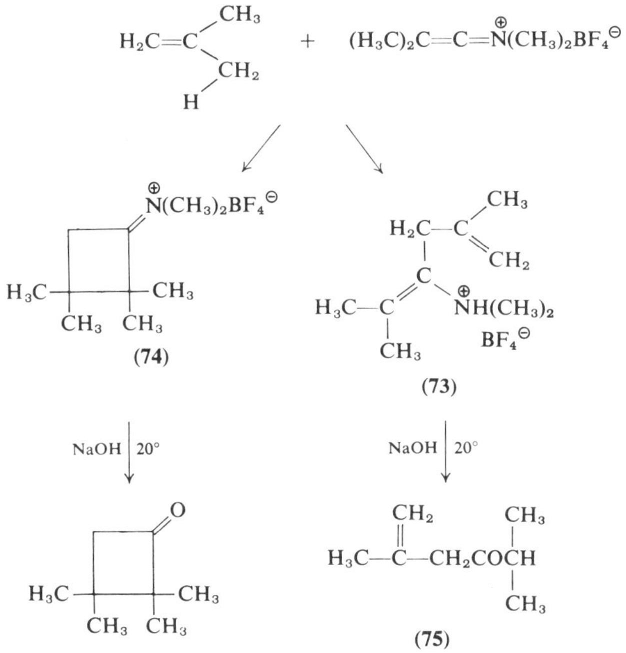

Steric factors have been found to modify the direction and even the course of the cycloadditions of TMK to olefins. Thus, the reaction of TMK with isobutene surprisingly gave an eneammonium salt (73) in addition to the expected cyclobutaniminium salt (74). Hydrolysis of the reaction mixture gave a linear ketone (75) as well as 2,3-tetramethylcyclobutanone (A. M. Leonard and L. Ghosez, unpublished results, 1976). These observations can be accounted for on the basis of a concerted [2 + 2] cycloaddition competing with an ene reaction. However, both products could also arise from a common cationic intermediate. On the other hand, no rearranged products have been obtained in the reaction of TMK with 2,2-dimethyl-1-methylenecyclopropane, but two regioisomeric spirohexanones have been isolated after hydrolysis of the reaction mixture (A. M. Leonard and L. Ghosez, unpublished results, 1976). This result can be readily accounted for on the basis of counterbalancing electronic and steric effects in a concerted reaction involving

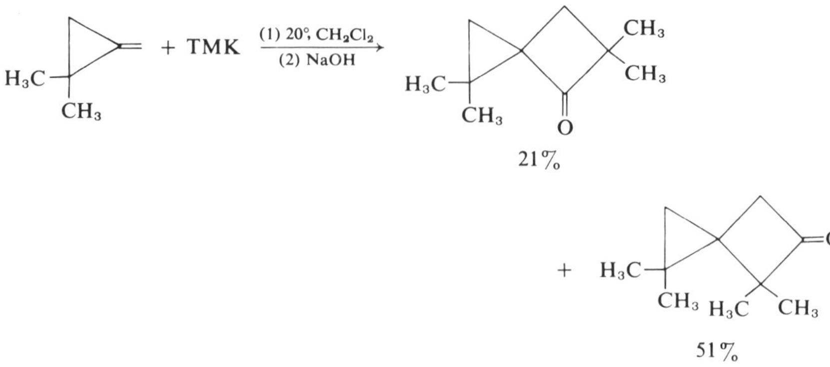

orthogonal approach of the reactants. Less readily explained is the isolation of regioisomeric adducts in the reaction of TMK with methyl- or *tert*-butylacetylene (Hoornaert *et al.*, 1975).

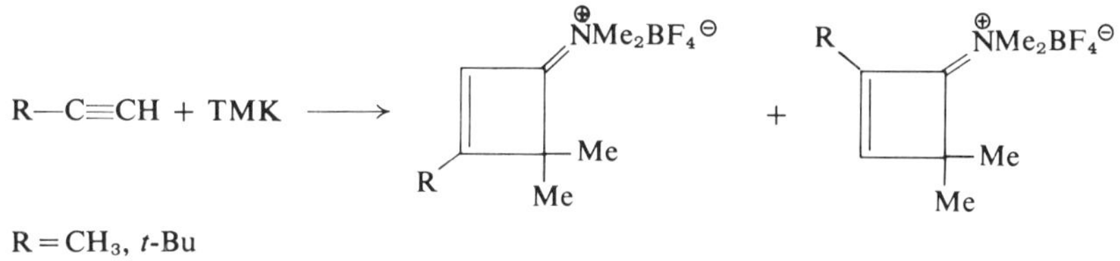

R = CH$_3$, *t*-Bu

The cycloadditions of tetramethylketeniminium tetrafluoroborate to olefins occur with retention of configuration for the double-bond substituents. Thus, *cis*- and *trans*-butenes or *cis*- and *trans*-cyclooctenes gave the corresponding *cis*- and *trans*-cyclobutaniminium tetrafluoroborates, respectively. However, hydrolysis of these adducts resulted in some epimerization.

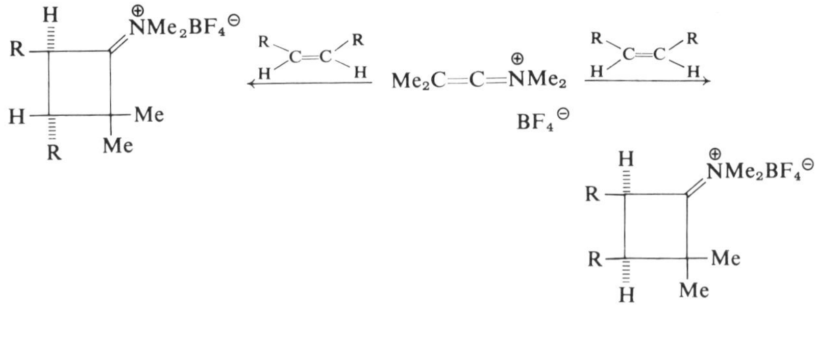

It is interesting that like ketenes (Ghosez et al., 1971; Dumont, 1972) and dichlorocarbene (Ghosez et al., 1964), tetramethyl keteniminium tetrafluoroborate adds preferentially to the less-strained double bond of dicyclopentadiene (Marchand-Brynaert, 1973).

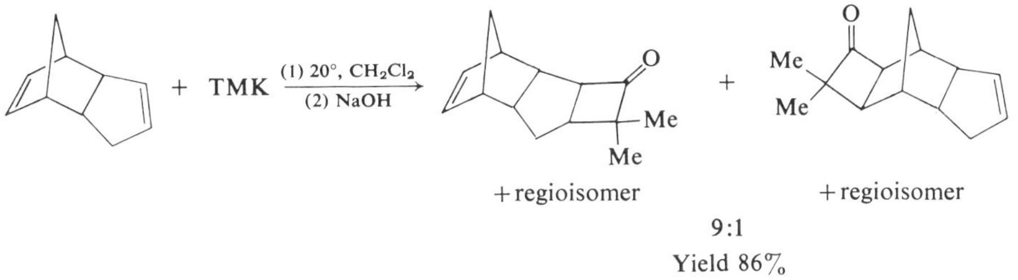

+ regioisomer   + regioisomer

9:1

Yield 86%

It would be rather difficult to explain this latter result on the basis of a stepwise mechanism involving successive formation of the two σ bonds. On the other hand, the data support the hypothesis of the formation of the two σ bonds in a single step involving an orthogonal interaction of the reactants. However, it is not necessary that bonding occur at both ends of the cumulene C=C bond as in the case of ketenes. The mechanism (Fig. 16) involving a three-membered ring intermediate resulting from the initial formation of both σ bonds at the central carbon atom of the cumulene has yet to be taken into consideration.

C. Reactions with Dienes

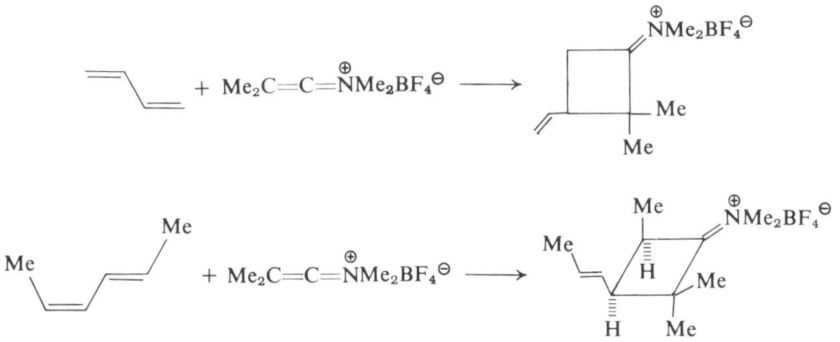

The reactions of tetramethylketeniminium salts with conjugated dienes are of particular interest. With transoid dienes such as butadiene, TMK reacts like ketenes and gives the [2 + 2] cycloadduct in 84% yield (Marchand-Brynaert and Ghosez, 1972). The reaction with cis,trans-2,4-hexadiene is particularly informative. It shows that these reactions are regio- and stereospecific and that a cis-1,2-disubstituted double bond reacts faster than the trans isomer. These observations are further arguments in favor of a multicenter process involving an orthogonal approach of the two reactants.

The behavior of tetramethylketeniminium salts toward cis-fixed dienes is totally different from that observed with ketenes. Thus, with cyclopentadiene, TMK reacts like an allene to give [4 + 2]-cycloadduct across the $C{=}\overset{\oplus}{N}{<}$ bond of the cumulene (Marchand-Brynaert and Ghosez, 1974).

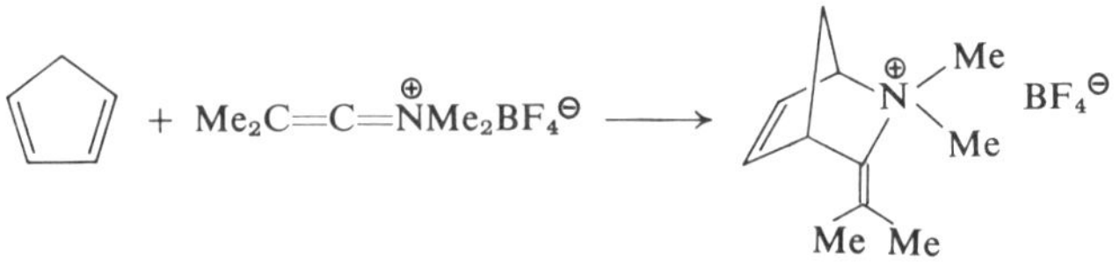

The importance of the population of cisoid and transoid conformations of the diene on the mode of cycloaddition is further demonstrated by the formation of both [2 + 2]- and [4 + 2]-adducts from 2,3-dimethylbutadiene and TMK (Marchand-Brynaert and Ghosez, 1974).

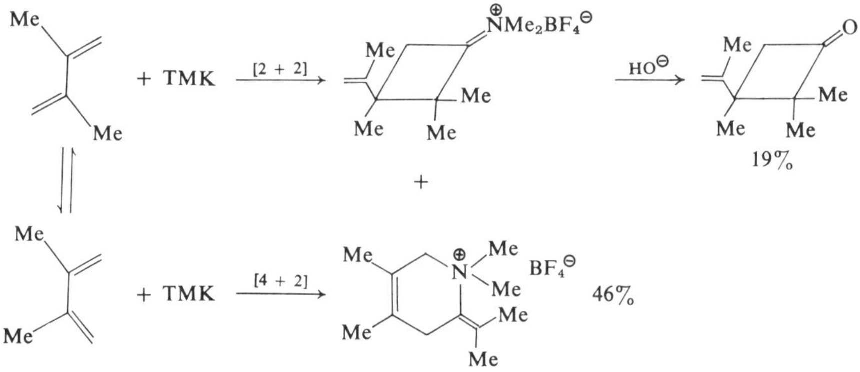

Only a few studies have been conducted concerning the mechanism of the cycloaddition reaction of keteniminium ions and allenes. Allene and TMK give a normal adduct whereas tetramethylallene reacts across the $C{=}\overset{\oplus}{N}{<}$ bond of TMK. 1,1-Dimethylallene gives both types of adducts: the unsubstituted double bond of the allene adds across the $C{=}C$ bond of TMK whereas the disubstituted double bond reacts at the $C{=}\overset{\oplus}{N}{<}$ bond (A. M. Leonard and L. Ghosez, unpublished results, 1976).

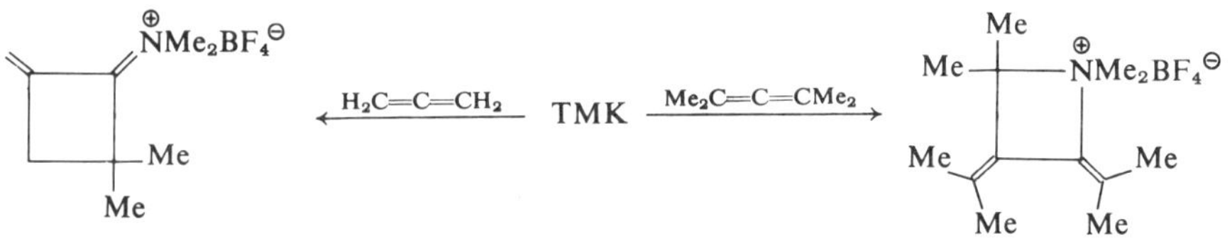

$Me_2C=C=CH_2$ + TMK ⟶ [structures shown] + [structure shown]

These data fit very well into the mechanistic picture discussed above for the reaction of TMK with olefins. The "normal" adducts are probably formed by a pericyclic reaction at the less-substituted double bond. On the other hand, addition at the 1,1-disubstituted double bond could involve an intermediate cation as shown in Fig. 17. The application of the Principle of

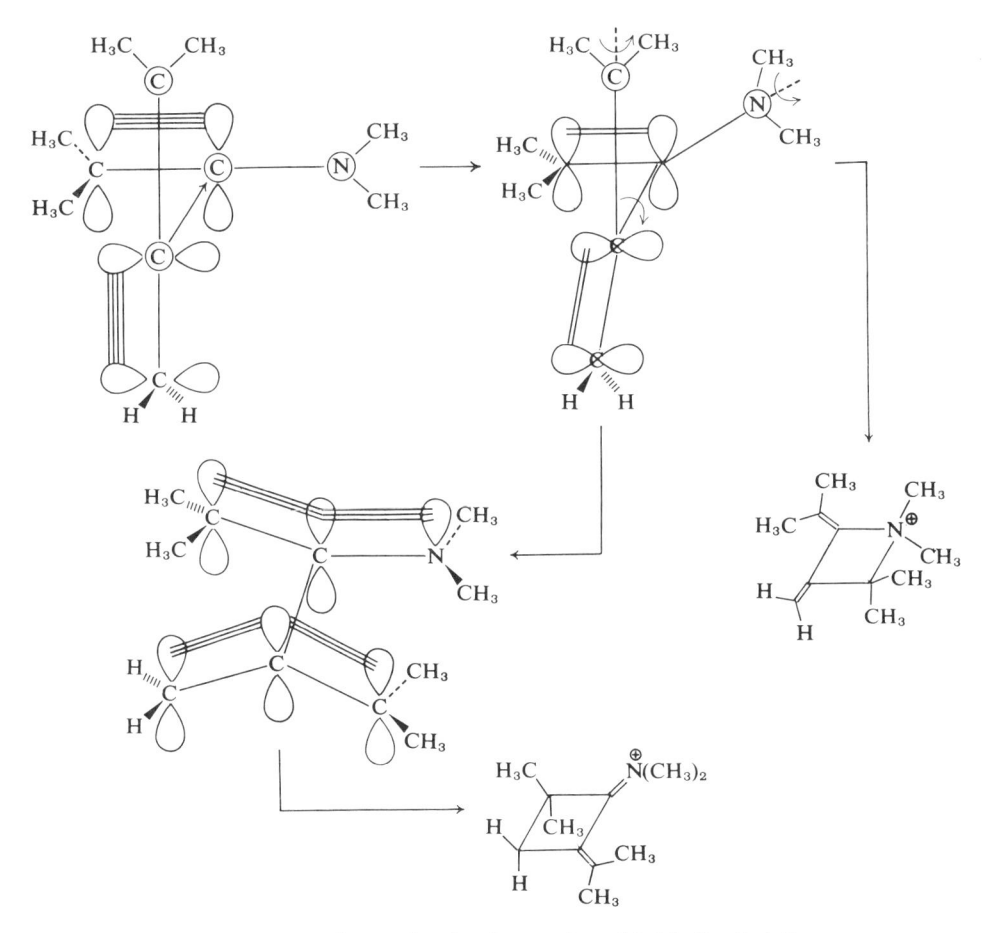

Fig. 17. Intermediate cation for the reaction with 1,1-dimethylallene.

Least Motion leads to the prediction that ring closure should occur preferentially between carbon and nitrogen. Carbon–carbon ring closure can only be achieved after a series of bond rotations.

IX. [2 + 4] Cycloadditions of Cumulenes

A. Structure and Dienophilic Reactivity

The cumulative arrangement of two carbon–carbon double bonds does not cause drastic changes in their dienophilic properties. Allene itself shows a weak dienophilic activity. With hexachlorocyclopentadiene the addition proceeds in satisfactory yield at 150° (Pledger, 1960). In accordance with the data for simple olefins, the dienophilic properties of the double bonds of

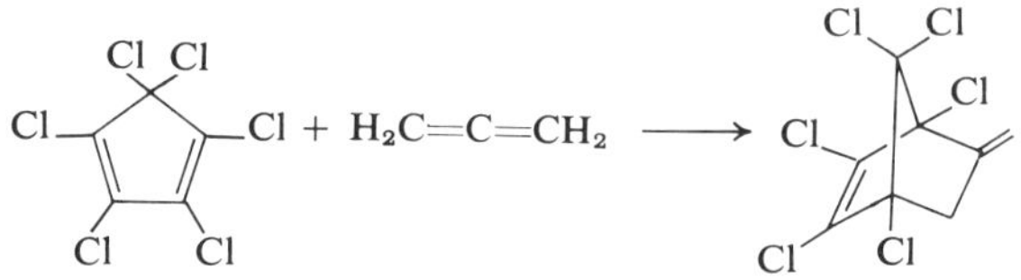

allenes increase when they are conjugated with electron-attracting groups. Cyanoallene and allene carboxylic and dicarboxylic acids undergo facile cycloadditions with cyclopentadiene (Jones et al., 1954; Kurtz et al., 1959; Agosta, 1962). Halogenated allenes are also potent dienophiles (Knoth and Coffman, 1960; Roedig et al., 1963) as shown, for instance, by the formation of a cycloadduct from tetrachloroallene and cyclopentadiene at 263°K.

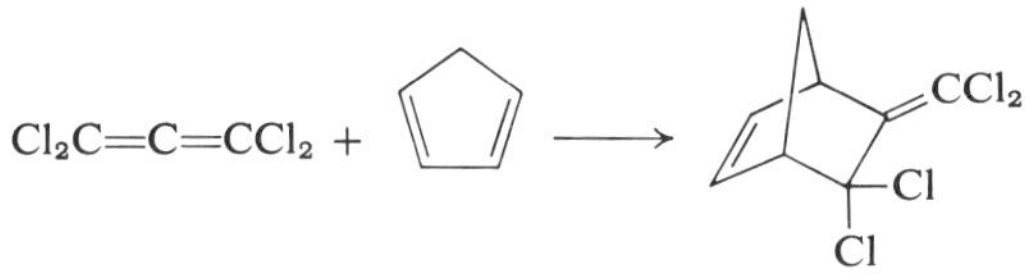

Replacement of the C=C bond in allene by the strongly electrophilic iminium functional group has already been shown (Section VIII) to increase the reactivity of the cumulene in [2 + 2] cycloadditions to such an extent that transoid dienes and keteniminium salts give four membered rings rather than Diels–Alder adducts. However, when the population of the cisoid conformation of the diene is increased or with cis-fixed dienes such as cyclopenta- or cyclohexadienes, Diels–Alder adducts are again formed (see Section VIII,C).

With ketenes which contain both the strong pull and push electronic systems favoring [2 + 2] cycloadditions, even the cis-constrained cyclopenta- and cyclohexadienes give cyclobutanones rather than the [4 + 2]-adducts. It is important to recall that, on the basis of electronic perturbations at the initial stages of the reaction, one should expect a [4 + 2] cycloaddition to occur at the C=O bond of the ketene rather than at the C=C bond. However, the introduction of electron-donating groups on the diene as well as the presence of anion-stabilizing substituents on the ketene are expected to increase the stability of a dipolar intermediate which could then lead, at least partially, to either of the two possible six-membered ring products. The formation of [2 + 4]-adducts in inter- and intramolecular reactions of ketenes with olefins are discussed in greater detail in Sections IX,B and IX,C.

B. Intermolecular [2 + 4] Cycloadditions

Introduction of the electron-donating alkoxy group at the 1-position of butadiene increases the reactivity of the diene and gives a mixture of [2 + 2]- and [2 + 4]-adducts with diphenyl ketene (Gouesnard, 1974). When a methoxy group is attached at the 2-position of the diene, only the 2,2-disubstituted double bond is activated. In this case, both diphenyl and butyl ethyl

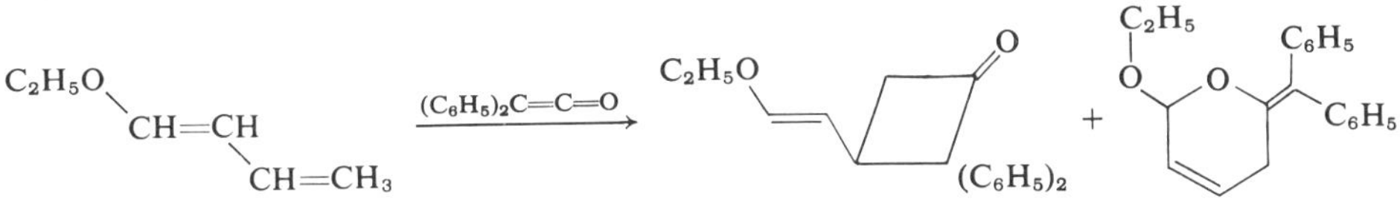

ketene yield [2 + 4]-adducts across the carbonyl double bond of the ketene (Martin et al., 1965).

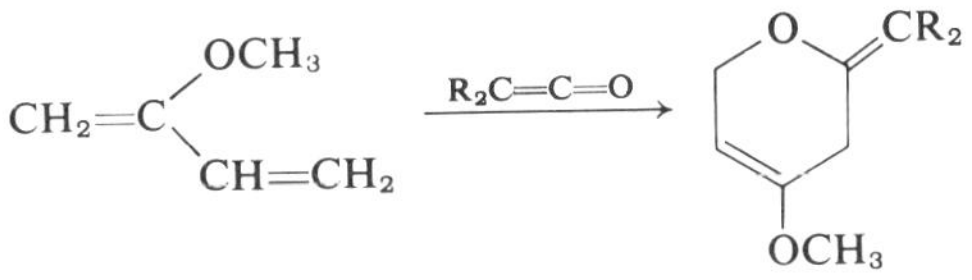

The anion stabilizing ability of the CF_3 group in bis(trifluoromethyl) ketene is again seen to alter the expected outcome of cycloaddition reactions with conjugated dienes. Thus, reaction with butadiene gives only a [2 + 4]-adduct resulting from addition of the carbonyl bond of the ketene across the 4_π diene system (England and Krespan, 1970a).

With *tert*-butylcyanoketene, the simultaneous presence of a good anion-stabilizing group and a bulky alkyl substituent is expected to favor a stepwise over a concerted pathway. Accordingly, both *tert*-butylcyanoketene (Brook and Hunt, 1974) and bis(trifluoromethyl) ketene (England and Krespan, 1970a) react anomalously with norbornadiene to give homo-Diels–Alder adducts, probably via a stepwise pathway.

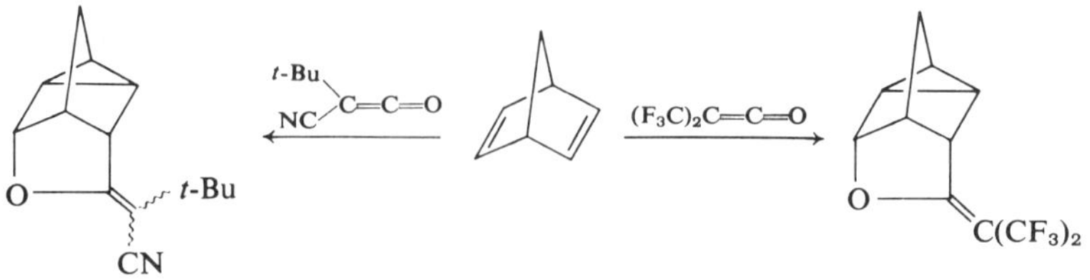

The [2 + 4] reactions of ketenes with activated dienes are most easily interpreted in terms of a two-step pathway involving a dipolar reaction intermediate which can cyclize to either four- or six-membered ring products. As the stability of such a dipolar species increase due to the substituent effects discussed above, it might be expected that the amount of six-membered ring product should increase. Thus, as the lifetime of **76** increases, the favorable conformation for the cyclization to the six-membered ring adduct can be obtained by bond rotation from the transoid zwitterionic intermediate first formed. This effect is well illustrated by the example given above. When R is phenyl, only the four-membered ring product is formed. If R is changed to $N(CH_3)_2$, which would delocalize the positive charge in **76** much more effectively than phenyl, only the six-membered ring adduct is observed (Gompper, 1969; Paul, 1969).

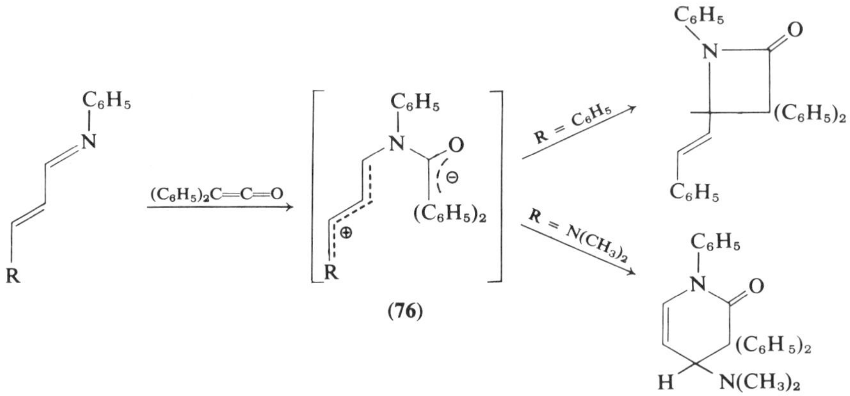

A further indication supporting the stepwise nature of ketene [2 + 4] cycloadditions is seen in the retro-Diels–Alder reaction of **77** (Murray and

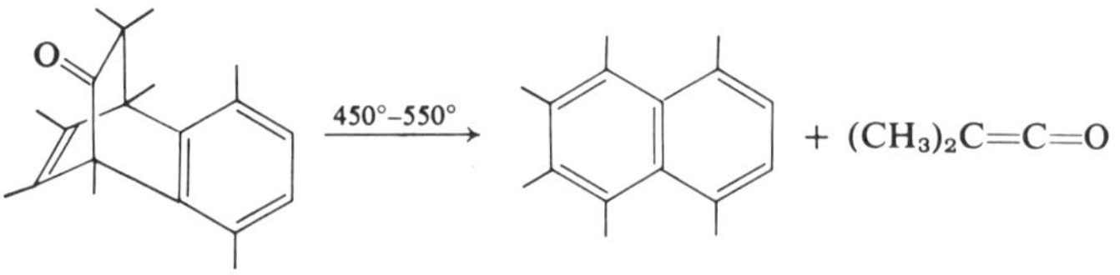

Hart, 1968). A concerted extrusion of dimethyl ketene to form an aromatic product might be expected to be easier than indicated by the extreme conditions necessary for the thermal reaction. The Principle of Microscopic Reversibility says that the presence of an intermediate for the reverse reaction necessarily implies a nonconcerted pathway for the forward reaction as well.

The activation parameters and the effect of a change in solvent polarity for ketene [2 + 4] cycloadditions have been determined (Gompper, 1969; Paul, 1969). Vinyl ether (**78**) reacts with diphenyl ketene to form a six-membered ring product. The activation parameters observed ($\Delta H^\ddagger = 10$ kcal mole^{-1}; $\Delta S^\ddagger = -47$ eu) as well as the small solvent effect ($k_{\text{acetonitrile}}/k_{\text{cyclohexane}} = 4.5$) can be interpreted in terms of either a concerted or a stepwise pathway. As discussed earlier, activation parameters and solvent effects must be considered with extreme care in the study of mechanistic pathways.

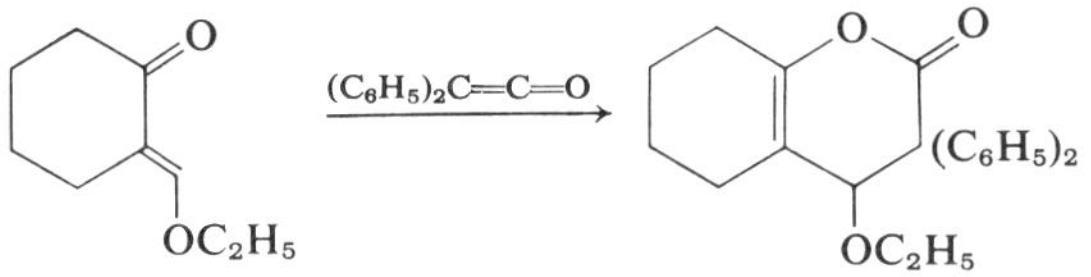

A system in which positive charge in the dipolar species is incorporated into an aromatic ring should be particularly well suited for a stepwise cycloaddition. This type of reaction is illustrated by the cycloaddition at room temperature of diphenyl ketene and tropone (Gompper *et al.*, 1968). The overall reaction is formally a cycloaddition in which tropone is utilized as a 1,3-dipolar species and the aromaticity of the tropylium cation can provide stabilization for a dipolar intermediate.

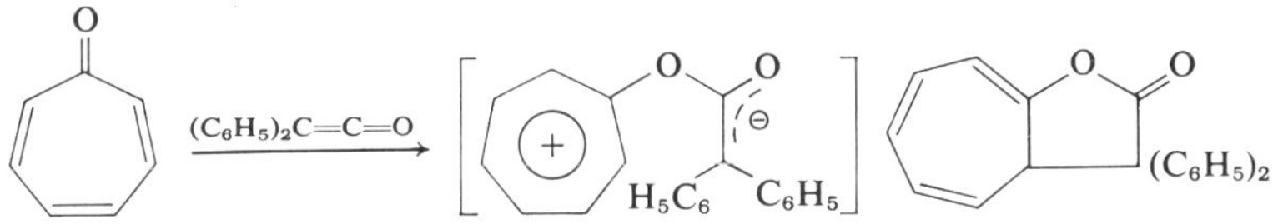

Ketene can be utilized as part of the 4π system in a [2 + 4] cycloaddition as shown below. The "*o*-quinoid" ketene (**79**) should benefit from a strong contribution of resonance form **80**. Reaction with acetone gives the six-membered adduct via addition across the carbonyl 2π system of the dienophile (Ziegler and Hanas, 1964).

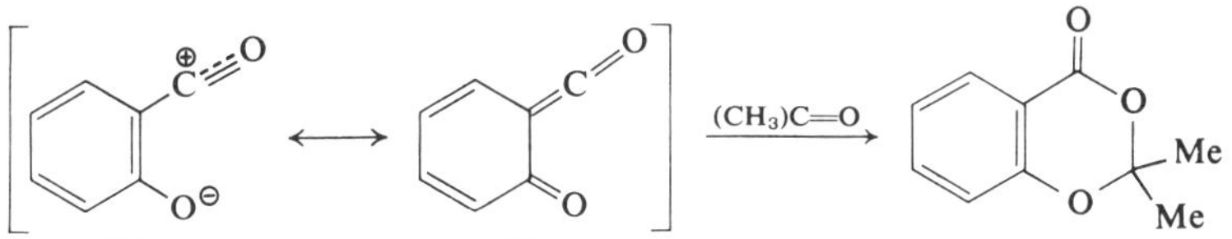

C. Intramolecular [2 + 4] Cycloadditions

Several interesting studies have appeared concerning the [2 + 4] intramolecular pericyclic reactions of ketenes (Hart *et al.*, 1966; Collins and Hart, 1967; Griffiths and Hart, 1968; Perst and Dimroth, 1968; Hart and Lankin, 1968; Chapman and Lassila, 1968; Hobson *et al.*, 1968; Chapman *et al.*, 1969; Kende *et al.*, 1969; Hart and Love, 1971; Goldschmidt *et al.*, 1973). The ketene, which is generated *in situ* either photochemically or thermally, can be trapped or observed spectroscopically at low temperature. Thermal reactions of these intramolecular ketenes often give interesting results.

When cyclohexadienone (**81**) was photolyzed at 77°K, an intense peak appeared below 250 nm (Morris and Waring, 1969). On warming this peak disappeared and the spectrum of the mixture of products **81** and **82** was observed. When the reaction was conducted in ether/isopentane/ethanol

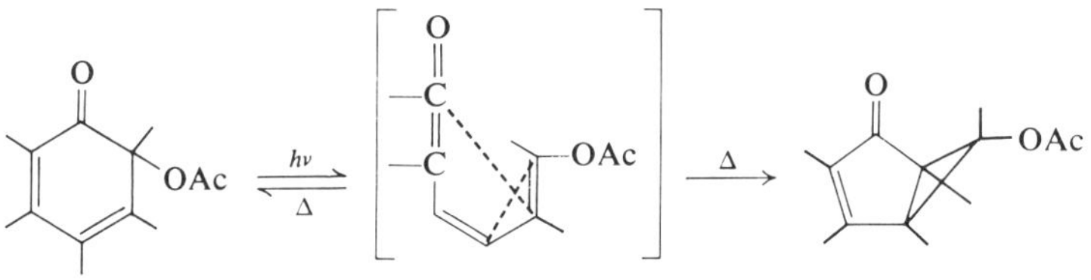

(5:5:2), the product distribution was 40% (**81**) and 60% (**82**). With hexane as the reaction medium, **81** was formed predominantly. The ketene (**83**) generated by photolysis of the starting cyclohexadienone (**81**) could recyclize thermally to starting material by a normal Diels–Alder type reaction or to the bicyclic product **82** by a thermally allowed ($_\pi 4_a + {_\pi}2_a$] process, a trans, trans Diels–Alder reaction. The solvent polarity effect on product composition was interpreted in terms of an asymmetrical transition state (or perhaps intermediate) along the pathway to **82**, which would favor its formation in more polar solvents.

X. Cycloadditions of Vinylcumulenes

The presence of a double bond conjugated with the cumulene system offers interesting possibilities for pericyclic cycloadditions.

Cycloadditions of vinylallene (**84**) with typical dienophiles have been reported by Jones *et al.* (1960), Fedorova and Petroc (1962), and Bertrand *et al.* (1968b, 1971). Tetracyanoethylene and **84** gave a crystalline adduct (**85**) in 88% yield after 3 hr at 0° in acetonitrile. With dimethylacetylenedicarboxylate the reaction was slower and gave the isomerized adduct (**86**).

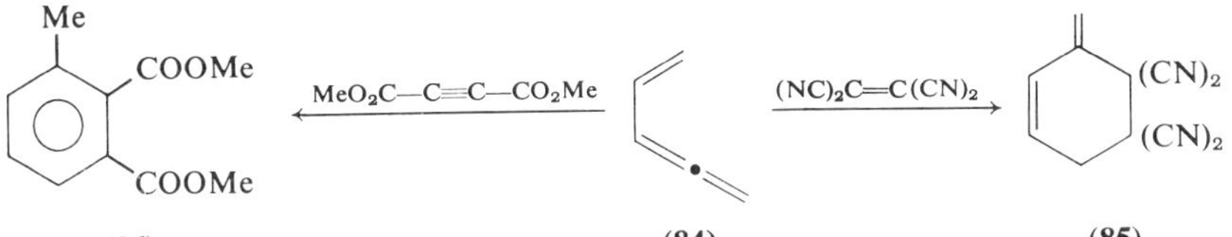

The cycloadditions of various substituted vinylallenes to methyl vinyl ketone have been found to be equally perispecific, yielding exclusively the [4 + 2]-cycloadducts (Bertrand et al., 1971). Significantly, *cis*-propenylallene, which cannot attain the cisoid conformation necessary for the Diels–Alder reaction, formed neither a [4 + 2]- nor a [2 + 2]-cycloadduct when exposed to methyl vinyl ketone.

The regiochemistry of the cycloadditions has been studied in detail (Bertrand et al., 1971). Vinylallene and its 3-alkyl-substituted derivatives reacted with methyl vinyl ketone to give a mixture of regioisomers **87** and **88**.

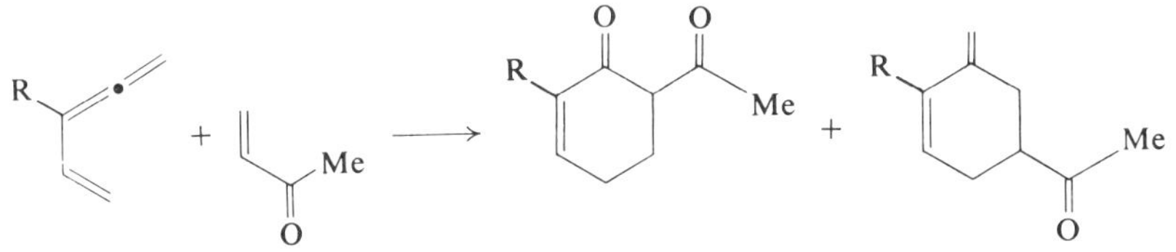

	87	88
R = H	28 : 72
R = CH$_3$	20 : 80
R = C$_3$H$_7$	15 : 85

The presence of an alkyl substitutent at the 5-position has a marked effect on the orientation of the cycloaddition. *trans*-Propenylallene and methyl vinyl ketone added in a single direction to yield the cis and trans cycloadducts **89** and **90** in a 3:1 ratio. These data are easily explained by a concerted pathway involving the vinylallene and the vinyl ketone as ($_\pi 4_s$ and $_\pi 2_s$] components, respectively.

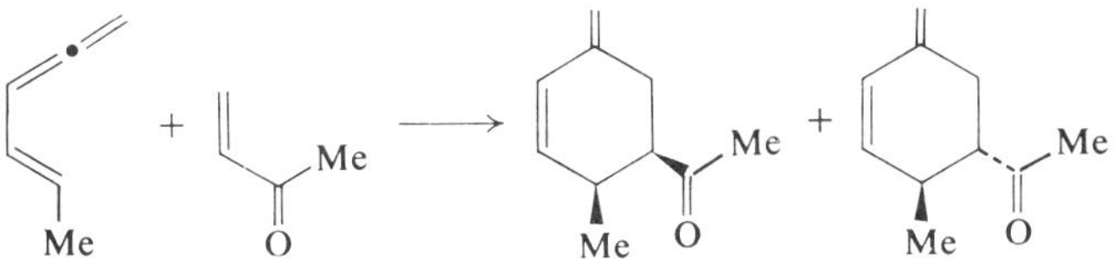

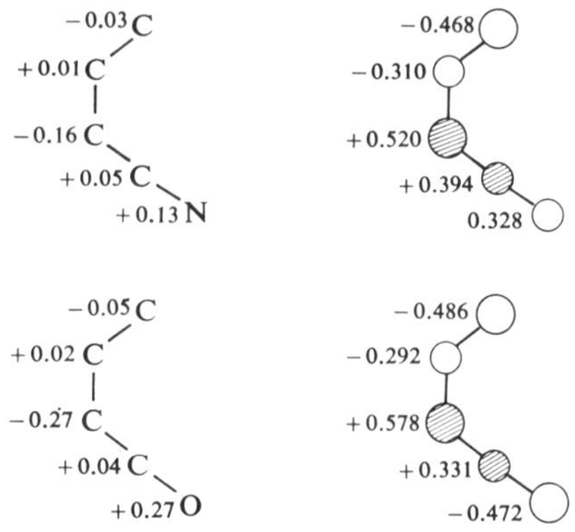

```
         −0.03 C                −0.468 ○
  +0.01 C                −0.310 ○
         |
  −0.16 C                +0.520 ◐
         +0.05 C         +0.394 ◐
                +0.13 N        0.328 ○


         −0.05 C                −0.486 ○
  +0.02 C                −0.292 ○
         |
  −0.27 C                +0.578 ◐
         +0.04 C         +0.331 ◐
                +0.27 O        −0.472 ○
```

Fig. 18. Electronic populations and coefficients for the structures of vinylheterocumulenes.

The diene systems of vinylketene and vinylketenimine are markedly influenced by the lone pair of the heteroatom as shown by the electronic populations and the coefficients of the HOMO's (Fig. 18) computed by E. Sonveaux, L. Ghosez, J. M. André, and A. Delhalle (unpublished results, 1975) for the cis planar geometry of both vinylheterocumulenes. These vinylheterocumulenes are electron-rich dienes which should react readily in Diels–Alder cycloadditions with electron-poor dienophiles. This has been experimentally verified by Sonveaux and Ghosez (1973; unpublished results,

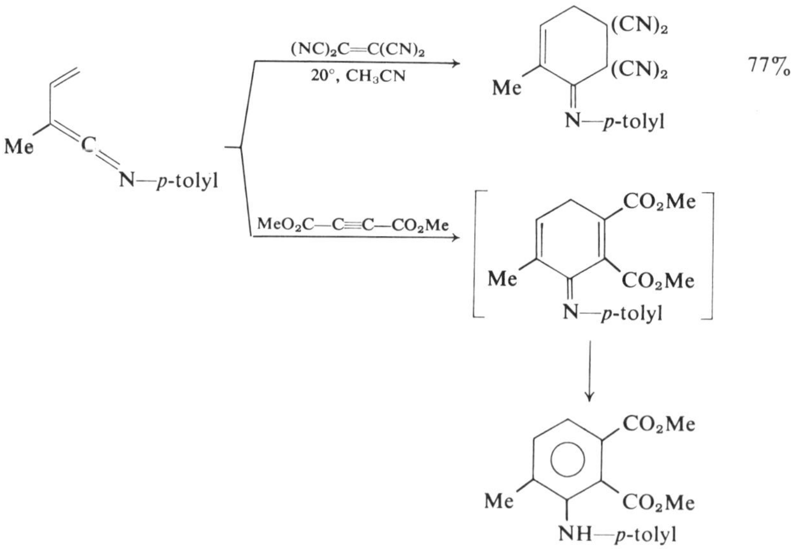

1974) in the preparation of the first representatives of the class of alkenylketenimines and the study of their facile cycloaddition to electron-deficient olefins and acetylenes, e.g., tetracyanoethylene and dimethylacetylenedicarboxylate. These cycloadditions occur in a single direction (Sonveaux and Ghosez, 1973; unpublished results, 1974) which can be predicted on the basis of the polarizability of the reacting π systems or of the coefficients of the interacting frontier molecular orbitals (HOMO of the vinylcumulene, LUMO of the dienophile) (Houk, 1973a,b; Epiotis, 1973c). The nature of the substituent R on nitrogen has a marked influence on the rate of the reaction. With 1,1-dicyanostyrene the reaction was completed in 35 min when R = cyclohexyl, whereas 30 hr were required when R = p-tolyl. This observation further demonstrates the important role of the nitrogen lone pair on the reactivity of the diene system in these pericyclic reactions.

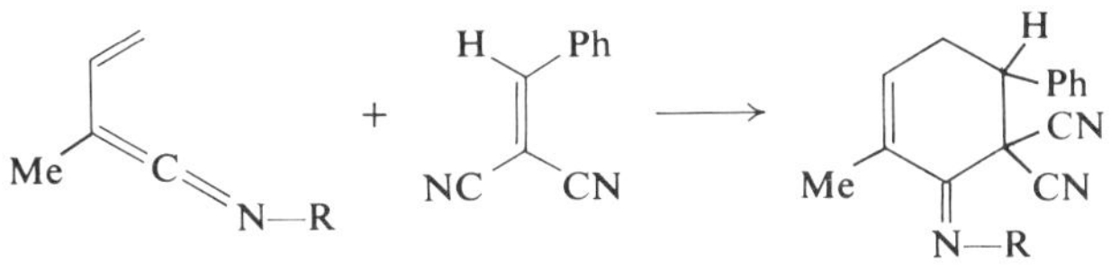

R = p-tolyl (100%)
R = cyclohexyl (96%)

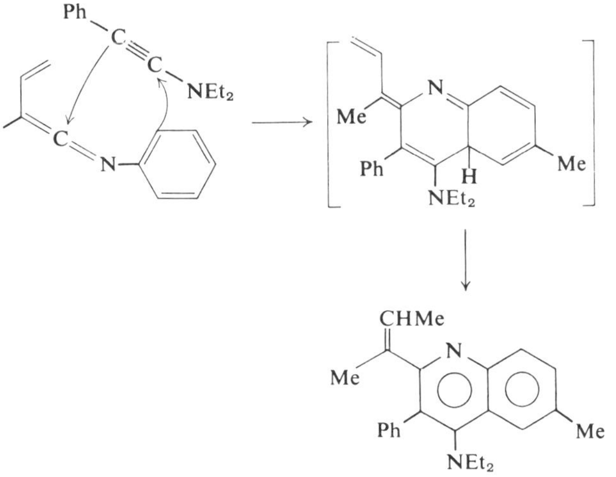

With electron-rich olefins and acetylenes, the dominant initial perturbation is expected to result from an interaction of the olefin HOMO and the vinylketenimine LUMO ($\pi_{C=N}-$). Interestingly, an electron-rich acetylene such as N-diethylaminophenylacetylene reacted with methyl vinyl ketone-p-tolylimine across the diene system formed by the C=N bond and one of the double bonds of the phenyl group to give a quinoline derivative (Sonveaux and Ghosez, 1973).

Few data are available concerning the behavior of the related vinyl ketones, probably because of their relative instability. However, it is interesting to mention here an observation by Payne (1966): a [2 + 2]-cycloadduct was obtained from reaction of isopropenyl ketene generated *in situ* and an electron-rich olefin such as ethyl vinyl ether.

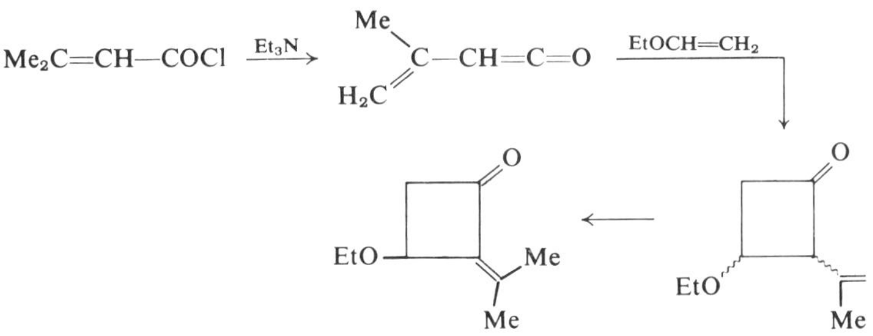

$$Me_2C=CH-COCl \xrightarrow{Et_3N} \underset{H_2C}{\overset{Me}{\diagdown}}C-CH=C=O \xrightarrow{EtOCH=CH_2}$$

These data on vinylcumulenes are good illustrations of the predictive power of simple and qualitative theoretical models like those used throughout this review. They also demonstrate the high chemical versatility of these cumulene systems. It can be safely predicted that our knowledge of the chemistry of these fascinating reactive organic molecules will continue to expand to the benefit of theoretical and synthetic organic chemistry.

Acknowledgments

The results which have been obtained in our laboratory at the University of Louvain come from the skilled efforts and ideas of Dr. R. Binamé, Dr. E. Cossement, J. P. Dejonghe, M. Delaunois-Job, M. Demuylder, Dr. W. Dumont, Dr. A. M. Hesbain-Frisque, C. Hoornaert, Dr. P. Laroche, Dr. A. M. Leonard, Dr. J. Marchand-Brynaert, Dr. R. Montaigne, C. de Perez, Dr. A. Roussel, Dr. A. Sidani, Dr. E. Sonveaux and Dr. P. Vannes.

We also wish to acknowledge the able collaboration of several young students and the technical assistance of A. Malengraux-Dekoker and H. Vanlierde.

We are grateful for the financial support of "Institut pour l'Encouragement de la Recherche Scientifique dans l'Industrie et l'Agriculture," "Fonds de la Recherche Fondamentale Collective," "Fonds National de la Recherche Scientifique," and Smith Kline and French Laboratories. One of us (MJO) thanks the University of Louvain for funds for a postdoctoral fellowship (1973–1975).

We also thank Mrs. P. De Ridder, Miss H. Vanlierde, Miss L. Coopman, and Mrs. Kathryn Krupa for their capable assistance in the preparation of the manuscript.

References

Agosta, W. (1962). *J. Am. Chem. Soc.* **84**, 110.
Alder, K., and Stein, G. (1931). *Justus Liebigs Ann. Chem.* **485**, 223.
Allred, E. L., Grant, D. M., and Goodlett, W. (1965). *J. Am. Chem. Soc.* **87**, 673.
Almenningen, A., Bastiansen, O., and Traetteberg, M. (1959). *Acta Chem. Scand.* **13**, 1699.
André, J. M., André, M. C., Leroy, G., and Weiler, J. (1969). *Int. J. Quantum Chem.* **3**, 1013.
Arendale, W. F., and Fletcher, W. H. (1953). *J. Chem. Phys.* **21**, 1898.
Baldwin, J. E., and Fleming, R. H. (1970). *Fortschr. Chem. Forsch.* **15**, 281.
Baldwin, J. E., and Fleming, R. H. (1972). *J. Am. Chem. Soc.* **94**, 2140.
Baldwin, J. E., and Fleming, R. H. (1973a) *J. Am. Chem. Soc.* **95**, 5249.
Baldwin, J. E., and Fleming, R. H. (1973b) *J. Am. Chem. Soc.* **95**, 5256.
Baldwin, J. E., and Fleming, R. H. (1973c) *J. Am. Chem. Soc.* **95**, 5261.
Baldwin, J. E., and Kapecki, J. A. (1969). *J. Am. Chem. Soc.* **91**, 3106.
Baldwin, J. E., and Kapecki, J. A. (1970a). *J. Am. Chem. Soc.* **92**, 4868.
Baldwin, J. E., and Kapecki, J. A. (1970b). *J. Am. Chem. Soc.* **92**, 4874.
Baldwin, J. E., and Roy, U. V. (1969). *Chem. Commun.* p. 1225.
Bampfield, H. A., and Brook, P. R. (1974a). *Chem. Commun.* p. 171.
Bampfield, H. A., and Brook, P. R. (1974b). *Chem. Commun.* p. 172.
Bampfield, H. A., Brook, P. R., and McDonald, W. S. (1975). *Chem. Commun.* p. 132.
Bartlett, P. D. (1971). *Q. Rev., Chem. Soc.* **25**, 473.
Barton, T. J., Rogido, R., and Clarady, J. C. (1970). *Tetrahedron Lett.* p. 2081.
Bauld, N. L., and Chang, C. S. (1972). *J. Am. Chem. Soc.* **94**, 7594.
Beetz, T., and Kellogg, R. M. (1973). *J. Am. Chem. Soc.* **95**, 7925.
Ben-Shoshan, R., and Pettit, R. (1968). *Chem. Commun.* p. 247.
Bertrand, M., and Le Gras, J. (1968). *Bull. Soc. Chim. Fr.* p. 3044.
Bertrand, M. Maurin, R., and Le Gras, J. (1968a). *C.R. Hebd. Seances Acad. Sci.* **267**, 417.
Bertrand, M., Grimaldi, J., and Waegell, B. (1968b). *Chem. Commun.* p. 1141.
Bertrand, M., Grimaldi, J., and Waegell, B. (1971). *Bull. Soc. Chim. Fr.* p. 962.
Bertrand, M., Le Gras, J., and Goré, J. (1972a). *Tetrahedron Lett.* p. 1189.
Bertrand, M., Le Gras, J., and Goré, J. (1972b). *Tetrahedron Lett.* p. 2499.
Bertrand, M., Le Gras, J., and Gil, G. (1974). *Tetrahedron Lett.* p. 37.
Bertrand, M., Maurin, R., Le Gras, J., and Gil, G. (1975a). *Tetrahedron* **31**, 849.
Bertrand, M., Le Gras, J., and Goré, J. (1975b). *Tetrahedron Lett.* **31**, 857.
Böhme, H., and Haake, M. (1976). *Adv. Org. Chem.* **9**, 107.
Borrmann, D. (1968). *In* "Methoden der organischen Chemie" (E. Mueller, ed.), Sauerstoff Verbindungen, II, Part 4, pp. 53–447. Thieme, Stuttgart.
Brady, W. T. (1971). *Synthesis* p. 415.
Brady, W. T., and Hoff, E. F., Jr. (1970). *J. Org. Chem.* **35**, 3733.
Brady, W. T., and Roe, R., Jr. (1970). *J. Am. Chem. Soc.* **92**, 4618.
Brady, W. T., Hoff, E. F., Roe, R., Jr., and Parry, F. H., Jr. (1969). *J. Am. Chem. Soc.* **91**, 5679.
Brady, W. T., Roe, R., Jr., Hoff, E. F., and Parry, F. H., III. (1970). *J. Am. Chem. Soc.* **92**, 146.
Braun, M., and Seebach, D. (1974). *Angew. Chem., Int. Ed. Engl.* **13**, 277.
Brook, P. R., and Hunt, K. (1974). *Chem. Commun.* p. 989.
Brook, P. R., Duke, A. J., and Duke, J. R. C. (1970a). *Chem. Commun.* p. 574.

Brook, P. R., Harrison, J. M., and Duke, A. J. (1970b). *Chem. Commun.* p. 589.
Brook, P. R., Duke, A. J., Harrison, J. M., and Hunt, K. (1974). *J. Chem. Soc., Perkin Trans. 1* p. 927.
Buenker, R. J. (1969). *J. Chem. Phys.* **48**, 1368.
Carless, H. A. J., and Lee, E. K. C. (1970). *J. Am. Chem. Soc.* **92**, 4482.
Chang, C. S., and Bauld, N. L. (1972). *J. Am. Chem. Soc.* **94**, 7593.
Chapman, O. L., and Lassila, J. D. (1968). *J. Am. Chem. Soc.* **90**, 2449.
Chapman, O. L., Kane, M., Lassila, J. D., Loeschen, R. L., and Wright, H. E. (1969). *J. Am. Chem. Soc.* **91**, 6856.
Collins, P. M., and Hart, H. (1967). *J. Chem. Soc. C* p. 895.
Cossement, E. (1973). Ph.D. Dissertation, University of Louvain.
Cossement, E., Binamé, R., and Ghosez, L. (1974). *Tetrahedron Lett.* p. 997.
Cox, A. P., Thomas, L. F., and Sheridan, J. (1959). *Spectrochim. Acta* **15**, 542.
Crandall, J. K., and Conover, W. W. (1974). *J. Org. Chem.* **39**, 63.
Crandall, J. K., and Sajka, S. A. (1972). *J. Am. Chem. Soc.* **94**, 5074.
Cripps, H. N., Williams, J. K., and Sharkey, W. H., Jr. (1959). *J. Am. Chem. Soc.* **81**, 2723.
Dai, S. H., and Dolbier, W. R., Jr. (1972). *J. Am. Chem. Soc.* **94**, 3946.
Davis, R. E. (1968). *Chem. Commun.* p. 248.
Delaunois, M., and Ghosez, L. (1969). *Angew. Chem., Int. Ed. Engl.* **8**, 72.
Del Bene, J. (1969). *J. Chem. Phys.* **51**, 2290.
Del Bene, J., and Jaffé, H. H. (1969). *J. Chem. Phys.* **50**, 1126.
Dewar, M. J. S. (1966). *Tetrahedron, Suppl.* **8**, Part I, 75.
Dewar, M. J. S. (1971). *Angew. Chem., Int. Ed. Engl.* **10**, 761.
Diels, O., and König, H. (1938). *Chem. Ber.* **71**, 1179.
Dolbier, W. R., Jr., and Dai, S. H. (1968). *J. Am. Chem. Soc.* **90**, 5028.
Dolbier, W. R., Jr., and Dai, S. H. (1970a), *J. Am. Chem. Soc.* **92**, 1774.
Dolbier, W. R., Jr., and Dai, S. H. (1970b). *Tetrahedron Lett.* p. 4645.
Do Minh, T., and Strausz, O. P. (1970). *J. Am. Chem. Soc.* **92**, 1766.
Dowd, P. (1970). *J. Am. Chem. Soc.* **92**, 1066.
Dumont, W. (1972). Ph.D. Dissertation, University of Louvain.
Duncan, W. G., Weyler, W., Jr., and Moore, H. W. (1973). *Tetrahedron Lett.* p. 4391.
Eaton, D. R., and Thompson, H. W. (1959). *Proc. R. Soc. London, Ser. A* **250**, 39.
Egger, K. W. (1973). *J. Am. Chem. Soc.* **95**, 1745.
England, D. C., and Krespan, C. G. (1970a). *J. Org. Chem.* **35**, 3300.
England, D. C., and Krespan, C. G. (1970b). *J. Org. Chem.* **35**, 3308.
England, D. C., and Krespan, C. G. (1970c). *J. Org. Chem.* **35**, 3312.
England, D. C., and Krespan, C. G. (1970d). *J. Org. Chem.* **35**, 3322.
Epiotis, N. D. (1972a). *J. Am. Chem. Soc.* **94**, 1924.
Epiotis, N. D. (1972b). *J. Am. Chem. Soc.* **94**, 1935.
Epiotis, N. D. (1973a). *J. Am. Chem. Soc.* **95**, 1191.
Epiotis, N. D. (1973b). *J. Am. Chem. Soc.* **95**, 1214.
Epiotis, N. D. (1973c). *J. Am. Chem. Soc.* **95**, 5624.
Epiotis, N. D. (1974). *Angew. Chem., Int. Ed. Engl.* **13**, 751.
Evans, M. G., and Warhurst, E. (1938). *Trans. Faraday Soc.* **34**, 614.
Farnum, D. G., Johnson, J. R., Hess, R. E., Marshall, T. B., and Webster, B. (1965). *J. Am. Chem. Soc.* **87**, 5191.
Fedorova, A. V., and Petrov, A. A. (1962). *Zh. Obshch. Khim.* **32**, 3537; *Chem. Abstr.* **58**, 12437c (1963).
Feiler, L. A., Huisgen, R., and Koppitz, P. (1973). *J. Am. Chem. Soc.* **96**, 2270.

Firl, J., and Runge, W. (1973), *Angew. Chem., Int. Ed. Engl.* **12**, 668.
Firl, J., and Runge, W. (1974). *Z. Naturforsch., Teil B* **29**, 393.
Fisher, H. (1964). *In* "The Chemistry of Alkenes" (S. Patai, ed.), pp. 1025–1159. Wiley (Interscience), New York.
Fletcher, V. R., and Hassner, A. (1970). *Tetrahedron Lett.* p. 1071.
Frey, H. M., and Hopf, H. (1973). *J. Chem. Soc., Perkin Trans. 2* p. 2016.
Frey, H. M., and Isaacs, N. S. (1970). *J. Chem. Soc. B* p. 830.
Fukui, K. (1971). *Acc. Chem. Res.* **4**, 57.
Gajewski, J. J., and Shih, C. N. (1967). *J. Am. Chem. Soc.* **89**, 4532.
Gajewski, J. J., and Shih, C. N. (1969). *J. Am. Chem. Soc.* **91**, 5900.
Gajewski, J. J., and Shih, C. N. (1972a). *J. Org. Chem.* **37**, 64.
Gajewski, J. J., and Shih, C. N. (1972b). *J. Am. Chem. Soc.* **94**, 1675.
Gheorgiu, M. D., Draghici, C., Stanescu, L., and Avram, M. (1973). *Tetrahedron Lett.* p. 9.
Ghosez, L., and de Perez, C. (1971). *Angew. Chem., Int. Ed. Engl.* **10**, 184.
Ghosez, L., and Marchand-Brynaert, J. (1976). *Adv. Org. Chem.* **9**, 421.
Ghosez, L., Laroche, P., and Bastens, L. (1964). *Tetrahedron Lett.* p. 3745.
Ghosez, L., Montaigne, R., and Mollet, P. (1966). *Tetrahedron Lett.* p. 135.
Ghosez, L., Haveaux, B., and Viehe, H. G. (1969). *Angew. Chem., Int. Ed. Engl.* **8**, 454.
Ghosez, L., Montaigne, R., Roussel, A., Vanlierde, H., and Mollet, P. (1971). *Tetrahedron* **27**, 615.
Ghosez, L., Cossement, E., Binamé, R., O'Donnell, M. J., and Dejonghe, J. P. (1974). "First International IUPAC Symposium on Organic Synthesis." Louvain-la-Neuve, Belgium.
Gilchrist, T. L., and Storr, R. C. (1972). "Organic Reactions and Orbital Symmetry." Cambridge Univ. Press, London and New York.
Gleiter, R. (1974). *Angew. Chem., Int. Ed. Engl.* **13**, 696.
Goldschmidt, Z., Gutman, U., Bakal, Y., and Worchel, A. (1973). *Tetrahedron Lett.* p. 3759.
Gompper, R. (1969). *Angew. Chem., Int. Ed. Engl.* **8**, 312.
Gompper, R., Studeneer, A., and Elser, W. (1968). *Tetrahedron Lett.* p. 1019.
Gouesnard, J. P. (1974). *Tetrahedron* **30**, 3113.
Griesbaum, K. (1966). *Angew. Chem., Int. Ed. Engl.* **5**, 933.
Griffiths, J., and Hart, H. (1968). *J. Am. Chem. Soc.* **90**, 3297.
Grimme, W., and Rother, H. J. (1973). *Angew. Chem., Int. Ed. Engl.* **12**, 505.
Grishin, Y. K., Ponomarev, S. V., and Lebedev, S. A. (1974). *Zh. Org. Khim.* **10**, 404.
Hanford, W. E., and Sauer, J. C. (1946). *Org. React.* **3**, 108.
Hart, H., and Lankin, D. C. (1968). *J. Org. Chem.* **33**, 4398.
Hart, H., and Love, G. M. (1971). *J. Am. Chem. Soc.* **93**, 6266.
Hart, H., Collins, P. M., and Waring, A. J. (1966). *J. Am. Chem. Soc.* **88**, 1005.
Hasek, R. H., Gott, P. G., and Martin, J. C. (1964). *J. Org. Chem.* **29**, 1239.
Hassner, A., Fletcher, V. R., and Hamon, D. P. G. (1971). *J. Am. Chem. Soc.* **93**, 264.
Hehre, W. J., Steward, R. F., and Pople, J. A. (1969). *J. Chem. Phys.* **51**, 2657.
Herndon, W. C. (1972). *Chem. Rev.* **72**, 157.
Herndon, W. C., Feuer, J., Giles, W. B., Otteson, D., and Silber, E. (1974). *In* "Chemical Reactivity and Reaction Paths" (G. Klopman, ed.), pp. 275–299. Wiley, New York.
Hewitt, T. G., and de Boer, J. J. (1971). *J. Chem. Soc. A* p. 817.
Hewitt, T. G., Anzenhofer, K., and de Boer, J. J. (1969). *Chem. Commun.* p. 312.
Hobson, J. D., Al Holly, M. M., and Malpass, J. R. (1968). *Chem. Commun.* p. 764.
Hoffmann, R. (1971). *Acc. Chem. Res.* **4**, 1.

Hoornaert, C., Hesbain-Frisque, A. M., and Ghosez, L. (1975). *Angew. Chem., Int. Ed. Engl.* **14**, 569.
Hopkinson, A. C. (1973). *J. Chem. Soc., Perkin Trans.* 2 p. 795.
Hopkinson, A. C., and Csizmadia, I. G. (1974). *Can. J. Chem.* **52**, 546.
Houk, K. N. (1972). *J. Am. Chem. Soc.* **94**, 8953.
Houk, K. N. (1973a). *J. Am. Chem. Soc.* **95**, 7287.
Houk, K. N. (1973b). *J. Am. Chem. Soc.* **95**, 7301.
Houk, K. N., Strozier, R. W., and Hall, J. A. (1974). *Tetrahedron Lett.* p. 897.
Hughes, R. P., and Powell, J. (1969). *J. Organomet. Chem.* **20**, 17.
Hughes, R. P., and Powell, J. (1973). *J. Organomet. Chem.* **60**, 409.
Huisgen, R., and Feiler, L. A. (1969). *Chem. Ber.* **102**, 3391.
Huisgen, R., and Mayr, H. (1975a). *Tetrahedron Lett.* p. 2965.
Huisgen, R., and Mayr, H. (1975b). *Tetrahedron Lett.* p. 2969.
Huisgen, R., and Otto, P. (1968). *J. Am. Chem. Soc.* **90**, 5342.
Huisgen, R., and Otto, P. (1969). *J. Am. Chem. Soc.* **91**, 5922.
Huisgen, R., and Steiner, G. (1973). *Tetrahedron Lett.* p. 3763.
Huisgen, R., Grashey, R., and Sauer, J. (1964). *In* "The Chemistry of Alkenes" (S. Patai, ed.), pp. 739–953. Wiley (Interscience), New York.
Huisgen, R., Feiler, L. A., and Otto, P. (1968). *Tetrahedron Lett.* p. 4485.
Huisgen, R., Feiler, L. A., and Binsch, G. (1969). *Chem. Ber.* **102**, 3460.
Inagaki, S., Minato, T., Yamak, S., Fujimoto, H., and Fukui, K. (1974). *Tetrahedron* **30**, 2165.
Isaacs, N. S., and Hatcher, G. B. (1974). *Chem. Commun.* p. 593.
Isaacs, N. S., and Stanbury, P. (1973). *J. Chem. Soc., Perkin Trans.* 2 p. 166.
Jacobs, T. L., and Kammerer, R. C. (1972). *J. Am. Chem. Soc.* **94**, 7190.
Janousek, Z., and Viehe, H. G. (1976). *Adv. Org. Chem.* **9**, 343.
Johns, J. W. C., Stone, J. M. R., and Winnerwisser, G. (1972). *J. Mol. Spectrosc.* **42**, 523.
Johnson, J. R., Strandberg, M. W. P. (1952). *J. Chem. Phys.* **20**, 687.
Jones, E. R. H., Mansfield, G. H., and Whitney, M. C. (1954). *J. Chem. Soc.* p. 3208.
Jones, E. R. H., Lee, H. H., and Whitney, M. C. (1960). *J. Chem. Soc.* p. 341.
Jutz, C. (1976). *Adv. Org. Chem.* **9**, 225.
Katz, T. J., and Dessau, R. (1963). *J. Am. Chem. Soc.* **85**, 2172.
Keim, W. (1968). *Angew. Chem., Int. Ed. Engl.* **7**, 879.
Kende, A. S., Goldschmidt, Z., and Izzo, P. T. (1969). *J. Am. Chem. Soc.* **91**, 6858.
Kiefer, E. F., and Okamura, M. Y. (1968). *J. Am. Chem. Soc.* **90**, 4187.
Kistiakowsky, G. B., Ruhoff, J. R., Smith, H. A., and Vaughan, W. E. (1936). *J. Am. Chem. Soc.* **58**, 146.
Knoth, W. H., and Coffman, D. D. (1960). *J. Am. Chem. Soc.* **82**, 3873.
Kraft, K., and Koltzenburg, G. (1967a). *Tetrahedron Lett.* p. 4357.
Kraft, K., and Koltzenburg, G. (1967b). *Tetrahedron Lett.* p. 4723.
Krow, G. R. (1971). *Angew. Chem., Int. Ed. Engl.* **10**, 435.
Kuehne, M. E., and Sheeran, P. J. (1968). *J. Org. Chem.* **33**, 4406.
Kurtz, P., Gold, H., and Disselnkötter, M., (1959). *Justus Liebigs Ann. Chem.* **624**, 1.
Lacey, R. N. (1960). *Adv. Org. Chem.* **2**, 213.
Lacey, R. N. (1964). *In* "The Chemistry of Alkenes" (S. Patai, ed.), pp. 1161–1227. Wiley (Interscience), New York.
Lawrence, R. B., and Strandberg, M. W. P. (1951). *Phys. Rev.* **83**, 363.
Letcher, J. H., Unland, M. L., and Van Wazer, J. R. (1969). *J. Chem. Phys.* **50**, 2185.
Luknitskii, F. I., and Vovsi, B. A. (1969). *Usp. Khim.* **38**, 1072; *Chem. Abstr.* **71**, 49141b (1969).

Marchand-Brynaert, J. (1973). Dissertation, University of Louvain.
Marchand-Brynaert, J., and Ghosez, L. (1972). *J. Am. Chem. Soc.* **94**, 2870.
Marchand-Brynaert, J., and Ghosez, L. (1974). *Tetrahedron Lett.* p. 377.
Martin, J. C., Gott, P. G., Goodlett, V. W., and Hasek, R. H. (1965). *J. Org. Chem.* **30**, 4175.
Maurin, R., and Bertrand, M. (1970). *Bull. Soc. Chim. Fr.* p. 998.
Mavrov, M. V., and Kucherov, V. F. (1967). *Usp. Khim.* **36**, 553; *Chem. Abstr.* **67**, 116478z (1967).
Metcalfe, J., Carless, H. A. J., and Lee, E. K. C. (1972). *J. Am. Chem. Soc.* **94**, 7235.
Montaigne, R., and Ghosez, L. (1968). *Angew. Chem., Int. Ed. Engl.* **7**, 221.
Montgomery, L. K., Schueller, K., and Bartlett, P. D. (1964). *J. Am. Chem. Soc.* **85**, 622.
Moore, W. R., Bach, R. D., and Ozretich, T. M. (1969). *J. Am. Chem. Soc.* **91**, 5918.
Moore, W. R., Mogoleska, P. D., and Traficante, D. D. (1972). *J. Am. Chem. Soc.* **94**, 4753.
Moriconi, E. J., and Crawford, W. C. (1968). *J. Org. Chem.* **33**, 370.
Morris, M. R., and Waring, A. J. (1969). *Chem. Commun.* p. 526.
Mulder, J. J. C., and Oosterhoff, L. J., (1970a). *Chem. Commun.* p. 305.
Mulder, J. J. C., and Oosterhoff, L. J. (1970b). *Chem. Commun.* p. 307.
Murray, R. K., Jr., and Hart, H. (1968). *Tetrahedron Lett.* p. 4995.
Nakamura, A. (1966). *Bull. Chem. Soc. Jpn.* **39**, 543.
Naser-Ud-Din, J., Riegl, J., and Skattebol, L. (1973). *Chem. Commun.* p. 271.
Nguyen Trong Anh (1970). "Les règles de Woodward-Hoffmann," pp. 156–157. Ediscience, Paris.
Okamoto, T. (1971). *Bull. Chem. Soc. Jpn.* **44**, 1353.
Okamoto, T. (1972). *Bull. Inst. Chem. Res., Kyoto Univ.* **50**, 450; *Chem. Abstr.* **78**, 96691k (1973).
Olah, G. A., and Westerman, P. W. (1973). *J. Am. Chem. Soc.* **95**, 3706.
Orville-Thomas, W. J., and Jones, W. J. (1960). *Z. Elektrochem.* **64**, 714.
Paquette, L. A., and Krow, G. R. (1969). *J. Am. Chem. Soc.* **91**, 6107.
Paul, K. P. (1969). Dissertation, University of Munich.
Payne, G. B. (1966). *J. Org. Chem.* **31**, 718.
Perst, H., and Dimroth, K. (1968). *Tetrahedron* **24**, 5385.
Petrov, A. A., and Fedorova, A. V. (1964). *Usp. Khim.* **33**, 3; *Chem. Abstr.* **60**, 7905c (1964).
Pitzer, K. S., and Strickler, S. J. (1964). *J. Chem. Phys.* **41**, 730.
Pledger, H. (1960). *J. Org. Chem.* **25**, 278.
Pople, J. A., and Gordon, M. (1967). *J. Am. Chem. Soc.* **89**, 4253.
Pullmann, B., and Berthier, G. (1949). *Bull. Soc. Chim. Fr.* p. 145.
Racanelli, P., Pautini, G., Immirzi, A., Allegra, G., and Porri, L. (1969). *Chem. Commun.* p. 361.
Reilly, J. L., and Krow, G. R. (1972). *J. Org. Chem.* **37**, 2364.
Rey, M., Roberts, S., Dieffenbacher, A., and Dreiding, A. S. (1970), *Helv. Chim. Acta* **53**, 417.
Roberts, J. D., and Sharts, C. M. (1962). *Org. React.* **12**, 1.
Roedig, A., Bischoff, F., Heinrich, B., and Märkl, G. (1963). *Justus Liebigs Ann. Chem.* **670**, 8.
Roth, W. R., and Erker, G. (1973a). *Angew. Chem., Int. Ed. Engl.* **12**, 503.
Roth, W. R., and Erker, G. (1973b). *Angew. Chem., Int. Ed. Engl.* **12**, 505.
Roth, W. R., Heiber, M., and Erker, G. (1973). *Angew. Chem., Int. Ed. Engl.* **12**, 504.
Roussel, A. (1970). Dissertation, University of Louvain.

Sabin, J. R., and Kim, H. (1972). *J. Chem. Phys.* **56**, 2195.
Salem, L. (1968). *J. Am. Chem. Soc.* **90**, 553.
Sidani, A., Marchand-Brynaert, J., and Ghosez, L. (1974). *Angew. Chem., Int. Ed. Engl.* **13**, 267.
Sonveaux, E., and Ghosez, L. (1973). *J. Am. Chem. Soc.* **95**, 5417.
Staudinger, H. (1912). "Die Ketene." Enke, Stuttgart.
Steur, R., van Dongen, J. P. C. M., de Brie, M. J. A., Drenth, W., de Haan, J. W., and van de Ven, L. J. M. (1971). *Tetrahedron Lett.* p. 3307.
Stevens, H. C., Reich, D. A., Brandt, D. R., Fountain, K. R., and Gaughan, E. J. (1965). *J. Am. Chem. Soc.* **87**, 5257.
Stoicheff, B. P. (1955). *Can. J. Phys.* **33**, 811.
Sustmann, R. (1971). *Tetrahedron Lett.* p. 2717.
Sustmann, R., Ansmann, A., and Vahrenholt, F. (1972). *J. Am. Chem. Soc.* **94**, 8099.
Taylor, D. R. (1967). *Chem. Rev.* **67**, 317.
Taylor, D. R., and Warburton, M. R. (1967). *Tetrahedron Lett.* p. 3277.
Taylor, D. R., Warburton, M. R., and Wright, D. B. (1972). *J. Chem. Soc., Perkin Trans. 1* p. 1365.
Trost, B. M. (1974). *Acc. Chem. Res.* **7**, 85.
Ulrich, H. (1967). "Cycloaddition Reactions of Heterocumulenes." Academic Press, New York.
van Dongen, J. P. C. M., de Brie, M. J. A., and Steur, R. (1973). *Tetrahedron Lett.* p. 1371.
Wagner, H. U., and Gompper, R. (1970). *Tetrahedron Lett.* p. 2819.
Weidler-Kubanek, A., and Litt, M. (1968). *J. Org. Chem.* **33**, 1844.
Weimann, L. J., and Christoffersen, R. E. (1973). *J. Am. Chem. Soc.* **95**, 2074.
Weingarten, H. (1970). *J. Org. Chem.* **35**, 3970.
Weiss, V. W., and Flygare, W. H. (1966). *J. Chem. Phys.* **45**, 3475.
Weyler, W., Jr., Byrd, L. R., Caserio, M. C., and Moore, H. W. (1972). *J. Am. Chem. Soc.* **94**, 1027.
Wilson, G., and Goldhamer, D. (1963). *J. Chem. Educ.* **40**, 599.
Woodward, R. B., and Hoffmann, R. (1969). *Angew. Chem., Int. Ed. Engl.* **8**, 871.
Ziegler, E., and Hanas, H. D. (1964). *Monatsh. Chem.* **95**, 1053.
Zimmerman, H. E. (1966). *J. Am. Chem. Soc.* **88**, 1564.
Zimmerman, H. E. (1969). *Angew. Chem., Int. Ed. Engl.* **8**, 1.
Zimmerman, H. E. (1971). *Acc. Chem. Res.* **4**, 272.

3

Cheletropic Reactions

WILLIAM L. MOCK

I.	Introduction	141
II.	Sulfolene Reactions	142
	A. Stereochemistry and Mechanism	143
	B. Sulfur Monoxide Extrusions and Additions	144
	C. Kinetics and Mechanism	146
	D. Cheletropic vs. 1,2-Addition	153
	E. Photochemistry	159
III.	Cheletropy in Other Systems	161
	A. Nitrous Oxide and Nitrogen Eliminations	162
	B. Phospholenes	165
	C. Sulfuranes	166
	D. Divalent Sulfur, Selenium, and Tellurium	168
	E. Silylene and Germylene	169
	F. Decarbonylations	171
	G. Miscellaneous Cheletropy	172
IV.	Conclusion	174
	References	174

I. Introduction

Cheletropic reactions were defined in the monumental article of Woodward and Hoffmann (1969a,b) as processes in which two σ bonds which terminate at a single atom are made, or broken, in concert. Provided were numerous examples of extrusions of CO, N_2, SO_2, and N_2O, together with a thorough stereochemical analysis of allowed reaction pathways. The latter are categorized as either *linear* (the least motion path) or *nonlinear* (nonleast motion, in which the process is contingent upon a sometimes difficultly visualized contortion of the reacting species). In this regard, Woodward and Hoffmann finally note mechanistic ambiguity for this class of pericyclic reactions. Since the cheletropic component involves scission (or formation) of a pair of bonds at one atom, in general for that moiety a stereochemical imprint of the transition state will be absent in the product. They conclude that more sophisticated scrutiny than examination of stereochemistry is required for verification of symmetry concepts in these reactions. Such an indication of a fruitful research area has not been ignored by the academic community.

We propose to examine developments since the formulation of the conceptual scheme of cheletropic reactions. Earlier work has been reviewed (Stark and Duke, 1967). Most intensively probed have been the cycloadditions and extrusions of sulfur dioxide. We shall review the accumulated evidence regarding the sulfolene reaction, its vinylogs and congeners, with a view to drawing such generalizations as are appropriate about the cheletropic process. In this exercise we shall draw most heavily upon our own work; limitations of space restrict comprehensive and equivalent coverage of all research on the subject. In a partial attempt to redress our slighting of much excellent work by others, this article concludes with an eclectic survey of recently reported cheletropic reactions which have caught this author's attention. It is his feeling that the true scope of synthetic possibilities of the cheletropic principle have not been widely enough appreciated. Perhaps the juxtaposition of a collection of such reactions connected by a unifying conceptual thread will contribute to that end.

II. Sulfolene Reactions

The addition of sulfur dioxide to butadiene is a venerable reaction [Eq. (1)] predating the Diels–Alder reaction in discovery and correct formulation (de Bruin, 1914a,b). Earlier work has been reviewed (Turk and Cobb, 1967), and

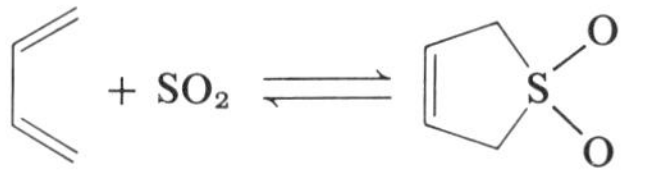

$$\text{butadiene} + SO_2 \rightleftharpoons \text{sulfolene} \quad (1)$$

the synthetic potential for the preparation and manipulation of dienes has been emphasized (Meyers, 1974). More recently an independent synthesis of 2,5-dihydrothiophenes has been recorded [Eq. (2)], much enhancing the

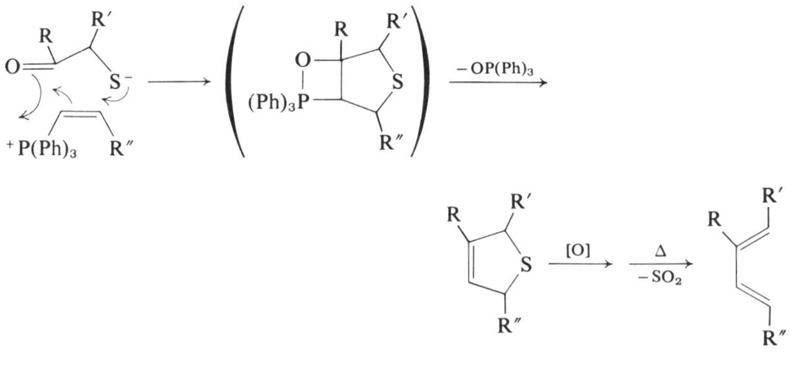

$$(2)$$

utility of this route to conjugated dienes (McIntosh *et al.*, 1974). In the following pages the term sulfolene will be used generically for sulfones which may be considered as cheletropic adducts of SO$_2$ and (poly)olefins, as well as for the pericyclic reactions so implied.

A. STEREOCHEMISTRY AND MECHANISM

Since complete stereochemical information pertaining to the cheletropic process is inaccessible by its nature, indirect evidence must be developed correlating structure with reactivity. One of the fundamental precepts of orbital symmetry conservation is reciprocal stereochemical alternation within a vinylogous series (e.g., [$_\pi 2_s + _\pi 2_a$], [$_\pi 4_s + _\pi 2_s$], [$_\pi 6_s + _\pi 2_a$]). Hence, the family of known heterocyclic fragmentations [Eqs. (3)–(5)] may reveal a pattern of reactivity which allows a stereochemical conclusion not sustainable by the behavior of a single member of the series. The facts are as presented.

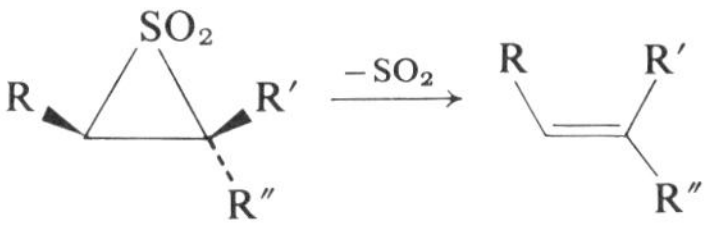

$$\underset{R''}{\underset{R}{\triangle}}\overset{SO_2}{\underset{}{}}R' \xrightarrow{-SO_2} \underset{R''}{\overset{R}{\diagdown}}=\underset{}{\overset{R'}{\diagup}} \qquad (3)$$

(1) R, R', R'' = Ar, alkyl or H in various combinations

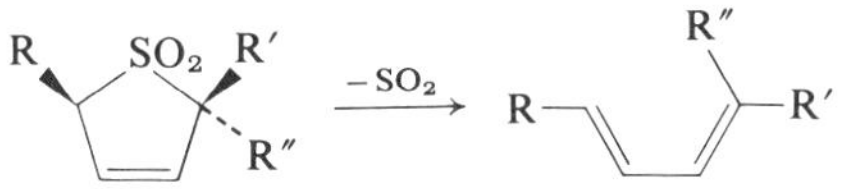

$$\underset{}{\overset{R\ SO_2\ R'}{\diagdown\!\!\diagup}}\ R'' \xrightarrow{-SO_2} R\diagdown\!\!\overset{R''}{\diagup}\!\!\diagdown R' \qquad (4)$$

(2) R, R' = CH$_3$; R'' = H
(3) R, R'' = CH$_3$; R' = H   > 99.9% stereospecific

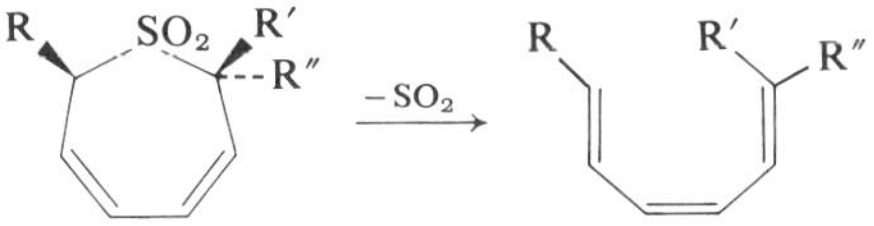

$$\underset{}{\overset{R\ SO_2\ R'}{\diagdown\!\!\!\diagup R''}} \xrightarrow{-SO_2} \underset{}{\overset{R\ \ R'\ R''}{\diagdown\!\!\diagup\!\!\diagdown}} \qquad (5)$$

(4) R, R' = CH$_3$; R'' = H
(5) R, R'' = CH$_3$; R' = H   > 97% stereospecific

In the case of **2** or **3** [Eq. (4)] SO$_2$ extrusion is cleanly *suprafacial* with respect to the incipient diene (Mock, 1966, 1975a; McGregor and Lemal, 1966; du Manoir *et al.*, 1972). On the other hand with **4** or **5** [Eq. (5)] the correlation between sulfone and triene indicates a unique *antarafacial* interaction for the hydrocarbon moiety (Mock, 1969, 1975a). These conclusions are summarized in T.S.-S and T.S.-A (where ··· means bond scission in the transition state). The observed stereochemistry is precisely that expected were the extrusion a *linear* concerted process; hence, these reactions may be tentatively described

T.S.-S   T.S.-A

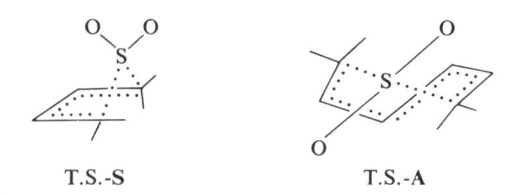

as retro-$[_\pi 4_s + _\omega 2_s]$ and $[_\pi 6_a + _\omega 2_s]$, respectively. However, this conclusion lacks firm support, for the following reasons:

1. In general, an experiment result which is in accord with a hypothesis only allows an *inference* that the hypothesis is true (Larrabee, 1974); many events in accord with the criteria for prediction are required before truth may be taken as established, according to the methodology of inductive reasoning.

2. In particular, in the examples cited [Eqs. (4) and (5)], an analysis of steric considerations (Mock, 1969, 1975a) reveals that bond-angle deformation and nonbonded interactions offer sufficient explanation [with the possible exception of the cis isomer (4)] for the observed stereospecificities, without the necessity of invoking orbital symmetry constraints.

3. The first member of the series (1) [Eq. (3)] is an obvious exception to the trend exhibited in the latter cases [Eqs. (4) and (5)]. Episulfone fragmentation is cleanly *suprafacial* in a large number of examples (Fischer, 1970; Carpino *et al.*, 1971). Accordingly, it must either be a *nonlinear* cheletropic reaction $[_\pi 2_s + _\omega 2_a]$, or it must be nonconcerted. In the latter case it is necessary and sufficient that dissociation of a dipolar (or diradical) intermediate be more rapid than internal rotation [Eq. (6)]. Whichever of these alternatives one

$$\text{(6)}$$

adopts (this question will be addressed subsequently), it must be concluded that the *magnitude* of orbital symmetry constraints in these systems is seriously in question, for even in the most highly selective example [Eq. (4)], the existence of a nonstereospecific pathway which is only 4 to 5 kcal/mole higher in activation energy may not be excluded by the observed results. The current veil of enthusiasm for the power of orbital symmetry considerations should not be allowed to obscure a critical uncertainty of their relevance in this particular case.

B. Sulfur Monoxide Extrusions and Additions

At this point it is convenient to digress to the analogous sulfoxides of the systems just considered. In the case of episulfoxides only partial stereospecificity is observed [Eq. (7)]. The reaction has been interpreted as proceeding through a short lived zwitterionic intermediate in which internal rotation partially occurs (Hartzell and Paige, 1967; Baldwin *et al.*, 1971a; Kondo *et al.*, 1972).

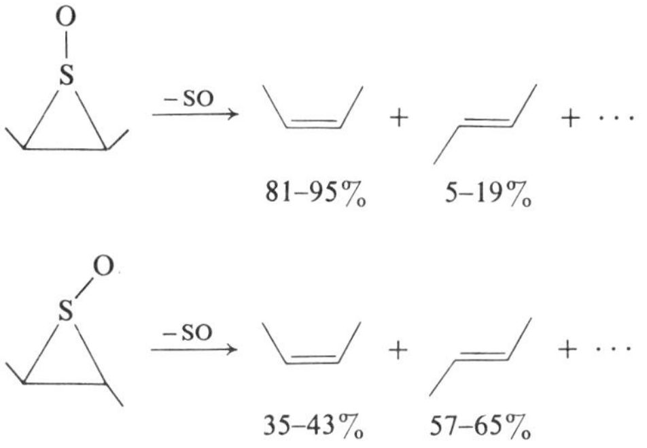

$$\underset{}{\text{(episulfoxide)}} \xrightarrow{-SO} \underset{81-95\%}{\diagup\!\!\!\diagdown} + \underset{5-19\%}{\diagup\!\!\!\diagdown} + \cdots$$

$$\underset{}{\text{(episulfoxide)}} \xrightarrow{-SO} \underset{35-43\%}{\diagup\!\!\!\diagdown} + \underset{57-65\%}{\diagup\!\!\!\diagdown} + \cdots \qquad (7)$$

Sulfur monoxide released from episulfoxides adds to dienes and trienes in manner analogous to sulfur dioxide, but in such cases it appears to do so with sequential bond formation (Dodson and Sauers, 1967; Dodson and Nelson, 1969). A puzzling pattern of relative reactivity and stereoselectivity [Eq. (8)]

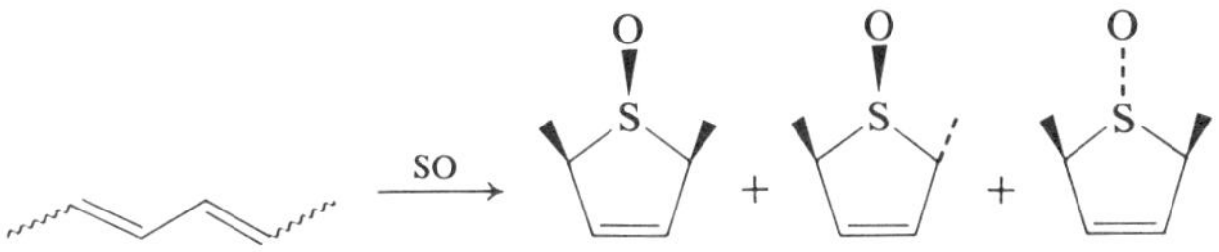

$$\text{diene} \xrightarrow{SO} \text{(sulfolene)} + \text{(sulfolene)} + \text{(sulfolene)} \qquad (8)$$

				Relative reactivity
E,E	0%	13%	87%	2.5
E,Z	trace	95%	5%	1.6
Z,Z	20%	61%	19%	1.0

results from the reversible addition of SO to the isomers of 2,4-hexadiene (Chao and Lemal, 1973; Lemal and Chao, 1973). It has been argued (on thermodynamic grounds) that *triplet* SO ($^3\Sigma^-$), the ground state species, is the reactive intermediate in these cycloadditions and cycloreversions, and that the *transoid* conformer of the diene must also be a major participant. (Attempts to isolate SO lead only to disproportionation products.) Accordingly, biradical intermediates are involved; therefore, these reactions do not meet the definition of cheletropic processes. However, the fact that substantial stereospecificities are observed in such cases should signal a note of caution in applying the steric criteria for concerted processes.

C. Kinetics and Mechanism

It was established (Sections I and II,A) that in the case of the sulfolene reactions, stereochemical correlations are inherently nonquantitative for the purpose of establishing the magnitude of orbital symmetry constraints. What is desired are relative activation energies for symmetry-allowed processes vs. their nonconcerted equivalents or (if possible) vs. disallowed pathways. This implies that rate measurements in suitably designed systems could reveal a substantial differential in reactivity which might be attributed to electronic factors.

Table I contains a number of sulfones for which the rate of extrusion of SO_2 has been measured. These will now be considered for the purpose indicated. The critical entries are n and o; the others in one way or another serve as controls which validate the conclusion to be derived from the former pair. However, the other sulfones do have some intrinsic interest, which will be developed first.

1. Five-Membered Ring

The first seven sulfolenes represent most of the possible simple substitution patterns for this five-membered ring. (Adducts with geminal α-substitution

TABLE I Rates and Free Energy of Activation for Sulfolene Dissociations

	Sulfone structure	Relative rate (125°)	$\Delta G^{\ddagger}$ (kcal/mole)	Reference
a.	 ⟨SO₂⟩	1.0	30.1 (125°)	(Grummitt et al., 1950)
b.	 ⟨SO₂⟩	0.49	30.6 (131°)	(Grummitt et al., 1950)
c.	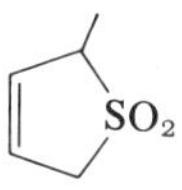 ⟨SO₂⟩	5	28.5 (105°)	(Grummitt et al., 1950)
d.	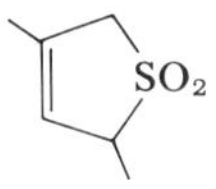 ⟨SO₂⟩	3.6	29.0 (111°)	(Grummitt et al., 1950)
e.	 ⟨SO₂⟩	0.1	31.9 (148°)	(Grummitt et al., 1950)

TABLE I (continued)

	Sulfone structure	Relative rate (125°)	$\Delta G^{\ddagger}$ (kcal/mole)	Reference
f.	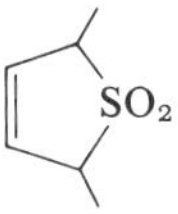 SO₂ (3-methyl cyclopentene sulfone)	57	26.5 (93°)	(Mock, 1975b)
g.	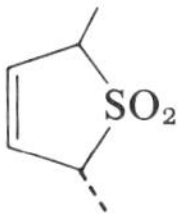 SO₂	0.62	30.3 (141°)	(Mock, 1975b)
h.	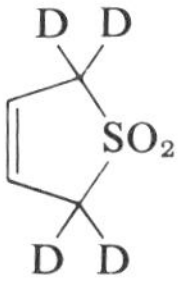 SO₂ (D,D,D,D)	0.92	—	(Ašperger et al., 1972)
i.	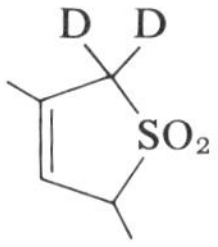 SO₂ (D,D)	3.40		(Ašperger et al., 1972)
j.	 ³⁴SO₂	0.99		(Ašperger et al., 1972)
k.	 SO₂	15	27.8 (93°)	(Mock, 1975b)
l.	 SO₂	1.2	29.7 (128°)	(Mock, 1975b)
m.	 SO₂	1.0	30.1 (125°)	(Mock, 1970b)
n.	 SO₂	5.9	28.6 (108°)	(Mock, 1970a, 1975b)
o.	 SO₂	0.0001 (est)	39.4 (256°)	(Mock, 1970a, 1975b)

appear generally to be thermodynamically disfavored with respect to their components; synthesis requires indirect means.) A pattern of reactivity is apparent, which generally parallels the corresponding dissociation constants for these sulfolenes. Substituents adjacent to the sulfone group are destabilizing, whereas β-substitution (on the sulfolene olefin) confers stability. Entry *g* (*trans*-2,5-dimethylsulfolene) is an exception; in this case the kinetic stability is attributed to methyl-hydrogen compression in the transition state for a concerted process; see T.S.-S (McGregor and Lemal, 1966; Mock, 1966, 1975a). In all cases the relative rates are surprisingly similar (compare $\Delta G^{\ddagger}$ values), indicating a mechanism with minimal susceptibility to substituent effects of the sort to be expected from ionic or radical mechanisms.

The secondary deuterium isotope effect (compare *a, h; d, i*) and the primary ^{34}S isotope effect (*a, j*) were determined (Ašperger *et al.*, 1972). In the former case $k_H k_D$ ratios of 1.094 (both methylenes deuterated) and 1.054 (one methylene deuterated) were taken to suggest that both C–S bonds in fact suffered rupture in the transition state (concerted reaction). The ^{34}S isotope effect (0.9%, compare with maximum expectation of 1.3% for a single C–S bond) was also consistent with a synchronous process with partial bond scission.

Quite recently, the rate of sulfolene *additions* has been measured (Isaacs and Laila, 1976).

2. Three-Membered Ring

Entry *k* in Table I, episulfone [thiirane dioxide (**1**)], possesses a kinetic stability which is surprisingly similar to the five-membered ring sulfolenes.

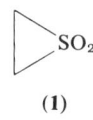

$$\triangleright\!\!\text{SO}_2$$

(1)

It will be recalled that the three-membered ring dissociation is also cleanly suprafacial with respect to the incipient olefin (Section II,A), and hence must either be nonlinear concerted or two-step (without intervening internal rotation). Previous experimental work on this ring has tended to favor the latter interpretation. In the case of phenyl-substituted thiirane dioxides, rates of dissociation have been observed leading to activation energies on the order of 17.5 to 18.5 kcal/mole (Bordwell *et al.*, 1968). The near identity of rates of decomposition of monophenyl-, *cis*-diphenyl-, and *trans*-diphenylepisulfones (inference: only one aryl group provides conjugative stabilization in the transition state) and rate acceleration accompanying increasing ionizing power of the medium (inference: "dipolar radical" mechanism) has led to a preference for sequential bond rupture (Bordwell *et al.*, 1968). However, it will be noted that the activation energy for the parent (*k*) is some *10 kcal/mole higher* than for its phenylated derivatives. This indicates a strong perturbation of the fragmentation mechanism for episulfone by an appended aryl group. Accordingly, while a stepwise mechanism may occur with thiirane rings bearing potential stabilizing substituents, the nonlinear concerted path may not be excluded for the parent. Rate studies of episulfones are inherently ambiguous because of the effects of ring strain. In the example **1** (*k*), there is certainly a thermodynamic destabilization due to bond angle deformation which would tend to accelerate fragmentation relative to sulfolenes *a–g*. Apparently a counterbalancing kinetic stabilization, which may be attributed to orbital symmetry constraints, neutralizes any potential rate acceleration. Since the magnitude of the expected destabilization cannot be known, no quantitative conclusion may be drawn about the importance of the symmetry constraint in *k*; for mechanistic evidence of cheletropy one must look to other substances. We have elsewhere provided further thoughts on the mechanism of SO_2 extrusion from episulfones (Mock, 1975b).

3. Seven-Membered Ring

Entry *l* in Table I, dihydrothiepin dioxide (**6**) represents a considerable expansion of the scope of the sulfolene reaction. To digress for a moment from questions of mechanism, the seven-membered ring sulfolene (**6**) may be prepared in good yield by 1,6-cycloaddition between *cis*-hexatriene and sulfur

(6)

dioxide, reverse of Eq. (5), R, R', R" = H (Mock, 1967). This is a general synthetic reaction; adducts have similarly been obtained from 2-acetoxymethyl-*cis*-hexatriene, 2-isopropyl-5-methyl-*cis*-hexatriene, and 2,4-dimethyl-*cis*-hexatriene, as well as other examples (Mock and McCausland, 1976). However, in at least one other case only a 1,4-adduct could be obtained from an apparently suitable triene (*cis*-1,2-dicyclohexenylethylene). That 1,4- and 1,6-additions are normally competing processes is borne out by the kinetic data in Table I. Although the near identity of rates (compare *a* and *l*) applies to extrusion, it may reasonably be inferred that the two modes of cycloaddition should be competitive, and that for any given triene case fairly subtle factors will dictate which path is favored. For the purpose of mechanistic investigation, the relevant point is that 1,6-extrusion in a *geometrically unconstrained* situation proceeds (antarafacially, see T.S.-A, Section II,A) at the same rate as 1,4-extrusion (suprafacially, T.S.-S). This conclusion is important to the comparison of *n* and *o* (see later, Section II,C,6). It might be noted in passing that the SO_2 adducts have good storage characteristics and may be recommended as protected forms of the highly reactive *cis*-hexatriene functionality, which may be liberated by mild thermolysis.

4. Homoconjugate Elimination

Entry *m* in Table I, 3-thiabicyclo[3.1.0]hexane 3,3-dioxide (**7**), constitutes a further salient example of the synthetic utility of the sulfolene reaction (Mock, 1970b). This substance and certain of its derivatives may be prepared by cycloaddition of diazomethane to sulfolenes followed by photolysis. Since thermolysis ($-SO_2$) readily yields 1,4-pentadiene, the overall scheme shown [Eq. (9)] allows insertion of a methylene group between conjugated double bonds.

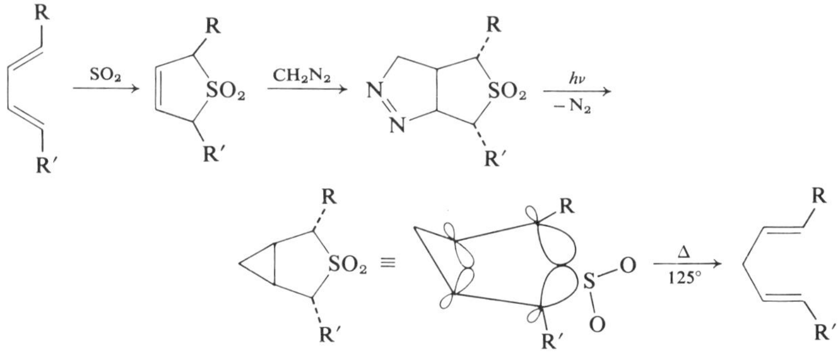

Considering that fragmentation of **7** is formally a $[_\sigma 2_s + _\sigma 2_s + _\sigma 2_s]$ process, the facility with which it takes place is surprising, for the rates of dissociation of *a* and *m* are virtually identical at 125°. Mechanistically, this was taken to indicate coupling of C–S bond scission with opening of the cyclopropane ring, i.e., a concerted reaction driven by relief of ring strain. (The extrusion is fully stereospecific.) This is supported by the inertness of the tricyclic structure (**8**), in which conformational and bridgehead constraints would force orthogonality (diminished π overlap) within the incipient double bonds of the potential homoconjugated product. The activation energy for dissociation of **8** [Eq. (10)] was observed to be more than 15 kcal ($\Delta\Delta G^\ddagger$) higher than that recorded for *a* or *m*.

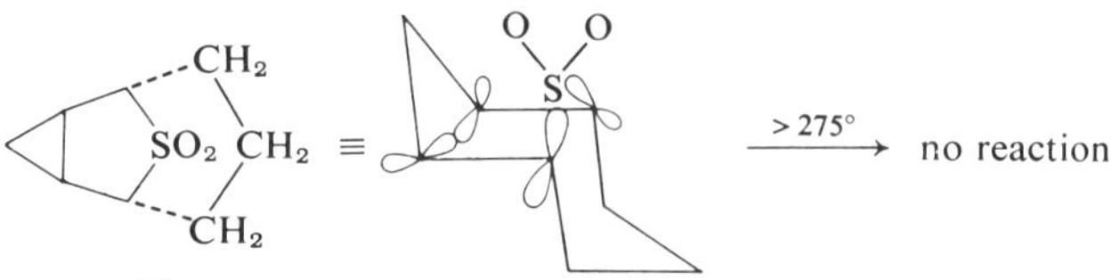

5. Sulfolane Fragmentation

In passing, it should be noted that the above described synthetic strategy is applicable to the stereospecific preparation of divinyl ethers and divinyl amines [Eq. (11)] (Mock, 1970b; Meyers and Takaya, 1971). Also, 3-thia-[3.2.2]propellane dioxides decompose similarly under very mild conditions

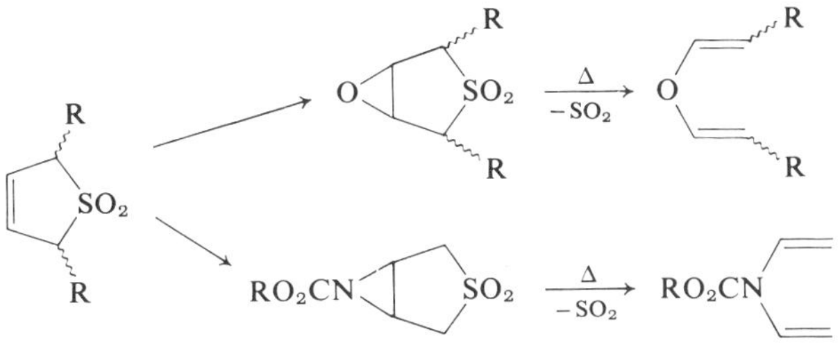

(11)

(Lantos and Ginsberg, 1972). Homoconjugate elimination, as in Section II,C,4, may be rendered even more facile in tricyclic systems by an appropriately situated double bond (van Tilborg et al., 1975).

Recently the stereochemical course of sulfolane fragmentation has been examined (Mock et al., 1975). Flash pyrolysis of the stereoisomers of 2,3-dimethyltetrahydrothiophene 1,1-dioxide under necessarily extremely vigorous conditions gave from either sulfone a mixture of 2-butenes (plus other products) in approximately the same ratio [Eq. (12)]. A sequential bond scission process is inferred.

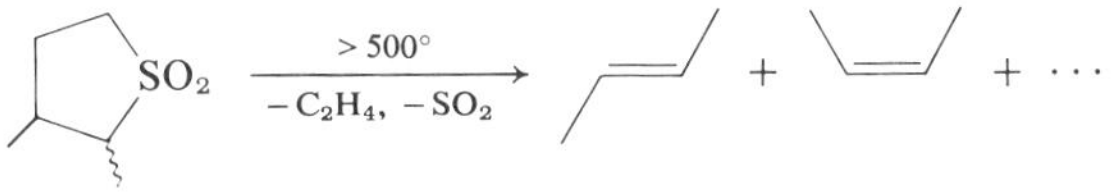

(12)

95% cis   36% + 53%
90% trans   51% + 40%

6. Steric Constraints

Finally we come to entries *n* and *o* in Table I, and are able to say something about the mechanism of the sulfolene reaction itself. The conformations of these molecules are depicted in **9** and **10**. The pertinent observation is that **10**

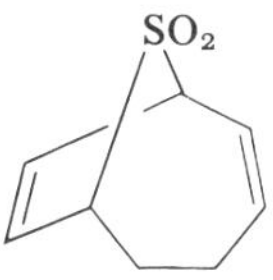

(9) 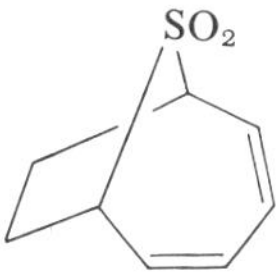   (10)

undergoes fragmentation some 60,000 times more slowly (*o* vs. *n*) than does **9** (Mock, 1970a, 1975b). It will be observed that **9** is a legitimate sulfolene, the product of an (allowed) suprafacial 1,4-addition (its mode of synthesis). The rate of dissociation (*n*) is typical for sulfolene decomposition (previous entries in Table I; but see Paquette *et al.*, 1971, 1973; Gasteiger and Huisgen, 1972). We therefore infer that **9** is a suitable reference model for an allowed cycloregression. In particular, the enthalpy ($\Delta H^{\ddagger} = 29.3$ kcal/mole) and entropy of activation ($\Delta S^{\ddagger} = +1.9$ eu) for **9** are representative; the other kinetically well-behaved five-ring sulfolenes in Table I have similar $\Delta H^{\ddagger}$ values and have $\Delta S^{\ddagger}$ in the range of -5 to $+10$ eu (Mock, 1975b). Furthermore, comparison of the equilibrium *dissociation constant* for **9** with that of other sulfolenes reveals the absence of ring strain as a significant rate accelerating factor in these bicyclic systems (as, e.g., in *k* and *m*). Also, there is no evidence that the extra double bond in the four-carbon bridge of **9** induces a perturbation of the reaction mechanism (as in phenylepisulfone, Section II,C,2).

The companion structure **10** is an isomer of **9** (from which it was indirectly prepared). Formally, **10** is a 1,6-adduct, which due to steric constraints of the bicyclic system may also undergo only *suprafacial* elimination with respect to the hydrocarbon (triene) moiety, instead of the *antarafacial* path normally observed (e.g., T.S.-A, Section II,A). As previously noted (see Table I), dissociation of **10** requires forcing conditions; its free energy of activation ($\Delta G^{\ddagger}$) is at least 10 kcal/mole higher than the other sulfolenes (particularly **9**). *This may be taken as a measure (minimal) of the magnitude of orbital symmetry constraints in these cycloregressions* (Mock, 1970a, 1975b.)

It is to be emphasized that the mechanism of fragmentation of **10** has not been established. Various circuitous schemes other than a simple one- or two-step extrusion may be postulated to meet the restriction of an allowed linear cheletropic process (Mock, 1975b). However, an independent estimate of the velocity of a direct *nonconcerted* process (sequential C–S bond scission) may be obtained by examining the rate of degenerate rearrangement of **11** (a deuterated intermediate in the conversion of **9** to **10**). In this molecule label scrambling between olefinic and bridgehead positions has been observed at elevated temperatures [Eq. (13)]. This requires formally a [1,3]-sigmatropic shift, and most likely occurs via a zwitterionic (or diradical) intermediate. The exact mechanism is not important; what matters is that only a single C–S bond participates. Accordingly, this rearrangement is a model for a *two-step* sulfolene dissociation in which the first bond scission is rate determining. The observed activation free energy (only crudely known due to experimental complications) for the rearrangement of **11** is of the same magnitude as for the decomposition of **10**. Consequently, a nonconcerted mechanism for the latter's dissociation is plausible.

However, the reservation must be made that the *nonlinear concerted* retro-[$_\pi 6_s + {_\omega}2_a$] alternative is in no way excluded for **10**. It is only required that such a pathway be at least 10 kcal/mole more endergonic than the *linear* process for **9**. We are obliged to give some consideration to the activation parameters for **10** ($\Delta H^{\ddagger} = 32$ kcal/mole, $\Delta S^{\ddagger} = -14$ eu). Although experimental accuracy is such as to render these determinations suspect, the apparent *enthalpy* contribution is no greater than for the other sulfolenes in Table I. The observation of a negative entropy factor which largely accounts for the inertness of **10** suggests a constrained transition state, and would appear more in accord with the nonlinear hypothesis than with a zwitterion or diradical (which should have additional degrees of freedom). However, it should be noted that the arylepisulfones, for which independent evidence suggested a nonconcerted transition state formulation (Section II,C,2), also exhibited negative activation entropies of exactly the same magnitude (Bordwell *et al.*, 1968). An explanation invoking a transition state solvation requirement seems adequate. (Limited volatility prevents vapor phase kinetics.) On the other hand, partitioning of the activation energy between $\Delta H^{\ddagger}$ and $\Delta S^{\ddagger}$ may be more reflective of whether the transition state comes early or late in the reaction coordinate than of whether the process is concerted (Skell and Cholod, 1969).

In summary, the significant conclusion is that the transition state for a *linear, allowed* sulfolene reaction is *stabilized* relative to that for a zwitterionic (or nonlinear) mechanism by at least 10 kcal/mole. Accordingly, delocalized bonding within the activated complex is indicated, and a concerted process may be inferred: retro-[$_\pi 4_s + {_\omega}2_s$] for Eq. (4), retro-[$_\pi 6_a + {_\omega}2_s$] for Eq. (5). Hence, we believe that the concept of the "aromatic" transition state here applies (Dewar and Kirschner, 1971).

D. CHELETROPIC VS. 1,2-ADDITION

The possibility of multiple modes of cycloaddition to sulfur dioxide may be appreciated by the reported spontaneous dissociation of 3,6-dihydro-1,2-oxathiin 2-oxides (prepared by oxidative cyclization of an appropriate unsaturated hydroxy sulfoxide) to dienes plus sulfur dioxide (Jung *et al.*, 1974) [Eq. (14)]. Although the cyclic sulfinate is too unstable for isolation in these instances, we have come across an actual case of competitive *cycloaddition* yielding both five- and six-membered ring products (W. L. Mock and L. M. Hanusa, unpublished results, 1972). In the course of exploring the scope of the sulfolene reaction, cycloadditions of conjugated *allenic* systems were

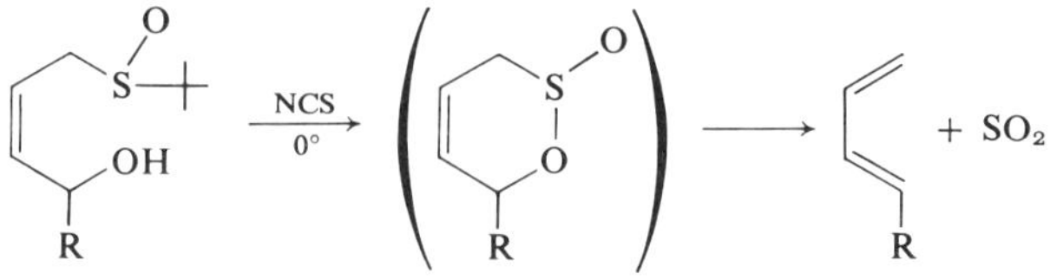

$$\text{[structure]} \xrightarrow[0°]{\text{NCS}} \left(\text{[structure]} \right) \longrightarrow \text{[structure]} + SO_2 \quad (14)$$

R = H, C$_6$H$_5$

examined [Eqs. (15)] (also see Kleveland and Skattebøl, 1973; Kleveland and Skattebøl, 1975). It was anticipated that the retro reaction in this series would provide alternative syntheses of vinylallenes, a difficultly accessible type of functionality. Whereas the addition reaction proved unusually facile, the latter hope (preparative fragmentation) was frustrated by the thermal stability of the products; all attempts at thermal cracking failed. By hap, in one

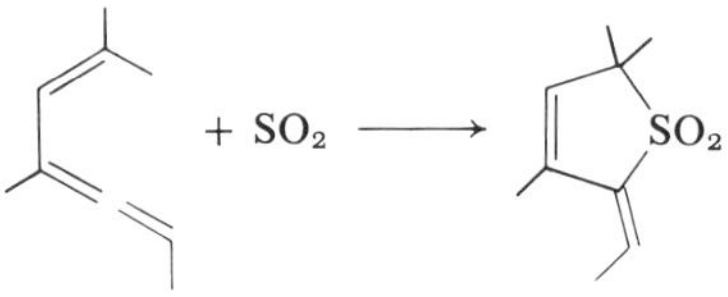

$$\text{[structure]} + SO_2 \longrightarrow \text{[structure]}SO_2 \quad (15a)$$

$$\text{[structure]} + SO_2 \longrightarrow \text{[structure]}SO_2 \quad (15b)$$

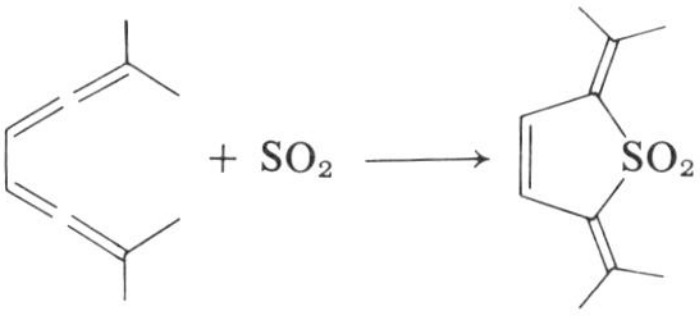

$$\text{[structure]} + SO_2 \longrightarrow \text{[structure]}SO_2 + \text{[structure]} \quad (15c)$$

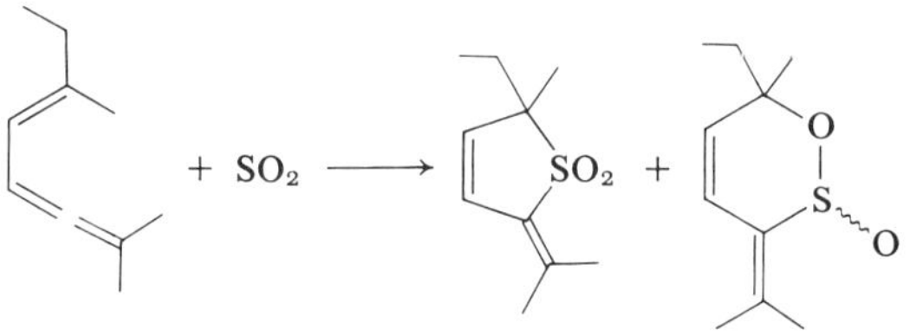

cycloaddition, epimeric cyclic sulfinates were in fact coproduced in good yield [Eq. (15c)]. An additional example of 1,2 addition has very recently been reported (Heldeweg and Hogeveen, 1976). Although the latter reaction does not qualify as cheletropic, the possibility is raised that mechanistically it might provide insight into the sulfolene reaction. This avenue has been pursued in the following analogous system.

1. Thiazine Oxide–Sulfoximine Correlation

At this juncture we make another slight digression to consider cycloadditions involving the imine analogs of sulfur dioxide. While N-sulfinylamines yield Diels–Alder adducts (**12**) [Eq. (16)] rather than the cheletropic sulfoximine analogs of sulfolenes (Kresze, 1967), an illuminating correlation has been established between these reaction manifolds (Mock and Nugent, 1975a,b).

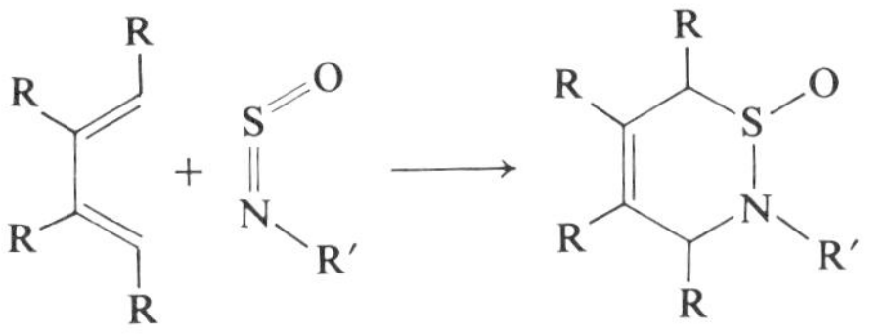

$$\text{(16)}$$

R′ = Ar, SO₂Ar, etc.

Ordinarily the thiazine oxides (**12a–c**) [Eq. (17)] are formed stereospecifically via a suprafacial process, i.e., (E,E)-hexadiene yields exo- and endo-cis adducts [Eq. (17a)] and (Z,E)-hexadiene likewise gives an (endo)-cis-addition product [Eq. (17b)]. However, (Z,Z)-hexadiene was an exception. In this case

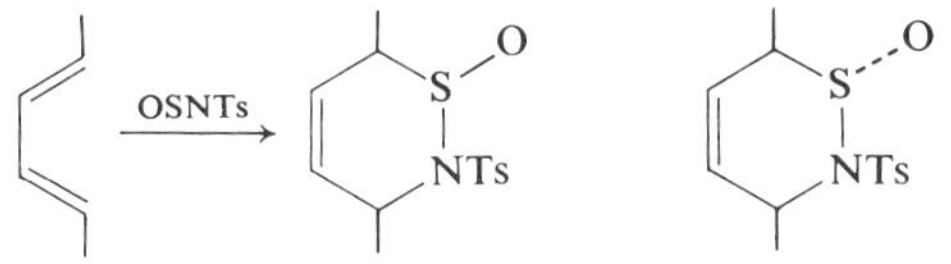

$$\text{(17a)}$$

(**12a**)　　(**12b**)

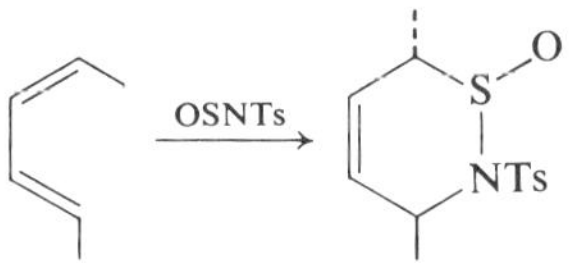

$$\text{(17b)}$$

(**12c**)

overall trans addition (**12d**) was observed exclusively [Eq. (18)]. The most plausible mechanism requires a *stepwise* addition, with initial C–S bond formation producing a zwitterionic intermediate which, due to severe methyl–methyl steric interference, must undergo a convoluting internal rotation as depicted before closure to the six-membered heterocycle (**12d**) can be achieved. The regioselective effects of substituents in other dienes is also consistent with such a sequential mechanism. Since the rate for Eq. (18) appeared similar to that for Eqs. (17), it was concluded that a common mechanism applied.

Therefore, thiazine oxide formation is probably in general *nonconcerted*, with the restriction that in most cases the lifetime of the intermediate is too brief to allow rotation about σ bonds (Mock and Nugent, 1975a).

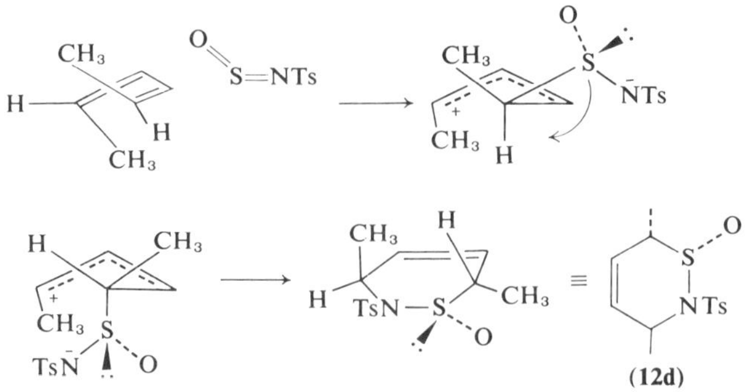

(18)

(12d)

In consequence of the preceding conclusion, there appears to be a marginal inconsistency between the mechanism postulated for the sulfolene reaction (synchronous σ-bond formation) and that for the thiazine oxide variant of the diene synthesis (nonconcerted), which involves an isoelectronic analog of SO_2. In order to rationalize this dichotomy in reactivity, a sulfoximine analog (13) of sulfolene was synthesized (indirectly, from tetramethylene sulfide; Mock and Nugent, 1975b). This substance was stable under conditions of

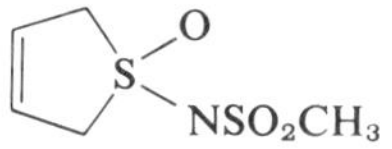

(13)

successful thiazine oxide formation (12e, from butadiene and $OSNSO_2CH_3$), and therefore such five-membered ring structures may not be intermediates in thiazine oxide formation. Upon thermolysis, 13 apparently rearranges to a thiazine oxide, demonstrating that 13 is thermodynamically unstable relative to 12e [Eq. (19)]. However, the mechanism of this isomerization was revealed

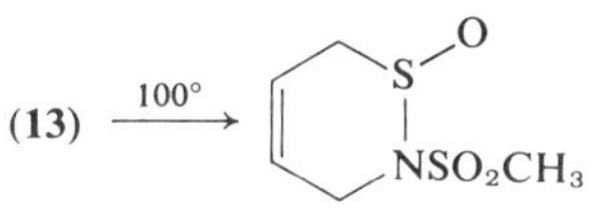

$$(13) \xrightarrow{100°} (12e) \qquad (19)$$

by attempting to carry out the rearrangement in sulfur dioxide solvent. In such case only sulfolene (and $OSNSO_2CH_3$) was produced, indicating most probably the capture of free *butadiene*. Hence, dissociation–recombination is the most probable path for the conversion of **13** to **12e** [Eq. (20)]. The significance of this finding is that it excludes a zwitterionic intermediate in the sulfolene dissociation of **13**. Formation of such an intermediate would require scission of only a single C–S bond; minimal electron redistribution appears to

$$\underset{(13)}{\text{sulfolene-NSO}_2\text{CH}_3} \xrightarrow{\times} \underset{\text{NSO}_2\text{CH}_3}{\text{zwitterion}^+_-} \longrightarrow \underset{(12e)}{\text{thiazine oxide-NSO}_2\text{CH}_3}$$

$$\downarrow \qquad \qquad \uparrow \text{CCl}_4$$

$$\diagdown\!\!\!\diagup\!\!\!\diagdown + \text{OSNSO}_2\text{CH}_3 \xrightarrow{\text{SO}_2} \text{sulfolene} \qquad (20)$$

be needed, and only slight atomic spatial reorganization would be required to shift the "ionic bond" from sulfur to nitrogen. However, since previous evidence indicated that closure of such an intermediate to a thiazine oxide would reasonably be expected to compete favorably with dissociation (note stereospecificity in **12d**, Mock and Nugent, 1975a), it must be concluded that *a dipolar intermediate does not occur in the fragmentation of* **13**; i.e., the $[_\sigma 2_s + {}_\sigma 2_s + {}_\pi 2_s]$ (retro sulfolene) reaction is in this case also a fully synchronous process. This point merits repeating: since there can be no *common* intermediate in the reactions of **12** and **13**, and since the formation of **12** plausibly involves a zwitterion, fragmentation of **13** must be concerted.

2. Theoretical Speculations

A different way to express the foregoing important conclusion is to say that the energy hypersurfaces for the sulfolene reaction and for thiazine oxide formation are nowhere contiguous. Why this *must* be so can be seen by examining orbital overlap within a *hypothetical* concerted isomerization of **13** to **12** [Eq. (21)]. In order for synchronous C–N bond formation and C–S bond rupture to occur, the nitrogen of the sulfoximine would have to swing toward the *backside* of the methylene bound to sulfur. (It should be obvious that a sigmatropic shift suprafacial at carbon is symmetry disallowed, since to be permitted it would require an antarafacial migration across the S–N bond to the distal p-lobe of nitrogen.) As the five-membered ring is carried into the six-membered ring, a pertinent point is revealed: *The penultimate precursor to* **12** *in this mechanism does not correspond to the transition state for a suprafacial Diels–Alder reaction*. In fact, by this analysis the $[_\pi 4_s + {}_\omega 2_s]$ transition state for the sulfolene reaction correlates with the $[_\pi 4_a + {}_\pi 2_a]$ variant of the diene synthesis rather than the normally observed $[_\pi 4_s + {}_\pi 2_s]$ process. The antarafacial geometry of the hypothetical diene–sulfinyl amide complex being exorbitantly strained [Eq. (21)], it follows that a sizable energy barrier should indeed exist between the two modes of cycloaddition (i.e., the latter structures depicted represent the ridge separating the accessible regions of the energy hypersurface for interaction of diene with sulfinylamide). Hence, dissociation–recombination is a lower energy path than the exothermic

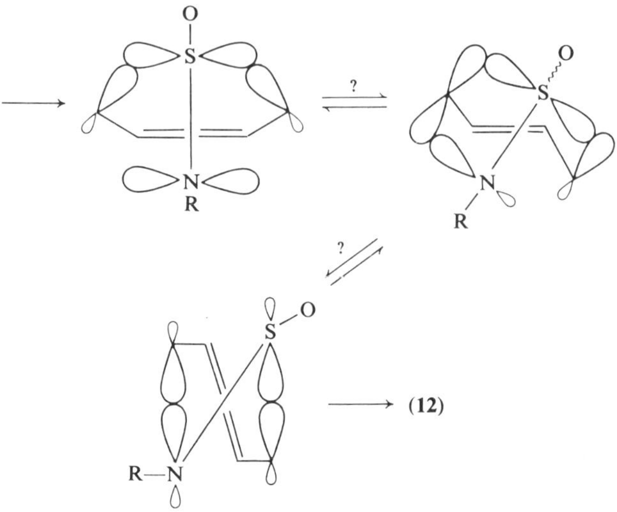

isomerization of **13** to **12**. There being no physical requirement for similar electron distributions in the two thusly separated reaction manifolds, it is fully plausible for the cheletropic sulfolene reaction to be a concerted process, forward = $[_\pi 4_s + {}_\omega 2_s]$, retro = $[_\sigma 2_s + {}_\sigma 2_s + {}_\pi 2_s]$, while thiazine oxide formation (Diels–Alder, nominally $[_\pi 4_s + {}_\pi 2_s]$), should in fact be a sequential (two-step) reaction.

However, it is a matter of some curiosity that the same two reactants should adopt such disparate mechanisms according to the manner in which they combine chemically (**12** or **13**). Although the evidence scarcely warrants further speculation, we should like tentatively to advance some thoughts on possible generalizations from this apparent dichotomy. The obvious feature of the sulfolene reaction, which is to be connected with its propensity to undergo synchronous cheletropy, is that electron pair interaction is facilitated (indeed, possibly forced) by the fact that the bonding interactions are focused at a *common* atom (in this case sulfur). With the six-membered ring, the loci of σ bonding are spatially separated and there are additional degrees of freedom (internal rotations and vibrations) to be sampled by the reacting system. The latter point, which has only intuitive support, recognizes that there is generally an entropic problem in achieving a synchronous transition state. Numbers of vibrational modes must be coordinated; the larger the ring the greater the difficulty. It would seem reasonable that the requisite coupling should be stronger in the *cheletropic* (geminal) situation. Related, but not identical to this is the Principle of Least Motion (PLM) (Hine, 1966; Tee *et al.*, 1974), an idea which we feel has not yet been given its due in conjunction with Orbital Symmetry Conservation (OSC). One does not need to resort to quasisophisticated computations to justify application of a principle which in fact is founded on common sense. We are not concerned with violations of symmetry conservation (where PLM will clearly lose in head-to-head competition with OSC). However, in comparing various *allowed* transformations or concerted vs. nonconcerted mechanisms, PLM has its place. It should be intuitively clear (and may be supported by back-of-an-envelope calculations) that less net interatom bond length adjustment is required in proceeding from the addends to a zwitterionic or diradical intermediate, than for a fully synchronous process in *macropericyclic* cases (i.e., multielectron pair cycloadditions). In such situations where an intermediate is likely to be stabilized by delocalization or other factors, overlap considerations (PLM) may dictate actual nonconcertedness. The crossover point from synchronous to stepwise mechanisms may occur with ring sizes as small as six or seven (e.g., **12**?). Subsequently, additional examples of PLM applicability will be mentioned. However, we wish to emphasize that the preceding speculations are offered more for the purpose of stimulating thought than as mechanistic explanation for a particular reaction (see also Bordwell, 1970, 1972).

E. PHOTOCHEMISTRY

Since there is an indicated reversal in selection rules for cycloadditions in the electronically excited state, the photochemical counterparts to the previously described thermal reactions are of interest. Triplet sensitized dissociation (via excited benzene) of *cis*- and *trans*-dimethylsulfolene (**2** and **3**) has been reported (Saltiel and Metts, 1967). In brief, a stereopreference was found toward the antarafacial mode with respect to the diene, in contradistinction to the thermal decomposition. However, the stereoselectivity was marginal, being 5:1 (antara:supra) for the cis isomer and 2.4:1 for the trans. [In each case some (Z,Z)-hexadiene was produced as well.] It was specifically pointed out that the major product forming step must be accompanied by intersystem crossing, since such stereospecificity as was observed would have been lost if triplet excited dienes were produced; (also, the triplet excitation energy of sulfur dioxide exceeds that of the dienes). However, we feel that the efficacy of the Woodward–Hoffmann rules is their universal applicability; to ascribe a partial result to symmetry factors is a minor abuse of the principles.

On the other hand, the results are also not in accord with expectation for a long-lived diradical (Kellogg and Prins, 1974). Without specifically disputing previous interpretations, we wish to suggest an alternative explanation in which the invocation of orbital symmetry selection rules is deliberately avoided.

Granting that the first step in the process is transfer of triplet excitation from benzene to sulfolene, one should inquire as to the consequence as regards geometrical *distortion* of the recipient olefin. It is an accepted canon of photochemistry that an orthogonal triplet alkene is of lower energy than its planar counterpart (Mulliken and Roothaan, 1947; Cundall, 1964; Coyle, 1974; Zimmerman *et al.*, 1974). Apparently some potential distortion about a formal double bond is feasible even in a five-membered ring* (Kropp and Krauss, 1967). As depicted (14) this places the sulfolene molecule in a *conrotatory* mode as regards eventual fragmentation; i.e., the ring methylene groups are displaced to opposite sides of the former π-bond nodal plane, and

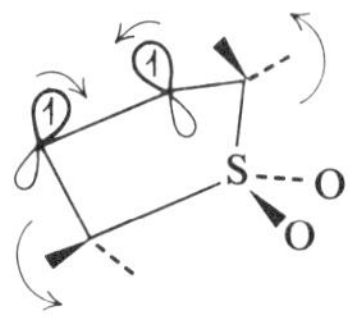

(14)

the *least motion* path leads to antarafacial cycloelimination from the incipient diene. Spin inversion may occur concurrently or immediately subsequently (i.e., a retro-$[_\pi 4_a + {}_\omega 2_s]$ process with some competing "diradical" crossover), or immediately prior to SO$_2$ extrusion (i.e., a retro-$[_\pi 4_a + {}_\omega 2_a]$ process, equivalent to fragmentation of a *vibrationally excited* ground state, specifically distorted as shown above, in which dissociation is competitive with relaxation back to the normal suprafacial mode). The *significant* photochemical act is placing of the sulfolene molecule in a conrotatory reaction mode, whence the observed geometrical results follow (by Least Motion; see also Section III,F).

Turning to the three- and seven-membered ring systems, it has been noted that ultraviolet irradiation induces decomposition in phenyl-substituted episulfones (Bordwell *et al.*, 1968). This is consistent with linear cheletropic fragmentation. Unpublished results with dihydrothiepin dioxide (6) and its bicyclic analog (10) give more insight into the photoprocesses which may take place (W. L. Mock, unpublished results, 1972). It was found that upon excita-

---

* It has been suggested that as much as 40° of twist are available to an excited state olefin in a cyclopentene (Kropp and Krauss, 1967). The longer C–S bond length should permit even greater torsion in sulfolene.

tion of the diene system of **6**, dissociation to give 1,*cis*-3,5-hexatriene (and presumably sulfur dioxide) did occur [Eq. (22)]. However, as minor primary coproducts 1,*trans*-3,5-hexatriene and an isomeric hydrocarbon (not 1,3-cyclohexadiene) were also detected. Therefore, it is surmised that electronically excited hexatriene was initially produced, and that in the process of demotion it was converted to the isomeric 1,3,5-hexatrienes and to 1,2,*cis*-4-hexatriene, which is suggested to be the other minor product (Srinivasan, 1963; Crowley, 1968). The bridged dihydrothiepin dioxide (**10**) was submitted to irradiation under approximately the same conditions as was **6**. Significantly, 1,3,5-cyclooctatriene was generated at about the *same rate* as were the products from **6** [Eq. (23)]. It follows that the departure of sulfur dioxide from both excited molecules *may be* occurring in linear suprafacial fashion, in accord with symmetry expectations. It should be noted that in contrast to the five-ring sulfolene dissociations, these reactions are likely occurring in the singlet manifold. Photoextrusions analogous to that of **10** have been reported (Paquette *et al.*, 1973).

$$\underset{(\mathbf{6})}{\diagup\!\!\diagdown\!\text{SO}_2} \xrightarrow[(-\text{SO}_2)]{h\nu} (\text{C}_6\text{H}_8)^* \longrightarrow \diagup\!\!\diagdown + \diagup\!\!\diagdown\!\!\diagup + \diagup\!\!\diagdown\!\!\diagup\!\!\diagdown \quad (22)$$

$$\underset{(\mathbf{10})}{\diagup\!\!\diagdown\!\text{SO}_2} \xrightarrow[(-\text{SO}_2)]{h\nu} \bigcirc \quad (23)$$

III. Cheletropy in Other Systems

This chapter concludes with consideration of a collection of recently reported reactions which either allow further insight into the cheletropic process or which are of intrinsic synthetic interest. We deliberately have omitted carbene (nitrene) reactions. In fact, they are one of the more consequential examples of cheletropy; ample evidence indicates that singlet methylene addition to double bonds is a *nonlinear* concerted process. However, it is our feeling that this topic is adequately covered elsewhere (Kirmse, 1971; Jones and Moss, 1973; see also Chapter 3 by Jones and Brinker in Vol. I of "Pericyclic Reactions") and that our allotted space should be used otherwise. Hence we must dismiss this subject with the preceding references (however, note Section III,F—carbon monoxide and isonitriles).

Of even potentially greater future importance is the area of metalloorganic "oxidative addition." There are numerous examples in transition metal chemistry of insertions into σ bonds (or π systems) with formal valency increase of the coordinating atom. Little mechanistic work appears to have been done, but it seems likely that at least some of these synthetically and catalytically promising reactions may be concerted (Cassar *et al.*, 1970; Mango and Schactschneider, 1971; van der Lugt, 1970; Mango, 1971, 1975; Noyori *et al.*, 1971; Fraser *et al.*, 1973; McDermott *et al.*, 1973; Dauben *et al.*, 1973). Regrettably, this subject cannot be developed here; the author does not feel equal to the task of critical selection.

A. Nitrous Oxide and Nitrogen Eliminations

One of the examples of cheletropy more extensively analyzed by Woodward and Hoffmann was the extrusion of NNO from nitrosoaziridines [Eq. (24)].

$$\triangleright\!\text{NNO} \xrightarrow[<0°]{\text{retro-}[_\pi 2_s + _\omega 2_a]} \| + \text{NNO} \qquad (24)$$

They described a disallowed "linear" path and distinguished between two allowed "nonlinear" paths (antarafacial interaction at departing nitrogen). An interpretational conundrum developed when the next higher vinylog of this system was examined. *N*-Nitrosopyrroline exhibited no tendency whatever to dissociate to butadiene plus nitrous oxide (McGregor and Lemal, 1966) [Eq. (25)]. Since this fragmentation could be shown to be exothermic,

$$\text{[pyrroline]}\text{NNO} \xrightarrow[>200°]{\times} \text{[butadiene]} + \text{NNO} \qquad (25)$$

and since a hypothetical allowed *linear* retro-$[_\pi 4_s + _\omega 2_s]$ pathway may be demonstrated, it was implicitly suggested that there is an intrinsic property of NNO, that it uniquely prefers nonlinear extrusion from nitrosamines over the least motion linear path (McGregor and Lemal, 1966; Woodward and Hoffmann, 1969a,b). This idea was tested by attempting a retro-$[_\pi 6_s + _\omega 2_a]$ extrusion, which should be relatively facile if the foregoing hypothesis were correct (Mock and Isaac, 1972). A bridged adduct of cyclooctatetraene (**15**) was indirectly prepared. It may be noted that the bicyclic nature of **15** so constrains the system that elimination of NNO would necessarily be suprafacial with respect to the hydrocarbon component (C_8H_8) [Eq. (26)]. In fact, two conceptually distinct fragmentations may be envisioned, both giving the same products. Based on the precedent of *N*-nitrosopyrroline, the linear retro-$[_\pi 4_s + _\omega 2_s]$ mode was not expected; the only function of the etheno bridge in **15** is to ensure that retro-$[_\pi 6_a + _\omega 2_s]$ cycloreversion may also not occur. In the event, **15** proved fully as disinclined to dissociate as *N*-nitrosopyrroline itself; it could be recovered unchanged after extensive exposure to elevated temperatures (200°, compared to $C_2H_4 \cdot NNO$, $\Delta\Delta G^{\ddagger} >$ 15 kcal/mole). It must be concluded that antarafacial cheletropic (nonlinear)

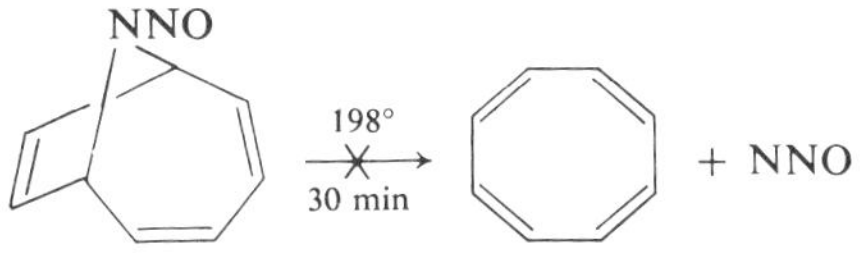

$$\text{(15)} \xrightarrow[30 \text{ min}]{198°} \bigcirc + \text{NNO} \qquad (26)$$

extrusion of NNO is not an intrinsic property of nitrosamines or of nitrous oxide, but more likely the facility of dissociation of *N*-nitrosoaziridines is to be associated with some other property of the three-membered ring.

We believe that the correct clue to successful explanation of reactivity in this series is recognition of a *noncoplanar* nitrosamide functionality in the *ground state* for the aziridine ring. This requires some initial credulity, since unconstrained nitrosamines exhibit substantial N–N π bonding, as evidenced by a rotational barrier of ca. 23 kcal/mole. However, comparable delocalization is observed in (isoelectronic) carboxylic amides, yet *N*-(*p*-bromobenzoyl)-aziridine was found to possess pyramidal nitrogen in the crystal (Shibaeva *et al.*, 1967). This behavior may be conventionally explained as the result of competing hybridization and conjugative effects. The enhanced *p* character of the carbon–nitrogen (ring) bonds destabilizes a potentially planar (sp^2) configuration of the ring nitrogen more than any energy gain that might be realized from amide-type resonance; the result is that normal pyramidal configuration of amine nitrogen more nearly obtains. The chemical consequence will be that the conformation depicted (**16**) should be either the ground state or a readily accessible rotamer thereof. Compared to unconstrained (i.e., coplanar) nitrosamines, the *N*-nitrosoaziridine has been promoted well along the reaction coordinate for a concerted retro-$[_\pi 2_s + {_\omega}2_a]$ cycloreversion [Eq. (27)]. The critical electron correlations are marked (*, †), and a diagram illustrating all of the symmetry-directed electron pair shifts has been presented elsewhere (Mock and Isaac, 1972).

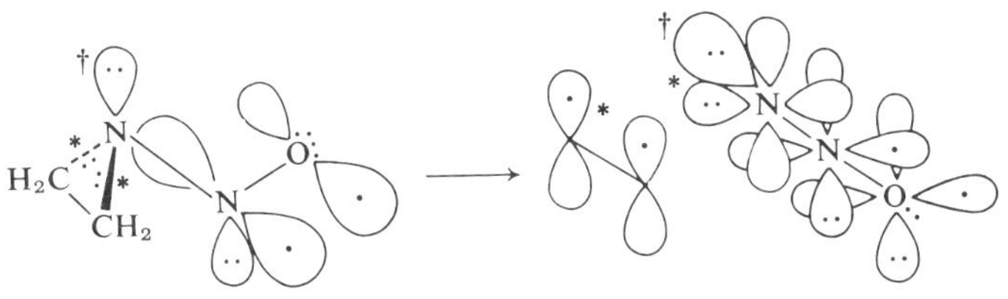

$$(27)$$

(**16**)

By this analysis we do not mean to ignore the influence of ring strain, which certainly contributes to the exothermicity and, therefore, likely adds to the velocity of NNO extrusion. What we have attempted to show is that *Least Motion* analysis of various *symmetry allowed* reaction paths, in conjunction with accurate structural information, can lead to useful interpretations (and perhaps predictions) of chemical reactivity. In the particular case at hand, Woodward and Hoffmann overlooked the propensity for rehybridization of aziridinyl nitrogen, and in consequence favored an alternative "nonlinear" pathway. The cheletropic process we describe (which in fact was previously considered and discounted—Freeman and Graham, 1967), ought properly to be categorized as *linear*, since it corresponds to the least motion process for a stable conformation of this ring.

While the above explanation is both adequate and insightful regarding orbital symmetry constraints on NNO extrusion, some attention should be drawn to Snyder's theoretical investigation comparing the ready extrusion of —N≡N— from azo compounds with the (analogous but difficultly observable) —N≡N(O)— fragmentation of *azoxy* compounds. A plausible explanation is offered based upon perturbation of the orbital energies of the azo linkage by the ⩾N → O dative bond of the azoxy structure (Snyder *et al.*, 1972a). An activation energy differential exceeding 26 kcal/mole has been attributed to this factor (Olsen and Snyder, 1974), which appears unrelated to the nitrosamine cases. Since this is not a cheletropic reaction, we shall omit an analysis of this interpretation. However, it should be candidly admitted that concerted NNO extrusion (except for aziridines and certain puzzling polycyclic azoxy cases) appears to be only a *virtual* reaction. In selecting a particular explanation as to why any given reaction does *not* proceed, one should be aware that the truly controlling factor may be obscured by the observer's ignorance. (This is the fundamental lesson which should be learned from the history of Orbital Symmetry Conservation.) This comment may well also apply to Sections II,C,6 and II,D,2.

The stereospecific molecular nitrogen expulsion counterparts to the previously discussed NNO extrusions have also received attention (Lemal and McGregor, 1966). Although the diazenes necessary for examination of such reactions are metastable at best, recent evidence has been presented to support preferred *linear* departure of N_2. In the bicyclic series shown (**17–19**)

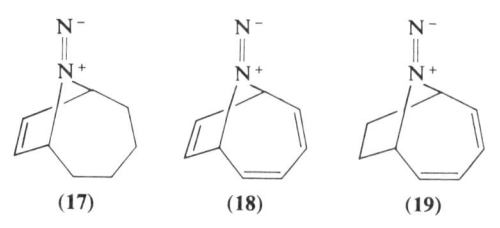

the first two molecules readily lost nitrogen (retro-[$_\pi 4_s + {_\omega}2_s$]), whereas **19** did not (forced retro-[$_\pi 6_s + {_\omega}2_a$]). It was concluded that nonlinear extrusion in the latter case was a relatively unfavored pathway (Anastassiou and Yamamoto, 1973). It might be noted that at least one example of nonstereospecificity in a three-membered ring analog (diphenylethyleneazamine) of the above systems has been recorded (Carpino and Kirkley, 1970). A novel olefin synthesis is based upon the fragmentation of the sulfoximine of N-aminooxazolidones [Eq. (28)]. The reaction plausibly is concerted (White and Kim, 1974; Kim and White, 1975). In contemplating the extrusion reactions of

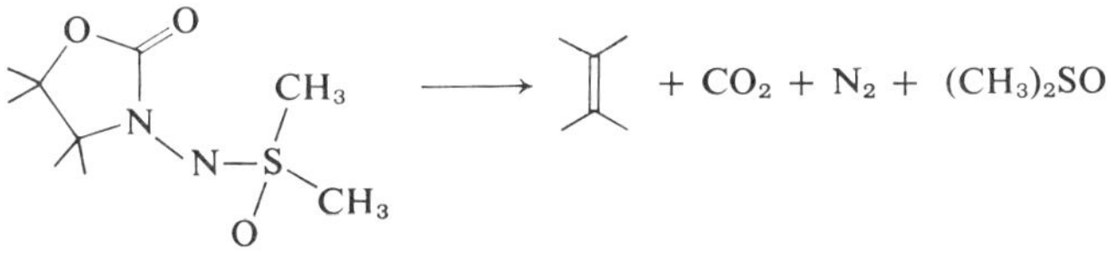

$$\text{(structure)} \longrightarrow \text{(structure)} + CO_2 + N_2 + (CH_3)_2SO \qquad (28)$$

nitrosamines and azoxy compounds, consideration might properly be given to the nitrosoalkane–azo dioxide equilibrium. Although this is debatably a cheletropic reaction, a nonlinear pathway is to be anticipated (Hoffmann et al., 1970); recent experimental evidence is consistent with this conclusion (Snyder et al., 1975; Greene and Gilbert, 1975).

B. PHOSPHOLENES

There is a well-known trivalent phosphorus analog of the sulfolene reaction (Quin, 1967). Of most interest to us is the question of mechanism. Unfortunately, to date no adduct of a (Z,E)-diene has been prepared, and therefore a rigorous test of stereospecificity has not been demonstrated. Reactions with (E,E)-hexadiene at least are highly stereoselective (Quin and Barket, 1970; Bond et al., 1968) [Eq. (29)]. However, the occasional occurrence of 2-phospholenes as initial products in these cycloadditions is sufficient

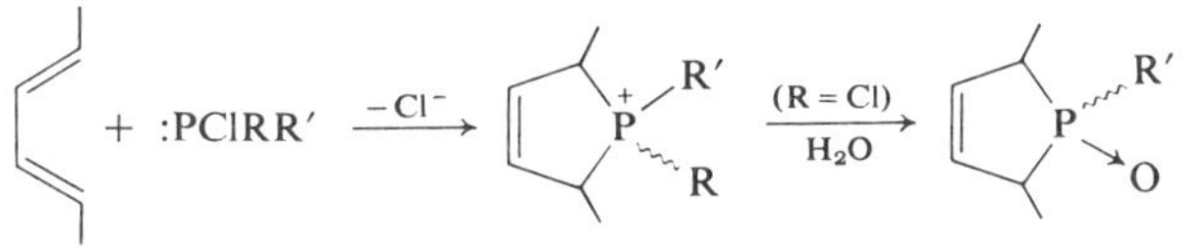

$$\text{(diene)} + :PClRR' \xrightarrow{-Cl^-} \text{(structure)} \xrightarrow[H_2O]{(R = Cl)} \text{(structure)} \qquad (29)$$

to alert one to the possibility of an ionic mechanism (Quin et al., 1968). The retro reaction has also been examined, in a novel phosphorus extrusion (Hall et al., 1972) [Eq. (30)]. Kinetic studies of the addition have been

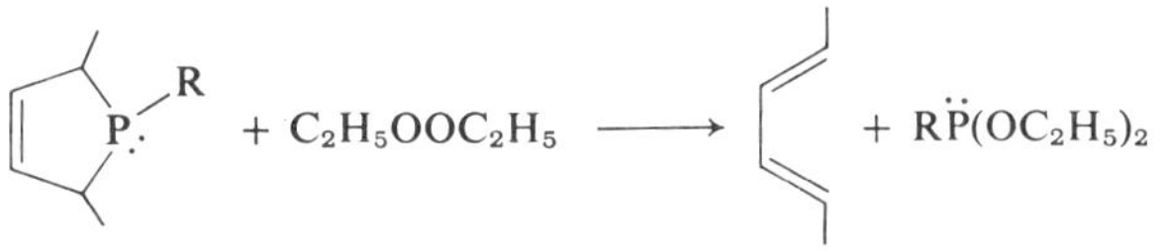

$$\text{(structure)} + C_2H_5OOC_2H_5 \longrightarrow \text{(diene)} + R\ddot{P}(OC_2H_5)_2 \qquad (30)$$

reported (Zubtsova et al., 1971) and theoretical considerations regarding the stereochemistry of the valence shell expansion of phosphorus have been presented (Hoffmann et al., 1972). Extrusion of phenylphosphinidine oxide proceeds with difficulty [Eq. (31)], but evidence indicating cheletropy in its

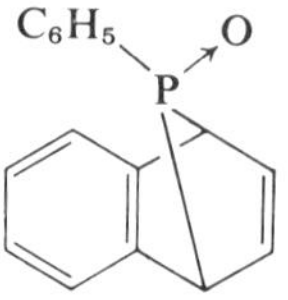

$$\text{[benzonorbornene-P(=O)C}_6\text{H}_5\text{]} \xrightarrow{\Delta} C_{10}H_8 + (C_6H_5-\ddot{P}=O) \longrightarrow \cdots \quad (31)$$

reactions has been advanced (Stille et al., 1972). Reactivity of phosphinothioylidene (Ph—P=S) has been analogously examined (Nakayama et al., 1971). The photoinduced dissociation of 3-phospholenes has been reported (Tomioka et al., 1974a,b). Several interesting examples of pentavalent phosphorane fragmentations in strained systems have been noted (Turnblom and Katz, 1973) [Eq. (32)]. Possible cases of cheletropic homoconjugate addition

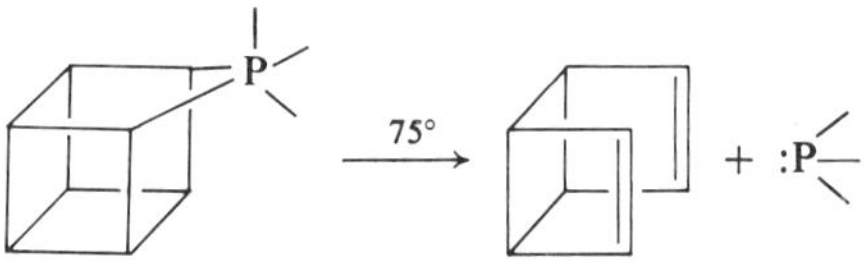

$$\text{[cubane-P]} \xrightarrow{75°} \text{[product]} + :P\!\!\!<\quad (32)$$

have been reported (Green, 1963; Kashman et al., 1973). The ozone–phosphite adduct (**20**) represents a formal example of cheletropy of a novel sort (Stephenson and McClure, 1973).

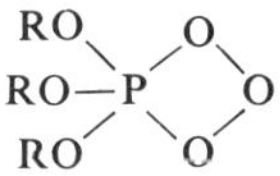

$$\begin{array}{c}\text{RO}\diagdown\diagup\text{O}\\ \text{RO}-\text{P}\text{O}\\ \text{RO}\diagup\diagdown\text{O}\end{array}$$
(**20**)

C. Sulfuranes

Reaction of the isomeric dihydrothiophenium salts **21** and **22** with *n*-butyllithium yields stereospecific formation of hexadienes, according to the scheme shown [Eq. (33)]. The results clearly imply concerted cycloelimination, with the unusual result in the latter instance [predominant formation of (Z,Z)-hexadiene] possibly explicable by steric hindrance between ring methyls and the butyl residue in the fragmentation (Trost and Ziman, 1971). Similar treatment of thietonium salts with butyllithium yields cyclopropanes,

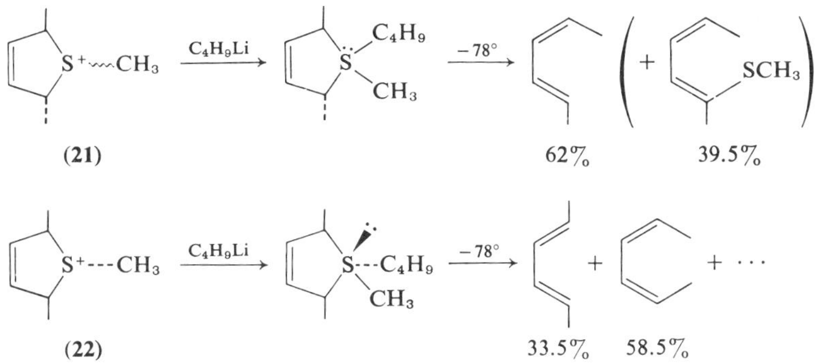

3. CHELETROPIC REACTIONS   167

evidently through a nonsynchronous process (Trost *et al.*, 1967, 1971) [Eq. (34)]. Thiiranes are apparently decomposed stereospecifically by butyl-

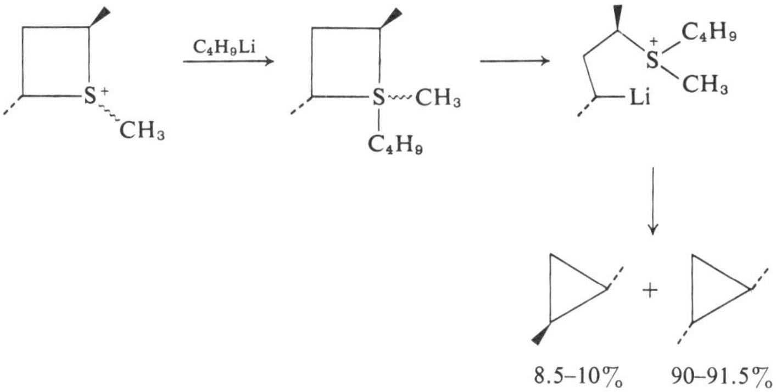

lithium [Eq. (35)]. A trigonal bipyramid intermediate (**23**, expanded sulfur valence shell) was postulated, inasmuch as similar treatment of the diastereomers of 2-bromo-3-ethylthiobutane afforded the same products with considerable crossover. However, detection of a slight amount of *trans*-butene from *cis*-episulfide was taken to indicate a competing nonconcerted pathway of only slightly higher energy (Trost and Ziman, 1969).

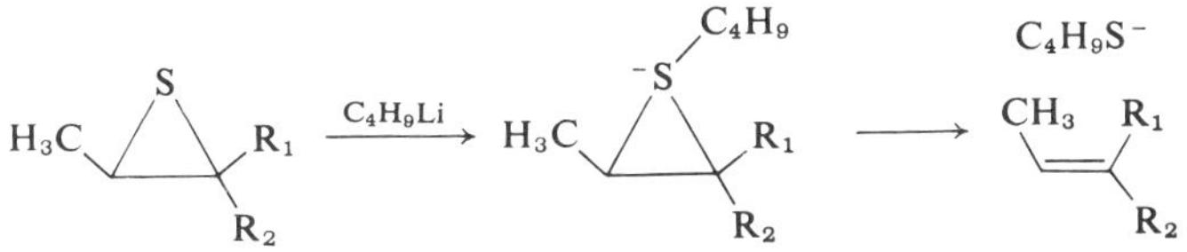

$R_1 = H, R_2 = CH_3$
$R_1 = CH_3, R_2 = H$

D. Divalent Sulfur, Selenium, and Tellurium

A variety of reagents are capable of inducing the cheletropic abstraction of sulfur from episulfides (Meyers, 1974; see also Section III,C on sulfuranes). To these may be added atomic carbon (Klabunde and Skell, 1971), as well as other carbenes [Eq. (36)]. Thiirane desulfurization is integral to a valuable new olefin synthesis (Barton and Willis, 1972; Buter *et al.*, 1972) [Eq. (37)]. The same theme has been extended to a novel alkyne synthesis (Mitchell, 1973) [Eq. (38)].

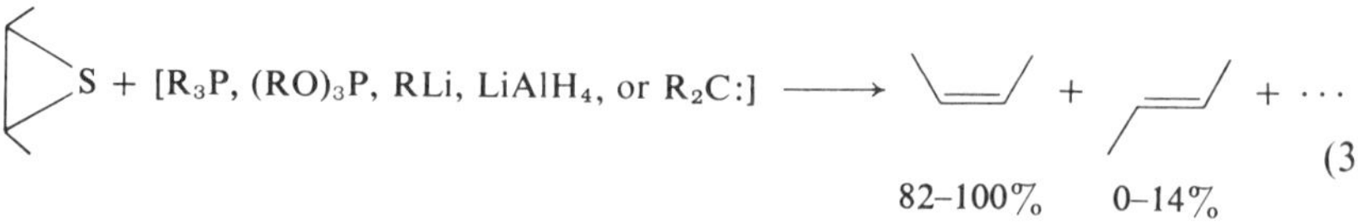

$$\triangleright\!S + [R_3P, (RO)_3P, RLi, LiAlH_4, \text{ or } R_2C:] \longrightarrow \text{(cis)} + \text{(trans)} + \cdots \quad (36)$$
$$\qquad\qquad\qquad\qquad\qquad\qquad\qquad\qquad\qquad 82\text{--}100\% \quad 0\text{--}14\%$$

$$2\ R_2CO \longrightarrow \longrightarrow R_2C\underset{N=N}{\overset{S}{\diagup\!\!\diagdown}}CR_2 \xrightarrow[-N_2]{R'_3P} R_2C=CR_2 + R'_3PS \quad (37)$$

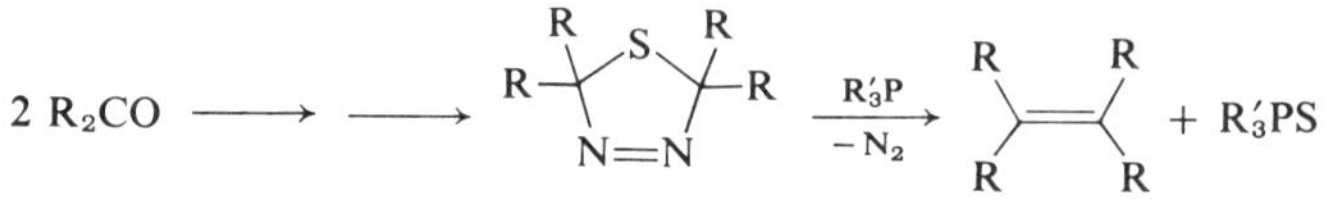

$$ArCH_2SC(Cl)_2Ar \xrightarrow[\text{KO-}t\text{-Bu}]{Ph_3P} ArC\equiv CAr \quad (38)$$

A special case of 1,4-sulfur extrusion from dienes is the thermolysis of thiophene–acetylene adducts (Helder and Wynberg, 1972; Kuhn and Gollnick, 1972; see, however, Kellogg and Prins, 1974) [Eq. (39)]. Analogous

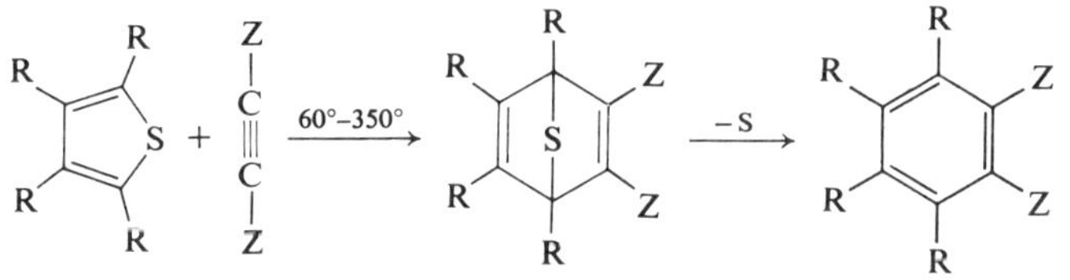

$$\text{(thiophene)} + \text{(alkyne ZC}\equiv\text{CZ)} \xrightarrow{60°\text{--}350°} \text{(bicyclic S adduct)} \xrightarrow{-S} \text{(arene)} \quad (39)$$

aromatizations of thiepins are documented (Wynberg and Helder, 1972; Barton *et al.*, 1972a; Traynelis *et al.*, 1973; Vogel *et al.*, 1974). Atomic sulfur extrusion involving as many as ten and twelve synchronous electron pair shifts have been hypothesized (Broadhurst *et al.*, 1972).

A valuable epoxide deoxygenation is based on stereospecific selenium extrusion (Clive and Denyer, 1973; Chan and Finkenbine, 1974) [Eq. (40)].

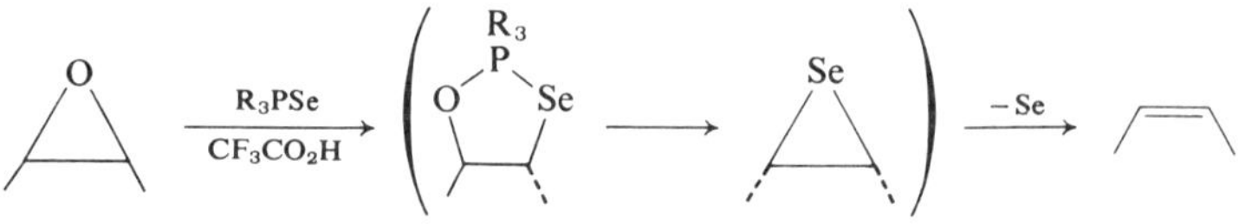

$$\triangle\!O \xrightarrow[CF_3CO_2H]{R_3PSe} \left(\overset{R_3P}{\underset{O}{\diagdown}}\!\!\diagup\!Se \longrightarrow \triangle\!Se \right) \xrightarrow{-Se} \diagup\!\!=\!\!\diagdown \quad (40)$$

Fragmentation of 1,2,3-selenadiazoles constitutes an exceptionally mild synthesis of acetylenes, and plausibly is concerted (Lalezari et al., 1970) [Eq. (41)]. An analogous sulfur extrusion is known (Martin, 1964). Seleno-

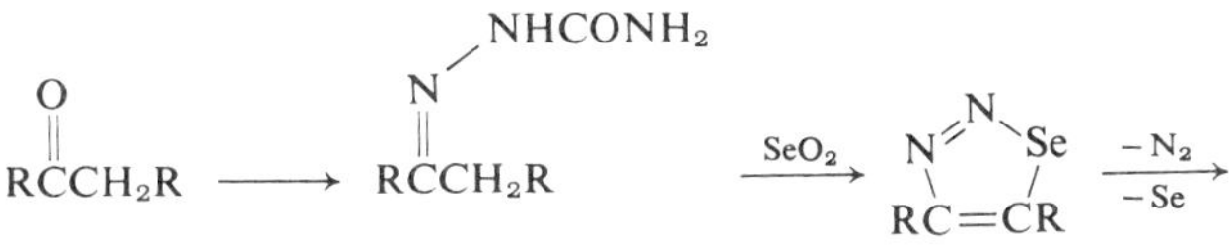

$$\text{RCCH}_2\text{R (O)} \longrightarrow \text{RCCH}_2\text{R (NNHCONH}_2\text{)} \xrightarrow{\text{SeO}_2} \underset{\text{RC=CR}}{\overset{\text{N=N}}{\text{Se}}} \xrightarrow[-\text{Se}]{-\text{N}_2} \text{RC}\equiv\text{CR} \quad (41)$$

phene yields a maleic anhydride bisadduct which probably is mediated by a labile 1:1 Diels–Alder adduct as in the thiophene reaction just cited (Arbuzov and Konovalov, 1959; Magdesieva, 1970) [Eq. (42)].

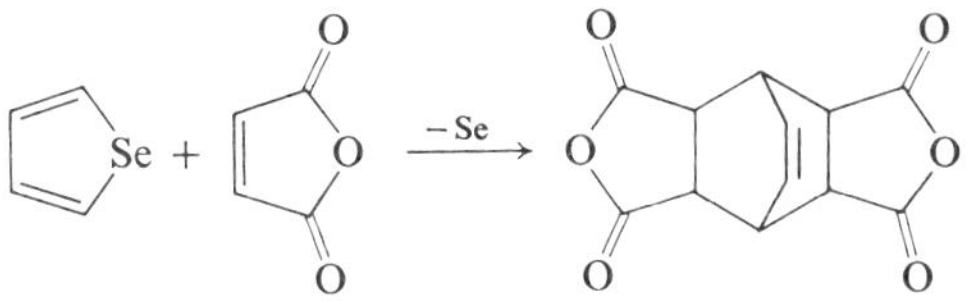

$$\text{(selenophene)} + \text{(maleic anhydride)} \xrightarrow{-\text{Se}} \text{(bisadduct)} \quad (42)$$

It should be noted in passing that the selenium dioxide adducts of dienes are not five-membered ring selenones as originally reported, but rather possess the six-membered ring structure in analogy to the thiazine oxides (Mock and McCausland, 1968) [Eq. (43)]. An example of "SeO" extrusion has been

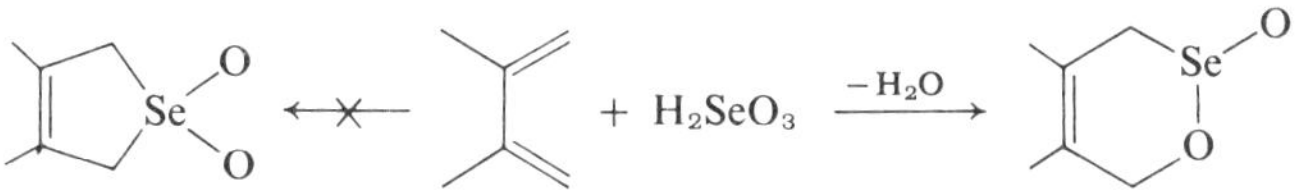

$$\text{(5-ring SeO}_2\text{)} \xleftarrow{\times} \text{(diene)} + \text{H}_2\text{SeO}_3 \xrightarrow{-\text{H}_2\text{O}} \text{(6-ring SeO}_2\text{)} \quad (43)$$

noted, although it may be neither general nor concerted (Arigoni et al., 1973).

An intriguing report on tellurium extrusion has appeared. The bridged structure shown cleanly yielded the 7,3-fused ring system to the exclusion of any 5,5-fused bicyclooctadiene [Eq. (44)]. Since the latter is known to result

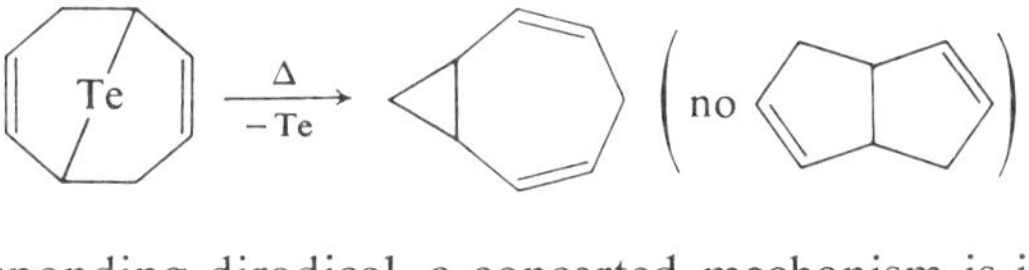

$$\text{(Te-bridged)} \xrightarrow[-\text{Te}]{\Delta} \text{(7,3-fused)} \quad \left(\text{no } \text{(5,5-fused)}\right) \quad (44)$$

from the corresponding diradical, a concerted mechanism is implied (Cuthbertson and MacNicol, 1974).

E. Silylene and Germylene

Although we have specifically excluded carbenes from this review, we shall cite several reports on congeners. Both silylene and its dialkyl derivatives (generated in several ways) react readily with unsaturated systems (Skell and Goldstein, 1964; Atwell and Weyenberg, 1969; Manuel et al., 1974; Gaspar et al., 1974; Childs and Weber, 1974; Jenkins et al., 1975). Of most relevance is the recent report that H_2Si: adds nonstereospecifically to (E,E)-hexadiene; apparently a concerted mechanism is excluded (Gaspar and Hwang, 1974) [Eq. (45)]. The allylic substitution products of certain silylenes have been formulated as arising from a preliminary cheletropic addition [Eq. (46)]

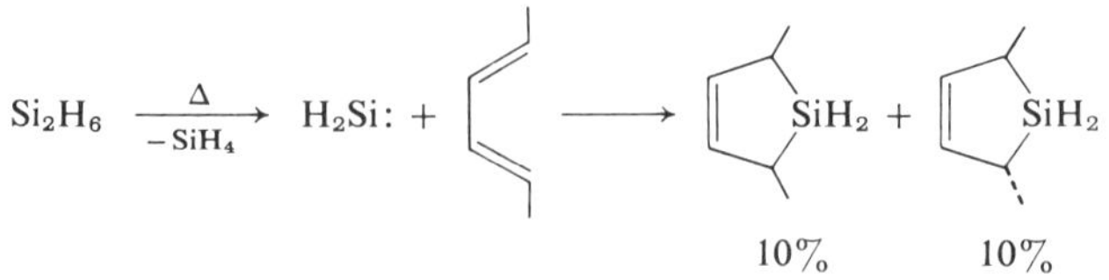

$$Si_2H_6 \xrightarrow[-SiH_4]{\Delta} H_2Si: + \text{[diene]} \longrightarrow \text{[ring-SiH}_2\text{]} + \text{[ring-SiH}_2\text{]} \quad (45)$$

$$10\% \quad 10\%$$

$$(CH_3)_3Si(CH_3PhSi)_2Si(CH_3)_3 \xrightarrow{h\nu} PhCH_3Si: \xrightarrow{\text{[cyclohexadiene]}}$$

$$\left(\text{[bicyclic-SiCH}_3Ph\text{]}\right) \longrightarrow \text{[cyclohexenyl-SiHCH}_3Ph\text{]} \quad (46)$$

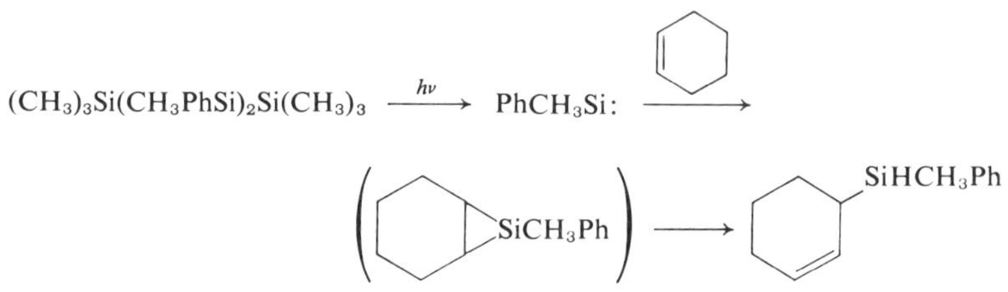

(Ishikawa et al., 1973; see also Childs and Weber, 1974; Barton and Juvet, 1975). The difficulty of divalent silicon preparation by extrusion is illustrated by thermolysis of the following silanorbornadiene [Eq. (47)], although

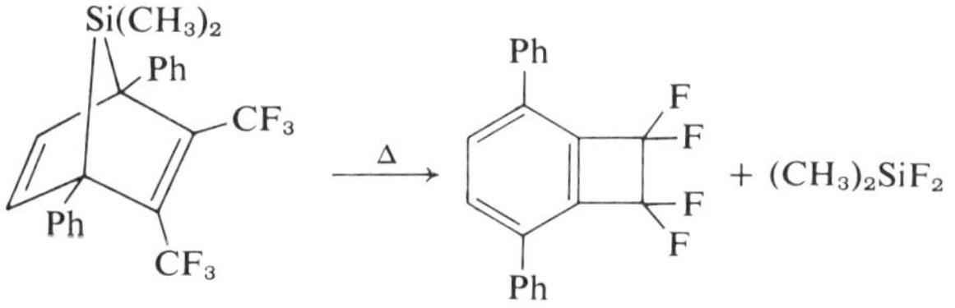

$$\text{[silanorbornadiene]} \xrightarrow{\Delta} \text{[product]} + (CH_3)_2SiF_2 \quad (47)$$

some evidence for dimethylsilylene liberation was presented (Barton et al., 1972b). Extrusion from silepins has also been reported (Barton et al., 1974).

Generation of divalent germanium appears somewhat easier. Five- and seven-membered ring syntheses by addition of germylenes to dienes and trienes [Eq. (48)], has been extensively reviewed (Satgé et al., 1973; Pant, 1974), although mechanistic examination has been limited.

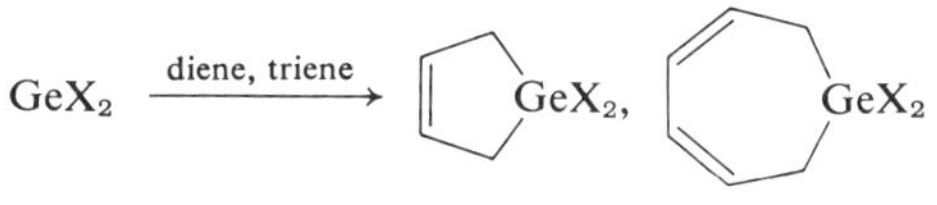

$$GeX_2 \xrightarrow{\text{diene, triene}} \text{[ring-GeX}_2\text{]}, \text{[ring-GeX}_2\text{]} \quad (48)$$

F. Decarbonylations

An apparently genuine case of concerted extrusion of CO is the fragmentation of the *endo*-cyclopropane isomer shown (Halton *et al.*, 1967; Clarke and Johnson, 1971) [Eq. (49)]. The rate is 10^7 times faster than for its *exo*-cyclopropane isomer, for which stereochemistry does not allow coupling of strain

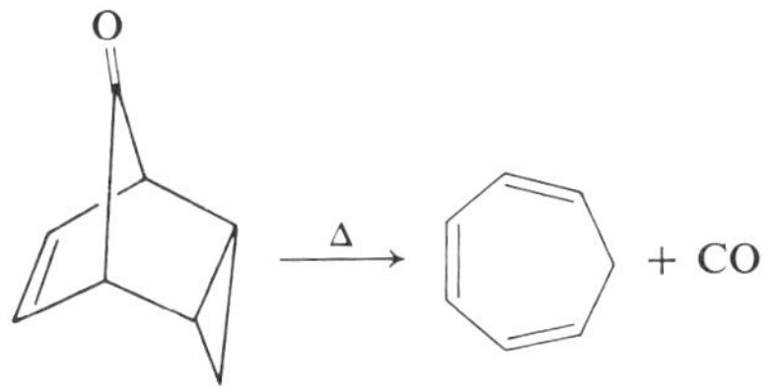

$$\text{(structure)} \xrightarrow{\Delta} \text{(cycloheptatriene)} + \text{CO} \tag{49}$$

release from the three-membered ring in a concerted transition state (Section II,C,4). An analogous cyclobutane-promoted decarbonylation allegedly does not involve scission of a bent σ bond (Kretschmer *et al.*, 1975). Rates for related, less strain-activated norbornen-7-ones have been reported (Baldwin, 1966; Sakai, 1973). The rate of extrusion of CO from 2,3-di-*tert*-butylcyclopropanone, $k_{150°} = 2 \times 10^{-4}$ sec^{-1} (Pazos *et al.*, 1974), is somewhat faster than from minimally strained cyclopentenones, $k_{100°} = 3$–12.5×10^{-9} sec^{-1} (extrapolated from higher temperatures, Sakai, 1973). Loss of CO from tricyclo[2.1.0.0^{2,5}]pentan-3-one derivatives occurs readily at 105° (Ona *et al.*, 1970). The influence of symmetry factors and the ordering of energy levels on the extrusion of CO from vinylene carbonates has been considered in a general way (Fleischauer and Scharf, 1972).

Recently, the *photo*decarbonylation of 3-cyclopentenones has been interpreted in terms of dual linear and nonlinear concerted pathways for decomposition (Darling *et al.*, 1974). Of relevance to the earlier discussion of the photosulfolene reaction (Section II,E) is the rationalization that spin-orbital coupling may relax spin inversion restrictions so that a concerted triplet reaction may lead directly to singlet products (see also Richardson *et al.*, 1974; Dewar and Kirschner, 1974; Barnett, 1974). Light-induced extrusions of CO from cycloheptadienones is well documented (Schuster and Kim, 1974) as is that for cyclopropanones (Turro *et al.*, 1965).

In an apparent cheletropic process, an *isonitrile adds* to a strained alkyne (Krebs and Kimling, 1971) [Eq. (50)]. Other cases of isonitrile addition and extrusion have been reported (Sheehan and Beeson, 1967; Deyrup, 1971; Burger and Fehn, 1972; Atkinson and Harger, 1974).

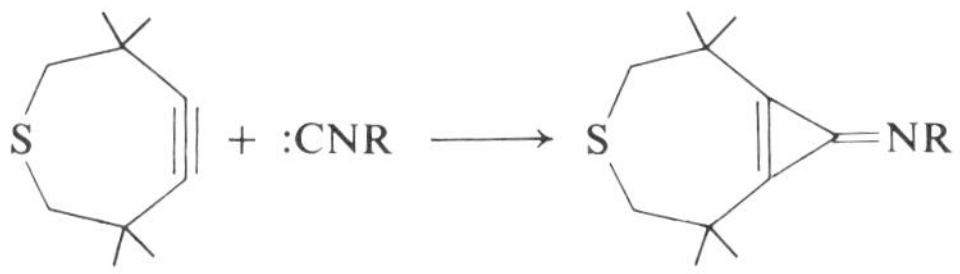

$$\text{(structure)} + :\text{CNR} \longrightarrow \text{(structure)}=\text{NR} \tag{50}$$

G. Miscellaneous Cheletropy

The thermal fragmentation of diazirine to N_2 plus singlet H_2C: has been treated theoretically by the SCF–MO–CNDO procedure (Snyder *et al.*, 1972b). Also, the fragmentation of cyclopropylmethylene (i.e., C_3H_5–CH:) to acetylene and ethylene has been treated by the MO following method with novel insight (Zimmerman and Sousa, 1972). A nitrene analog of this transformation is known (Wulfman and Steinheimer, 1972; Szeimies and Harnisch, 1973).

Another SO_2 extrusion which possibly proceeds through a pericyclic mechanism is the thermolysis of allyl sulfones (La Combe and Stewart, 1961; Hendrickson and Bergeron, 1973; Fisher and Lin, 1973; King *et al.*, 1974) [Eq. (51)]. Thiaziridine dioxides extrude SO_2 (Quast and Kees, 1974) [Eq. (52)], as do 1,2,5-thiadiazole 1,1-dioxides (Ege and Beisiegel, 1974) [Eq. (53)].

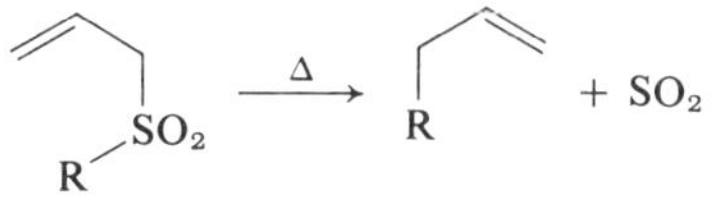

$$\text{R-allyl sulfone} \xrightarrow{\Delta} \text{alkene} + SO_2 \tag{51}$$

$$\text{thiaziridine dioxide} \longrightarrow SO_2 + \text{R}_2\text{C=N-R} \tag{52}$$

$$\text{1,2,5-thiadiazole 1,1-dioxide} \longrightarrow SO_2 + 2\,RC\equiv N \tag{53}$$

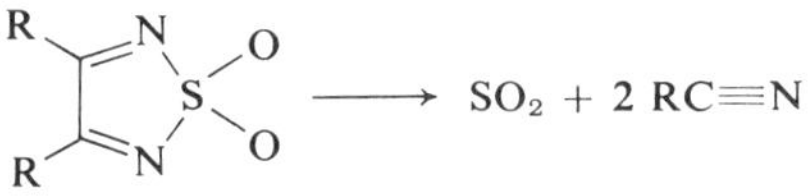

Thermolysis of 2-ethoxy-1,3-dioxolanes in the presence of benzoic acid has been reported (Josan and Eastwood, 1968; Hiyama and Nozaki, 1973); we suggest a complex cheletropic process as shown [Eq. (54)]. This bears a

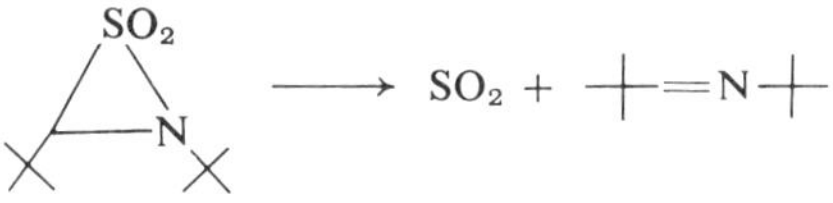

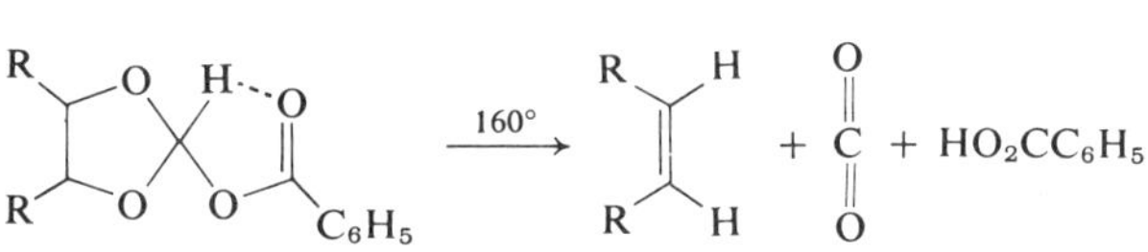

$$\text{dioxolane} \xrightarrow{160°} \text{alkene} + \text{C} + HO_2CC_6H_5 \tag{54}$$

certain formal resemblance to the long-accepted mechanism of percarboxylic acid epoxidation (Bartlett, 1950), which is revealed as a double cheletropic transformation [Eq. (55)].

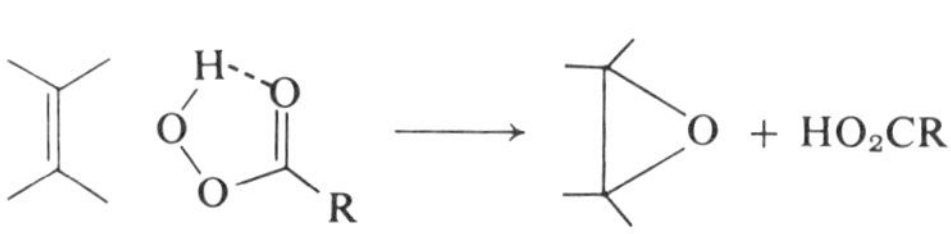

$$\text{alkene} + \text{percarboxylic acid} \longrightarrow \text{epoxide} + HO_2CR \tag{55}$$

Arguments supporting perepoxides as intermediates in the $^1\Delta_g$ O$_2$-olefin "ene" addition [Eq. (56)] were discounted (Foote et al., 1972; Hasty et al.,

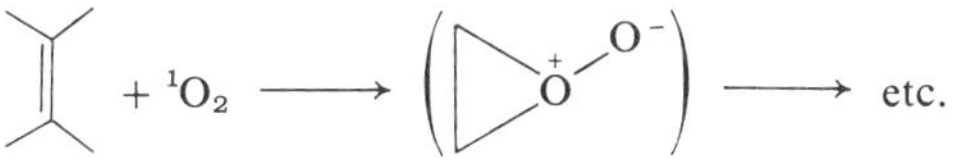

$$\ce{>=<} + {}^1O_2 \longrightarrow \left(\triangle\!\!\!\overset{+}{\underset{O}{}}\!\!\!\overset{O^-}{} \right) \longrightarrow \text{etc.} \tag{56}$$

1972), but have subsequently been resurrected (Hasty and Kearns, 1973; Schaap and Faler, 1973; Stephenson et al., 1973). Similar transient adducts have been postulated in the ozonolysis of olefins, as explanation of the occasional production of epoxides (Story et al., 1971; Murray and Suzui, 1973) [Eq. (57)].

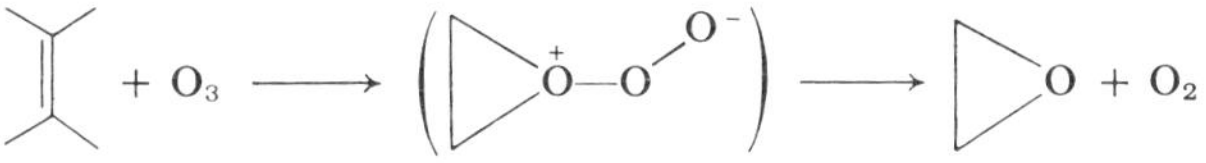

$$\ce{>=<} + O_3 \longrightarrow \left(\triangle\!\!\!\overset{+}{\underset{O-O}{}}\!\!\!\overset{O^-}{} \right) \longrightarrow \triangle\!\text{O} + O_2 \tag{57}$$

Examples of cheletropic interaction at nitrogen were provided previously (Section III,A). Mechanistically related are pericyclic fragmentations of intermediate ylids in the reaction of carbenes with aziridines (Hata and Watanabe, 1972) [Eq. (58)]. Recently, 2H-pyrrole has been advocated as a useful cheletropic departing moiety (Hata and Watanabe, 1973) [Eq. (59)]. Also docu-

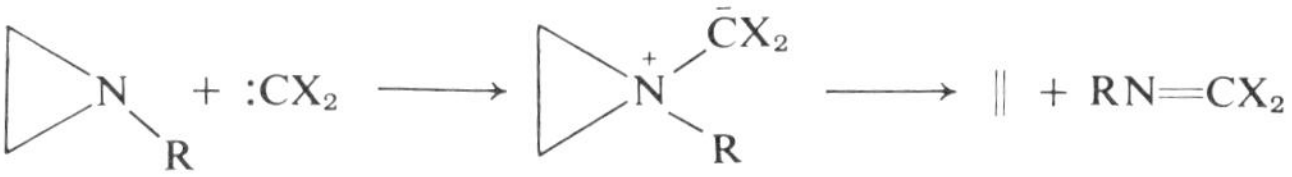

$$\triangle\!\text{N}\!-\!\text{R} + :CX_2 \longrightarrow \triangle\!\overset{+}{\text{N}}\!(\text{R})\!-\!\bar{C}X_2 \longrightarrow \| + RN\!\!=\!\!CX_2 \tag{58}$$

$$\text{(equation 59 scheme)}$$

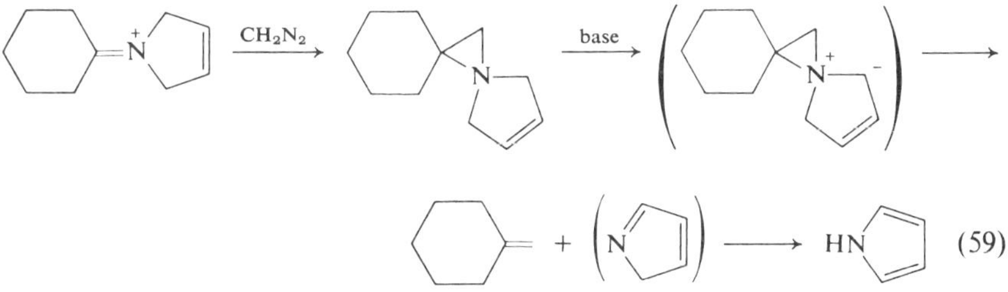

$$\tag{59}$$

mented is the apparently facile dissociation of aziridine and oxaziridine N-oxides (Emmons, 1957; Baldwin et al., 1971b) [Eq. (60)]. Likely related is the formation of diimide from 1-amino-2,2-diphenylaziridine (Annunziata et al., 1974).

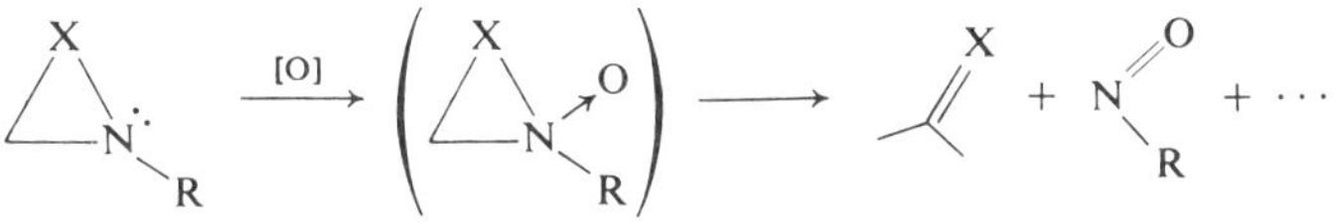

$$\overset{X}{\triangle}\!\!-\!\!\text{N}\!-\!\text{R} \xrightarrow{[O]} \left(\overset{X}{\triangle}\!\!-\!\!\overset{\nearrow O}{\text{N}\!-\!\text{R}} \right) \longrightarrow \overset{X}{\diagup\!\!\diagdown} + \text{N}(R)\!\!=\!\!O + \cdots \tag{60}$$

(X = O, CH$_2$)

We shall end this survey with what we regard as the most elegantly arcane cheletropic process yet reported (Felix *et al.*, 1972). The useful and synthetically ingenious fragmentation of *N*-aminoaziridine monohydrazones [Eq. (61)] releases four components (although it may in fact proceed stepwise). The occurrence of such an improbable reaction persuasively argues for the potency of the cheletropic concept.

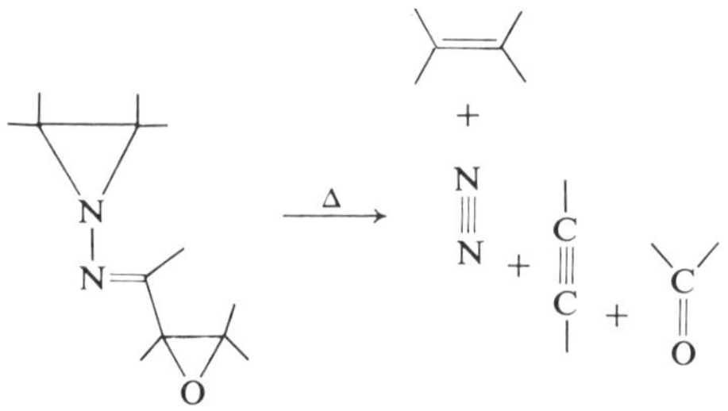

(61)

IV. Conclusion

We have attempted to show the scope of the cheletropic principle; that numerous useful synthetic transformations may be realized by such extrusion reactions. It has been suggested that mechanistically there is a special facility intrinsic to such processes, which is rooted in entropic and least motion considerations. Accordingly, conjecturing on new cheletropric reactions may be an especially rewarding way for discovery of novel chemistry. We have further attempted to demonstrate that least motion analysis of reactions, as exemplified by nitrosoaziridines, photodissociation of sulfolenes, and possibly the thiazine oxide–sulfolene cycloaddition manifold, constitutes a principle which ought to be more widely applied in attempting to understand qualitatively the nature of chemical reactivity.

References

Anastassiou, A. G., and Yamamoto, H. (1973). *J. Chem. Soc., Chem. Commun.* p. 840.
Annunziata, R., Fornasier, R., and Montanari, F. (1974). *J. Org. Chem.* **39**, 3195.
Arbuzov, B. A., and Konovalov, A. I. (1959). *Izv. Akad. Nauk SSSR, Otd. Khim. Nauk* p. 2130; *Chem. Abstr.* **54**, 10813 (1960).
Arigoni, D., Vasella, A., Sharpless, K. B., and Jensen, H. P. (1973). *J. Am. Chem. Soc.* **95**, 7917.
Ašperger, S., Hegedić, D., Pavlović, D., and Borčić, S. (1972). *J. Org. Chem.* **37**, 1745.
Atkinson, R. S., and Harger, M. J. P. (1974). *J. Chem. Soc., Perkin Trans. 1* p. 2619.

Atwell, W. H., and Weyenberg, D. R. (1969). *Angew. Chem., Int. Ed. Engl.* **8**, 469.
Baldwin, J. E. (1966). *Can. J. Chem.* **44**, 2051.
Baldwin, J. E., Höfle, G., and Choi, S. C. (1971a). *J. Am. Chem. Soc.* **93**, 2810.
Baldwin, J. E., Bhatnagar, A. K., Choi, S. C., and Shortridge, T. J. (1971b). *J. Am. Chem. Soc.* **93**, 4082.
Barnett, G. (1974). *Can. J. Chem.* **52**, 3837.
Bartlett, P. D. (1950). *Rec. Chem. Prog.* **11**, 47.
Barton, D. H. R., and Willis, B. J. (1972). *J. Chem. Soc., Perkin Trans. 1* p. 305.
Barton, T. J., and Juvet, M. (1975). *Tetrahedron Lett.* p. 3893.
Barton, T. J., Martz, M. D., and Zika, R. G. (1972a). *J. Org. Chem.* **37**, 552.
Barton, T. J., Witiak, J. L., and McIntosh, C. L. (1972b). *J. Am. Chem. Soc.* **94**, 6229.
Barton, T. J., Kippenhan, R. C., Jr., and Nelson, A. J. (1974). *J. Am. Chem. Soc.* **96**, 2272.
Bond, A., Green, M., and Pearson, S. C. (1968). *J. Chem. Soc. B* p. 929.
Bordwell, F. G. (1970). *Acc. Chem. Res.* **3**, 281.
Bordwell, F. G. (1972). *Acc. Chem. Res.* **5**, 374.
Bordwell, F. G., Williams, J. M., Jr., Hoyt, E. B., Jr., and Jarvis, B. B. (1968). *J. Am. Chem. Soc.* **90**, 429.
Broadhurst, M. J., Grigg, R., and Johnson, A. W. (1972). *J. Chem. Soc., Perkin Trans. 1* pp. 1124 and 2111.
Burger, K., and Fehn, J. (1972). *Angew. Chem., Int. Ed. Engl.* **11**, 47.
Buter, J., Wassenaar, S., and Kellogg, R. M. (1972). *J. Org. Chem.* **37**, 4045.
Carpino, L. A., and Kirkley, R. K. (1970). *J. Am. Chem. Soc.* **92**, 1784.
Carpino, L. A., McAdams, L. V., III, Rynbrandt, R. H., and Spiewak, J. W. (1971). *J. Am. Chem. Soc.* **93**, 476.
Cassar, L., Eaton, P. E., and Halpern, J. (1970). *J. Am. Chem. Soc.* **92**, 3515.
Chan, T. H., and Finkenbine, J. R. (1974). *Tetrahedron Lett.* p. 2091.
Chao, P., and Lemal, D. M. (1973). *J. Am. Chem. Soc.* **95**, 920.
Childs, M. E., and Weber, W. P. (1974). *Tetrahedron Lett.* p. 4033.
Clarke, S. C., and Johnson, B. L. (1971). *Tetrahedron* **27**, 3555.
Clive, D. L. J., and Denyer, C. V. (1973). *J. Chem. Soc., Chem. Commun.* p. 253.
Coyle, J. D. (1974). *Chem. Soc. Rev.* **3**, 329.
Crowley, K. J. (1968). *J. Org. Chem.* **33**, 3679.
Cundall, R. B. (1964). *Prog. React. Kinet.* **2**, 165.
Cuthbertson, E., and MacNicol, D. D. (1974). *J. Chem. Soc., Chem. Commun.* p. 498.
Darling, T. R., Pouliquen, J., and Turro, N. J. (1974). *J. Am. Chem. Soc.* **96**, 1247.
Dauben, W. G., Kielbania, A. J., Jr., and Raymond, K. N. (1973). *J. Am. Chem. Soc.* **95**, 7166.
de Bruin, G. (1914a). *Proc. K. Akad. Wetensch. Amsterdam* **17**, 585; *J. Chem. Soc. Abstr.* **106**, 1197 (1914).
de Bruin, G. (1914b). *Versl. Akad. Wet.* **23**, 445; *Chem. Abstr.* **9**, 623 (1915).
Dewar, M. J. S., and Kirschner, S. (1971). *J. Am. Chem. Soc.* **93**, 4292.
Dewar, M. J. S., and Kirschner, S. (1974). *J. Am. Chem. Soc.* **96**, 7578.
Deyrup, J. A. (1971). *Tetrahedron Lett.* p. 2191.
Dodson, R. M., and Nelson, J. P. (1969), *J. Chem. Soc., Chem. Commun.* p. 1159.
Dodson, R. M., and Sauers, R. F. (1967). *Chem. Commun.* p. 1189.
du Manoir, J. R., King, J. F., and Fraser, R. R. (1972). *J. Chem. Soc., Chem. Commun.* p. 541.
Ege, G., and Beisiegel, E. (1974). *Synthesis* p. 22.
Emmons, W. D. (1957). *J. Am. Chem. Soc.* **79**, 6522.

Felix, D., Müller, R. K., Horn, U., Joos, R., Schreiber, J., and Eschenmoser, A. (1972). *Helv. Chim. Acta* **55**, 1276.
Fischer, N. H. (1970). *Synthesis* p. 393.
Fischer, N. H., and Lin, H.-N. (1973). *J. Org. Chem.* **38**, 3073.
Fleischhauer, J., and Scharf, H. D. (1972). *Tetrahedron Lett.* p. 1119.
Foote, C. S., Fujimoto, T. T., and Chang, Y. C. (1972). *Tetrahedron Lett.* p. 45.
Fraser, A. R., Bird, P. H., Bezman, S. A., Shapley, J. R., White, R., and Osborn, J. A. (1973). *J. Am. Chem. Soc.* **95**, 597.
Freeman, J. P., and Graham, W. H. (1967). *J. Am. Chem. Soc.* **89**, 1761.
Gaspar, P. P., and Hwang, R.-J. (1974). *J. Am. Chem. Soc.* **96**, 6198.
Gaspar, P. P., Hwang, R.-J., and Eckelman, W. C. (1974). *J. Chem. Soc., Chem. Commun.* p. 242.
Gasteiger, J., and Huisgen, R. (1972). *J. Am. Chem. Soc.* **94**, 6541.
Green, M. (1963). *Proc. Chem. Soc., London* p. 177.
Greene, F. D., and Gilbert, K. E. (1975). *J. Org. Chem.* **40**, 1409.
Grummitt, O., Ardis, A. E., and Fick, J. (1950). *J. Am. Chem. Soc.* **72**, 5167.
Hall, C. D., Bramblett, J. D., and Lin, F. F. S. (1972). *J. Am. Chem. Soc.* **94**, 9264.
Halton, B., Battiste, M. A., Rehberg, R., Deyrup, C. L., and Brennan, M. E. (1967). *J. Am. Chem. Soc.* **89**, 5964.
Hartzell, G. E., and Paige, J. N. (1967). *J. Org. Chem.* **32**, 459.
Hasty, N. M., and Kearns, D. R. (1973). *J. Am. Chem. Soc.* **95**, 3380.
Hasty, N., Merkel, P. B., Radlick, P., and Kearns, D. R. (1972). *Tetrahedron Lett.* p. 49.
Hata, Y., and Watanabe, M. (1972). *Tetrahedron Lett.* pp. 3827 and 4659.
Hata, Y., and Watanabe, M. (1973). *J. Am. Chem. Soc.* **95**, 8451.
Helder, R., and Wynberg, H. (1972). *Tetrahedron Lett.* p. 605.
Heldeweg, R. F., and Hogeveen, H. (1976). *J. Am. Chem. Soc.* **98**, 2341.
Hendrickson, J. B., and Bergeron, R. (1973). *Tetrahedron Lett.* p. 3609.
Hine, J. (1966). *J. Amer. Chem. Soc.* **88**, 5525.
Hiyama, T., and Nozaki, H. (1973). *Bull. Chem. Soc. Jpn.* **46**, 2248.
Hoffmann, R., Gleiter, R., and Mallory, F. B. (1970). *J. Am. Chem. Soc.* **92**, 1460.
Hoffmann, R., Howell, J. M., and Muetterties, E. L. (1972). *J. Am. Chem. Soc.* **94**, 3047.
Isaacs, N. S., and Laila, A. A. R. (1976). *Tetrahedron Lett.* p. 715.
Ishikawa, M., Ishiguro, M., and Kumada, M. (1973). *J. Organomet. Chem.* **49**, C71.
Jenkins, R. L., Kedrowski, R. A., Elliott, L. E., Tappen, D. C., Schlyer, D. J., and Ring, M. A. (1975). *J. Organomet. Chem.* **86**, 347.
Jones, M., Jr., and Moss, R. A. (1973). "Carbenes." Wiley, New York.
Josan, J. S., and Eastwood, F. W. (1968). *Aust. J. Chem.* **21**, 2013.
Jung, F., Molin, M., Van Den Elzen, R., and Durst, T. (1974). *J. Am. Chem. Soc.* **96**, 935.
Kashman, Y., Menachem, Y., and Benary, E. (1973). *Tetrahedron* **29**, 4279.
Kellogg, R. M., and Prins, W. L. (1974). *J. Org. Chem.* **39**, 2366.
Kim, M., and White, J. D. (1975). *J. Am. Chem. Soc.* **97**, 451.
King, J. F., Enanoza, R. M., and Lewars, E. G., (1974). *Can. J. Chem.* **52**, 2409.
Kirmse, W. (1971). "Carbene Chemistry," 2nd ed. Academic Press, New York.
Klabunde, K. J., and Skell, P. S. (1971). *J. Am. Chem. Soc.* **93**, 3807 and 5315.
Kleveland, K., and Skattebøl, L. (1973). *J. Chem. Soc., Chem. Commun.* p. 432.
Kleveland, K., and Skattebøl, L. (1975). *Acta Chem. Scand.* **B29**, 827.
Kondo, K., Matsumoto, M., and Negishi, A. (1972). *Tetrahedron Lett.* p. 2131.
Krebs, A., and Kimling, H. (1971). *Angew. Chem., Int. Ed. Engl.* **10**, 409.

Kresze, G. (1967). In "1,4-Cycloaddition Reactions" (J. Hamer, ed.), p. 453. Academic Press, New York.
Kretschmer, G., McCay, I. W., Padden-Row, M. N., and Warrener, R. N. (1975). *Tetrahedron Lett.* p. 1339.
Kropp, P. J., and Krauss, H. J. (1967). *J. Am. Chem. Soc.* **89**, 5199.
Kuhn, H. J., and Gollnick, K. (1972). *Tetrahedron Lett.* p. 1909.
La Combe, E. M., and Stewart, B. (1961). *J. Am. Chem. Soc.* **83**, 3457.
Lalezari, I., Shafiee, A., and Yalpani, M. (1970). *Angew. Chem., Int. Ed. Engl.* **9**, 464.
Lantos, I., and Ginsburg, D. (1972). *Tetrahedron* **28**, 2507.
Larrabee, R. B. (1974). *J. Organomet. Chem.* **74**, 313.
Lemal, D. M., and Chao, P. (1973). *J. Am. Chem. Soc.* **95**, 922.
Lemal, D. M., and McGregor, S. D. (1966). *J. Am. Chem. Soc.* **88**, 1335.
McDermott, J. X., White, J. F., and Whitesides, G. M. (1973). *J. Am. Chem. Soc.* **95**, 4451.
McGregor, S. D., and Lemal, D. M. (1966). *J. Am. Chem. Soc.* **88**, 2858.
McIntosh, J. M., Goodbrand, H. B., and Masse, G. M. (1974). *J. Org. Chem.* **39**, 202.
Magdesieva, N. N. (1970). *Adv. Heterocycl. Chem.* **12**, 1.
Mango, F. D. (1971). *Tetrahedron Lett.* p. 505.
Mango, F. D. (1975). *Coord. Chem. Rev.* **15**, 109.
Mango, F. D., and Schachtschneider, J. H. (1971). *J. Am. Chem. Soc.* **93**, 1123.
Manuel, G., Mazerolles, P., and Cauquy, G. (1974). *Synth. React. Inorg. Met.-Org. Chem.* **4**, 133.
Martin, D. (1964). *Angew. Chem., Int. Ed. Engl.* **3**, 311.
Meyers, A. I. (1974). "Heterocycles in Organic Synthesis." Wiley, New York.
Meyers, A. I., and Takaya, T. (1971). *Tetrahedron Lett.* p. 2609.
Mitchell, R. H. (1973). *J. Chem. Soc., Chem. Commun.* p. 955.
Mock, W. L. (1966). *J. Am. Chem. Soc.* **88**, 2857.
Mock, W. L. (1967). *J. Am. Chem. Soc.* **89**, 1281.
Mock, W. L. (1969). *J. Am. Chem. Soc.* **91**, 5682.
Mock, W. L. (1970a). *J. Am. Chem. Soc.* **92**, 3807.
Mock, W. L. (1970b). *J. Am. Chem. Soc.* **92**, 6918; **95**, 4472.
Mock, W. L. (1975a). *J. Am. Chem. Soc.* **97**, 3666.
Mock, W. L. (1975b). *J. Am. Chem. Soc.* **97**, 3673.
Mock, W. L., and Isaac, P. A. H. (1972). *J. Am. Chem. Soc.* **94**, 2749.
Mock, W. L., and McCausland, J. H. (1968). *Tetrahedron Lett.* p. 391.
Mock, W. L., and McCausland, J. H. (1976). *J. Org. Chem.* **41**, 242.
Mock, W. L., and Nugent, R. M. (1975a). *J. Am. Chem. Soc.* **97**, 6521.
Mock, W. L., and Nugent, R. M. (1975b). *J. Am. Chem. Soc.* **97**, 6526.
Mock, W. L., Mehrotra, I., and Anderko, J. A. (1975). *J. Org. Chem.* **40**, 1842.
Mulliken, R. S., and Roothaan, C. C. J. (1947). *Chem. Rev.* **41**, 219.
Murray, R. W., and Suzui, A. (1973). *J. Am. Chem. Soc.* **95**, 3343.
Nakayama, S., Yoshifuji, M., Okazaki, R., and Inamoto, N. (1971). *J. Chem. Soc., Chem. Commun.* p. 1186.
Noyori, R., Suzuki, T., and Takaya, H. (1971). *J. Am. Chem. Soc.* **93**, 5896.
Ona, H., Yamaguchi, H., and Masamune, S. (1970). *J. Am. Chem. Soc.* **92**, 7495.
Olsen, H., and Snyder, J. P. (1974). *J. Am. Chem. Soc.* **96**, 7839.
Pant, B. C. (1974). *J. Organomet. Chem.* **66**, 321.
Paquette, L. A., Wingard, R. E., Jr., and Meisinger, R. H. (1971). *J. Am. Chem. Soc.* **93**, 1047.

Paquette, L. A., Meisinger, R. H., and Wingard, R. E., Jr. (1973). *J. Am. Chem. Soc.* **95**, 2230.
Pazos, J. F., Pacifici, J. G., Pierson, G. O., Sclove, D. B., and Greene, F. D. (1974). *J. Org. Chem.* **39**, 1990.
Quast, H., and Kees, F. (1974). *Angew. Chem., Int. Ed. Engl.* **13**, 742.
Quin, L. D. (1967). *In* "1,4-Cycloaddition Reactions" (J. Hamer, ed.), p. 47. Academic Press, New York.
Quin, L. D., and Barket, T. P. (1970). *J. Am. Chem. Soc.* **92**, 4303.
Quin, L. D., Gratz, J. P., and Barket, T. P. (1968). *J. Org. Chem.* **33**, 1034.
Richardson, W. H., Montgomery, F. C., Yelvington, M. B., and O'Neal, H. E. (1974). *J. Am. Chem. Soc.* **96**, 7525.
Sakai, M. (1973). *Tetrahedron Lett.* p. 2297.
Saltiel, J., and Metts, L. (1967). *J. Am. Chem. Soc.* **89**, 2232.
Satgé, J., Massol, M., and Rivière, P. (1973). *J. Organomet. Chem.* **56**, 1.
Schaap, A. P., and Faler, G. R. (1973). *J. Am. Chem. Soc.* **95**, 3381.
Schuster, D. I., and Kim, C. W. (1974). *J. Am. Chem. Soc.* **96**, 7437.
Sheehan, J. C., and Beeson, J. H. (1967). *J. Am. Chem. Soc.* **89**, 366.
Shibaeva, R. P., Atovmyan, L. O., and Kostyanovskii, R. G. (1967). *Dokl. Akad. Nauk SSSR* **175**, 586.
Skell, P. S., and Cholod, M. S. (1969). *J. Am. Chem. Soc.* **91**, 7131.
Skell, P. S., and Goldstein, E. J. (1964). *J. Am. Chem. Soc.* **86**, 1442.
Snyder, J. P., Lee, L., Bandurco, V. T., Yu, C. Y., and Boyd, R. J. (1972a). *J. Am. Chem. Soc.* **94**, 3260.
Snyder, J. P., Boyd, R. J., and Whitehead, M. A. (1972b). *Tetrahedron Lett.* p. 4347.
Snyder, J. P., Heyman, M. L., and Suciu, E. N. (1975). *J. Org. Chem.* **40**, 1395.
Srinivasan, R. (1963). *J. Chem. Phys.* **38**, 1039.
Stark, B. P., and Duke, A. J. (1967). "Extrusion Reactions." Pergamon, Oxford.
Stephenson, L. M., and McClure, D. E. (1973). *J. Am. Chem. Soc.* **95**, 3074.
Stephenson, L. M., McClure, D. E., and Sysak, P. K. (1973). *J. Am. Chem. Soc.* **95**, 7888.
Stille, J. K., Eichelberger, J. L., Higgins, J., and Freeburger, M. E. (1972). *J. Am. Chem. Soc.* **94**, 4761.
Story, P. R., Alford, J. A., Ray, W. C., and Burgess, J. R. (1971). *J. Am. Chem. Soc.* **93**, 3044.
Szeimies, G., and Harnisch, J. (1973). *J. Chem. Soc., Chem. Commun.* p. 739.
Tee, O. S., Altmann, J. A., and Yates, K. (1974). *J. Am. Chem. Soc.* **96**, 3141.
Tomioka, H., Hirano, Y., and Izawa, Y. (1974a). *Tetrahedron Lett.* p. 1865.
Tomioka, H., Yoshimasa, H., and Yasuji, I. (1974b). *Tetrahedron Lett.* p. 4477.
Traynelis, V. J., Yoshikawa, Y., Sih, J. C., Miller, L. J., and Livingston, J. R., Jr. (1973). *J. Org. Chem.* **38**, 3978.
Trost, B. M., and Ziman, S. (1969). *J. Chem. Soc., Chem. Commun.* p. 181.
Trost, B. M., and Ziman, S. D. (1971). *J. Am. Chem. Soc.* **93**, 3825.
Trost, B. M., Schinski, W. L., and Mantz, I. B. (1967). *J. Am. Chem. Soc.* **91**, 4320.
Trost, B. M., Schinski, W. L., Chen, F., and Mantz, I. B. (1971). *J. Am. Chem. Soc.* **93**, 676.
Turk, S. D., and Cobb, R. L. (1967). *In* "1,4-Cycloaddition Reactions" (J. Hamer, ed.), p. 13. Academic Press, New York.
Turnblom, E. W., and Katz, T. J. (1973). *J. Am. Chem. Soc.* **95**, 4292.
Turro, N. J., Leermakers, P. A., Wilson, H. R., Neckers, D. C., Byers, G. W., and Vesley, G. F. (1965). *J. Am. Chem. Soc.* **87**, 2613.
van der Lugt, W. T. A. M. (1970). *Tetrahedron Lett.* p. 2281.

van Tilborg, W. J. M., Smael, P., Visser, J. P., Kouwenhoven, C. G., and Reinhoudt, D. N. (1975). *Rec. Trav. Chim.* **94**, 85.
Vogel, E., Schmidbauer, E., and Altenbach, H.-J. (1974). *Angew. Chem., Int. Ed. Engl.* **13**, 736.
White, J. D., and Kim, M. (1974). *Tetrahedron Lett.* p. 3361.
Woodward, R. B., and Hoffmann, R. (1969a). *Angew. Chem., Int. Ed. Engl.* **8**, 781.
Woodward, R. B., and Hoffmann, R. (1969b). "The Conservation of Orbital Symmetry." Academic Press, New York.
Wulfman, D. S., and Steinheimer, T. R. (1972). *Tetrahedron Lett.* p. 3933.
Wynberg, H., and Helder, R. (1972). *Tetrahedron Lett.* p. 3647.
Zimmerman, H. E., and Sousa, L. R. (1972). *J. Am. Chem. Soc.* **94**, 834.
Zimmerman, H. E., Kamm, K. S., and Werthemann, D. P. (1974). *J. Amer. Chem. Soc.* **96**, 7821.
Zubtsova, L. I., Razumova, N. A., and Yakovleva, T. V. (1971). *J. Gen. Chem. USSR* **41**, 2450.

4

Applications of Frontier Molecular Orbital Theory to Pericyclic Reactions

K. N. HOUK*

I. Introduction	182
II. The Development of the Frontier Molecular Orbital Method	182
A. "Fundamental Postulates" and Early Applications	182
B. The Contributions of Woodward and Hoffmann	183
III. The Frontier Orbital Approximation	184
A. Perturbation Theories of Chemical Reactivity	184
B. Approximations Involved in Frontier Molecular Orbital Theory	191
C. Derivations in Terms of Charge-Transfer, or Configuration Interaction, Formalisms	191
D. The Bader–Salem–Pearson Formalism for Unimolecular Reactions	195
IV. Practical Applications of Frontier Molecular Orbital Theory	199
A. Calculated and Experimental Quantities Used in Frontier Orbital Applications	199
B. Frontier Molecular Orbitals	203
C. Charge-Controlled, Frontier-Energy-Controlled, and Frontier-Density-Controlled Reactions	206
V. Applications to Thermal Intermolecular Cycloadditions	208
VI. Applications to Thermal Intramolecular Reactions	240
A. Intramolecular Reactions as Cycloadditions	240
B. Electrocyclizations	241
C. Sigmatropic Shifts	248
D. Cheletropic Reactions and Eliminations	255
E. Miscellaneozations	257
VII. Photochemical Pericyclic Reactions	258
A. Cycloadditions	260
B. Electrocyclizations	263
C. Sigmatropic Shifts	265
D. Cheletropic Reactions	265
E. Miscellaneozations	265
VIII. Conclusion	266
References	266

* This article was prepared in part while the author was Visiting Professor at Princeton University.

I. Introduction

The explanation of chemical reactivity phenomena by quantum mechanical techniques has been a hotly pursued goal. Although quantum mechanical calculations of potential surfaces, coupled with classical trajectory calculations (see, for example, Wang and Karplus, 1973; Jean and Chapuisat, 1974), may eventually provide a complete understanding of chemical reactivity, the chemists' desires to explain the results of their experiments in a satisfactory, if qualitative, fashion, have led to the development of a number of approximate quantum mechanical theories of reactivity (Fukui, 1964). One of these approximate methods—frontier molecular orbital theory—has been particularly successful.

This chapter describes the ways in which frontier molecular orbital theory and more detailed perturbation approaches have enriched our understanding of one important class of reactions—pericyclic reactions—to the point that it is now possible in many cases to predict how substituents will affect the rates and mechanisms of reactions, to anticipate rates and the products of thermal or photochemical reactions of a variety of molecules, and to grasp the physical origins of these phenomena.

II. The Development of the Frontier Molecular Orbital Method

A. "Fundamental Postulates" and Early Applications

Soon after the development of a number of useful, but not infallible MO reactivity indices, many of which are cited by Fukui (1964), Fukui *et al.* (1954) made the empirical observation that the site of attack by electrophiles on benzenoid aromatics corresponded to the position of highest "frontier electron density" in the aromatic molecule. That is, the position of largest coefficient in the highest occupied molecular orbital (HOMO) on the substrate was the same as the position at which electrophilic substitution occurred most rapidly. A tentative explanation based on involvement of triplet states in the reaction was soon replaced by a more compelling suggestion that partial electron transfer was involved in chemical reactions, from the HOMO of the substrate to an electrophile, or, in the case of reactions with a nucleophile, from the reagent to the substrate lowest unoccupied molecular orbital (LUMO). For the reaction with a radical reagent, the HOMO of the substrate could donate charge to the radical center and LUMO could accept charge from the radical center (Fukui *et al.*, 1954). These "fundamental postulates" were shown to rationalize correctly several examples of reactions of each type.

R. D. Brown (1959) proposed a charge-transfer model for the transition state of electrophilic aromatic substitution reactions, and he showed that the HOMO electron density should be of major, but not exclusive importance. Charge transfer from the HOMO of the aromatic to a vacant electrophile orbital occurs more readily than other electron transfers because the lowest ionization potential involves removal of an electron from the HOMO of the aromatic compound.

Theoretical underpinnings for Fukui's frontier orbital postulates were provided by perturbation theory. Coulson and Longuet-Higgins (1947a,b, 1948a,b) had shown that the union of π systems with a reagent orbital could be treated as a perturbation of the isolated molecule orbitals, and Dewar (1952) explored the applications of perturbation theory to a variety of chemical phenomena, including chemical reactivity. A recent exposition of the principles of perturbation theory as applied to organic molecules can be found in Dewar's monograph (1969) and that by Dewar and Dougherty (1975).

Fukui et al. (1957) discussed the relationship between perturbation theory and the frontier molecular orbital postulates, and showed that a number of the theoretical reactivity indices proposed by others were interrelated. In ensuing papers, Fukui and his co-workers have shown that frontier molecular orbital theory is simply a first approximation to a perturbation treatment of chemical reactivity. Recently, Fujimoto and Fukui (1974) have reviewed the developments in the perturbation theory of chemical reactivity made by Fukui's group, as well as by many others. Some of these will be mentioned below.

B. The Contributions of Woodward and Hoffmann

The creation of a unifying theoretical concept with which to understand reactions which may occur in a concerted fashion began with Woodward and Hoffmann's 1965 article on the stereochemistry of electrocyclic reactions. This communication was soon followed by a series of communications and a general treatment of the whole field of pericyclic reactions (Woodward and Hoffmann, 1969). These papers, as well as some earlier works which forecast the eventual creation of this treatment have been collected by Simmons and Bunnett (1974). The initial Woodward–Hoffmann publication pointed out that electrocyclic reactions followed the stereochemistry dictated by the symmetry, or nodal properites of the HOMO of the polyene, and, later, the origin of this control in terms of correlation diagrams was pointed out (Longuet-Higgins and Abrahamson, 1965; Hoffmann and Woodward, 1965a). Woodward and Hoffmann realized that a great body of apparently unrelated chemical fact fits a simple pattern, and they were able to effect a revolution in the way organic chemists think about reaction mechanisms. Woodward and Hoffmann's specific applications of frontier molecular orbital theory are treated in later sections.

III. The Frontier Orbital Approximation

A. Perturbation Theories of Chemical Reactivity

A number of different perturbation expressions for calculation of the energy change which occurs upon the interaction of two molecules have been published. Before comparing the details of each of these expressions, a qualitative summary will be given, based primarily on the elegant descriptions given by Dewar (1969), Hoffmann (1971), and Jorgenson and Salem (1973).

The interaction of two orbitals, ϕ_a and ϕ_b, gives rise to a new set of orbitals, $\phi_a + \lambda\phi_b$ and $\phi_b - \lambda^*\phi_a$, whose shapes (determined by mixing coefficients, λ and λ^*) and energies are qualitatively described below and in Fig. 1. Full calculations on a variety of molecules, or applications of perturbation theory such as described below, lead to the following conclusions: The lower energy orbital (ϕ_a) will be stabilized by mixing in some of the higher energy orbital (ϕ_b) in a bonding fashion, while the upper energy orbital will be destabilized by mixing in some of the lower energy orbital in an antibonding fashion. The extent of interaction will depend on both the difference in energy between ϕ_a and ϕ_b—the closer these orbitals in energy, the more extensive the interaction and the larger is $\Delta\epsilon$—and also on the extent of overlap of ϕ_a with ϕ_b—the larger the overlap, the larger is $\Delta\epsilon$.

Adopting the simplest possible model in which the total energy of a system is a sum of energies of individual electrons, the interaction of two filled orbitals is destabilizing because $\Delta\epsilon$ is somewhat smaller than $\Delta\epsilon^*$. However, the interaction of one filled orbital (e.g., ϕ_a) with a vacant orbital (e.g., ϕ_b)

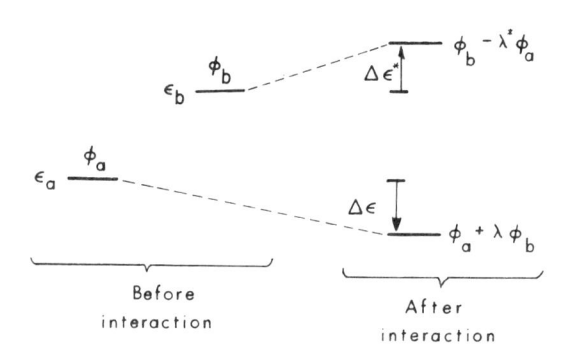

Fig. 1. Schematic representation of orbital interactions.

will result in a lower electronic energy after interaction than for isolated ϕ_a and ϕ_b, because the two electrons will both go into a lower energy orbital upon interaction. Similarly, interaction of a doubly occupied orbital (ϕ_a) with a singly occupied orbital (ϕ_b) or of a singly occupied (ϕ_a) with a vacant (ϕ_b) leads to stabilization, the former because $\Delta\epsilon^*$ is much less than twice $\Delta\epsilon$.

A calculation of the energy of a transition state for a bimolecular reaction may be made using an assumed geometry, but with the MO's of the reactants as the starting point. Stabilizing interactions will arise from interactions of occupied orbitals on one reactant and vacant on the other (and vice versa); of these stabilizing occupied–vacant interactions, the interactions of the HOMO of one reactant with the LUMO of the other (and vice versa) usually will be the largest terms, because these are the orbitals closest in energy. Thus, considerations of reactivity based on the frontier orbital theory are simply first approximations to actual calculations of transition state energies. However, as will be shown below, frontier orbital interactions may, in some cases, be less important than other interactions. As experience with the theory has increased, these cases have become relatively easy to recognize, so that there is little cause for apprehensions about the validity of the theory, such as expressed recently: "The frontier orbital method has little to recommend it" (Dewar, 1969, p. 332).

The various numerical formulations of the perturbation theory of chemical reactivity will now be discussed with reference to the qualitative model discussed above.

1. Dewar

Dewar (1952) outlined a general theory of organic reactivity using perturbation theory within the Hückel LCAO approximation. The interaction (or union) of two systems, A and B, which have MO's ϕ_{a_1}, ϕ_{a_2}, and ϕ_{b_1}, ϕ_{b_2}, respectively, gives rise to a composite system with different molecular orbitals, the energies and shapes of which may be approximated reasonably by perturbation theory. A second-order perturbation gives rise to mixing of the orbitals of A with those of B. For an individual orbital on A, for example, ϕ_a in Fig. 1, interaction with orbitals of B gives rise to two new orbitals, $\phi_a + \lambda\phi_b$ and $\phi_b - \lambda^*\phi_a$. The values of λ and λ^* calculated from second-order perturbation theory are (Dewar, 1952)

$$\lambda = -\frac{H_{ab}}{\epsilon_a - \epsilon_b}; \qquad \lambda^* = -\frac{H_{ab}}{\epsilon_b - \epsilon_a} \qquad (1)$$

Here H_{ab} is the resonance integral, $\int \phi_a H \phi_b d\tau$, for interaction between orbitals ϕ_a and ϕ_b. In Hückel theory, H_{ab} is the resonance integral β_{ab} which is often left as a parameter. For two orbitals overlapping at several sites i, ii, ..., and assuming that $\int \phi_x H \phi_y d\tau$ for atomic orbitals ϕ_x and ϕ_y is proportional to the overlap integral, $S_{ab} = \int \phi_a \phi_b d\tau$ (for example, $\int \phi_x H \phi_y d\tau = k S_{xy}$, where k is a constant), then H_{ab} will be equal to $\sum_i C_a{}^i C_b{}^i k S_{ab}$, where $C_a{}^i$ and $C_b{}^i$ are the AO coefficients at the points where orbital ϕ_a overlaps with orbital ϕ_b.

Thus, Eq. (1) can be replaced by the more complete expression

$$\lambda = -\frac{\sum_i C_a{}^i C_b{}^i k S_{ab}}{\epsilon_a - \epsilon_b}$$

In this approximation (overlap neglected) $\lambda = -\lambda^*$ and λ is >0 because $\epsilon_a < \epsilon_b$ and ϵ_a is a negative number, while H_{ab} is ordinarily negative. The expression discussed here is just a semiquantitative expression of the earlier statement that ϕ_a mixes in some of ϕ_2 in a bonding fashion, and ϕ_b mixes in some of ϕ_a in an antibonding fashion.

Turning to a quantity of more interest, the change in energy upon interaction, this is found, by a second-order perturbation theory, as follows:

$$\Delta\epsilon = \frac{-(H_{ab})^2}{\epsilon_a - \epsilon_b} = -\Delta\epsilon^*$$

where H_{ab} has the same meaning as before, and can be expanded in terms of atomic orbital coefficients and overlaps. The change in energy is still inversely proportional to the difference in the energies of ϕ_a and ϕ_b, but is proportional to the square of the resonance integral.

This derivation is based on the Hückel method and contains all of the Hückel assumptions: overlap is neglected, electron–electron repulsion is neglected, and the perturbation, $\beta_{rs} = \int \phi_r H \phi_s d\tau$, is a parameter which is not explicitly calculated. If overlap were included at this level of approximation, then $\Delta\epsilon < -\Delta\epsilon^*$, as discussed earlier.

For interaction between ϕ_a and ϕ_b of the same energy (degenerate) the change in energy calculated approximately by first-order perturbation theory is

$$\Delta\epsilon = H_{ab}$$

Up to now, only individual orbital energies have been considered. The total change in energy is, at this level of approximation, the sum of the one-electron energies, that is

$$\Delta E = 2 \sum_a \Delta\epsilon_a + 2 \sum_b \Delta\epsilon_b$$

if all orbitals are doubly occupied. By substituting the expressions given above the following formidable-looking expression is obtained.

$$\Delta E = 2 \sum_a^{occ} \sum_b^{vac} \frac{(\sum_i C_a{}^i C_b{}^i \beta_{ab}^i)^2}{\epsilon_a - \epsilon_b} + 2 \sum_a^{occ} \sum_b^{vac} \frac{(\sum_i C_a{}^i C_b{}^i \beta_{ab}^i)^2}{\epsilon_b - \epsilon_a} \qquad (2)$$

This is, however, just a sum of single electron energy changes.

The limits of accuracy of these perturbation expressions can be more easily seen from a derivation starting with the full calculation. A complete calculation would consist of solutions of the secular determinant.

$$\begin{vmatrix} H_{aa} - \epsilon & H_{ab} \\ H_{ab} & H_{bb} - \epsilon \end{vmatrix} = 0$$

The following are energies after interaction.

$$\epsilon = \frac{(\epsilon_a + \epsilon_b) \pm \sqrt{(\epsilon_a - \epsilon_b)^2 + 4H_{ab}^2}}{2} \tag{3}$$

where $\epsilon_a = H_{aa}$ and $\epsilon_b = H_{bb}$, the unperturbed orbital energies.

For the degenerate case ($\epsilon_a = \epsilon_b$), the full expression reduces to

$$\epsilon = \epsilon_a \pm H_{ab}$$

that is,

$$\Delta\epsilon = H_{ab}$$

The nondegenerate case is best treated by the following rearrangement of equation (3); as discussed recently by Jotham (1975).

$$\epsilon = \tfrac{1}{2}\left[(\epsilon_a + \epsilon_b) \pm (\epsilon_a - \epsilon_b)\sqrt{1 + \frac{4H_{ab}^2}{(\epsilon_a - \epsilon_b)^2}}\right] \tag{4}$$

Equation (4) is still an exact solution (overlap neglected), and for $4H_{ab}^2 \ll \epsilon_a - \epsilon_b$, a reasonable estimate to (4) is made by using the approximation, $\sqrt{1 + 4x} \approx (1 + 2x)$, which reduces (4) to

$$\epsilon = \tfrac{1}{2}\left[(\epsilon_a + \epsilon_b) \pm \left(\epsilon_a - \epsilon_b + \frac{2H_{ab}^2}{\epsilon_a - \epsilon_b}\right)\right]$$

that is,

$$\Delta\epsilon = -\Delta\epsilon^* = \frac{H_{ab}^2}{\epsilon_a - \epsilon_b}$$

which is identical to the perturbation expression. This approximation works well for occupied–vacant interactions, but overestimates $\Delta\epsilon$ as $\epsilon_a - \epsilon_b$ becomes small.

2. Salem

Salem (1968) derived the energy of interaction between two conjugated polyenes, neglecting electron–electron repulsion, but including overlap. Only π–π interactions between the isolated systems A and B were considered. The interaction between the two systems was assumed proportional to overlap, and the overlap between the π systems of two molecules approaching in parallel planes was shown to be small—that is, $2p\sigma, 2p\sigma$ overlap is maximum (0.270) at about 1.7 Å—so if the energy change is expressed in a series in powers of overlap, only the first few terms need be considered. This observation is of great importance, because it indicates that perturbation theory will provide an excellent approximation to the actual energy change, even at distances like those in the transition state.

The new feature obtained in this treatment arises from the inclusion of overlap, which means that the interaction of filled orbitals on the two molecules leads to net repulsion. For the interaction of two molecules, the energy change for interaction of occupied orbitals on the two molecules becomes:

$$\Delta E_{ooo} = -\sum_i (q_a{}^i + q_b{}^i) k S_{ab}^{i2}$$

In this equation, $q_a{}^i$ and $q_b{}^i$ are total electron densities on atomic orbitals $\phi_a{}^i$ and $\phi_b{}^i$ at sites of interaction, i. The interaction between the two systems has been assumed to be proportional to overlap. As two molecules approach each other, the interaction of filled orbitals leads to a destabilizing, closed shell repulsion term, ΔE_{occ}.

The stabilizing term, which arises from interactions of filled orbitals on one molecule and vacant on the second takes a somewhat different form than in Dewar's derivation, because of the inclusion of overlap, and the explicit dissection of H_{ab} into kinetic and potential energy terms. However, the largest term (about 90% of the total) is the same as Eq. (1). Salem applied these equations to the calculation of approximate surfaces for several thermal and photochemical reactions, and showed how the Woodward–Hoffmann rules were a natural result of the frontier orbital interactions.

Devaquet and Salem (1969) derived a new perturbation expression for the interaction of two molecules making assumptions similar to those in the preceding discussion, but including electron repulsion. The theory leads to the same expression as before, except that inclusion of electron repulsion leads to an additional term, represented approximately as

$$\Delta E(\text{polar}) = \sum_i \Delta q_a{}^i \Delta q_b{}^i \frac{1}{r_i}$$

which is the Coulombic interaction between the two molecules arising from charges $\Delta q_a{}^i$ and $\Delta q_b{}^i$ on atoms of A and B at interacting centers i separated by a distance r_i. Applications of the total expression given below [Eq. (5)] to several examples were presented, and found to correctly rationalize the selectivity observed in thermal and photochemical reactions.

$$\Delta E = \Delta E(\text{closed-shell repulsion}) + \Delta E(\text{occupied–vacant}) + \Delta E(\text{polar}) \quad (5)$$

Recently, Sustmann and Binsch (1971) have cast the interaction energy in an iterative SCF formalism in order to include electron repulsion more exactly than was done by Salem and Devaquet.

Epiotis (Yates *et al.*, 1975) has recently reemphasized the symmetry properties of closed-shell repulsion. That is, the full expression for the change in energy which occurs upon the interaction of two filled orbitals, ϕ_i and ϕ_j, is

$$\Delta\epsilon = \frac{4S_{ij}(\epsilon_{av} S_{ij} - H_{ij})}{1 - S_{ij}^2}$$

where S_{ij} and H_{ij} are the overlap and resonance integrals, respectively, between ϕ_i and ϕ_j, and ϵ_{av} is the average of the energies of ϕ_i and ϕ_j (Mulliken, 1955). This equation has the property that the higher the energy of the interacting orbitals, and the greater the overlap, the larger the closed-shell repulsion. Importantly, only orbitals of the same symmetry repel each other.

3. Fukui and Fujimoto

The early contributions of Fukui have been discussed. More recently, Fukui and Fujimoto (1968), as well as Murrell *et al.* (1965) have derived the expressions for interactions between two molecules in essentially a configuration interaction approach.

The total energy of the interacting molecules may be calculated starting with the MO's of the isolated molecules as a starting point, and allowing configuration interaction between the various possible electronic configurations of the complex. That is

$$E(\text{AB complex}) = E(A, B) + \sum_{\substack{\text{all CT} \\ \text{configurations}}} \lambda E(A^+, B^-) + \sum_{\substack{\text{all excited} \\ \text{configurations} \\ \text{of A or B}}} \lambda^* E(A^*, B)$$

Here $E(A,B)$ is the energy of the electronic configuration of lowest energy at the distance of interest, $E(A^+,B^-)$ is the energy of a charge transfer configuration in which an electron has been transferred from A to B (or vice versa), and $E(A^*,B)$ is the energy of a configuration with one molecule excited. The λ's are coefficients which measure how much of the higher energy configurations are mixed into the ground configuration to give the final energy of the ground state. Other configurations, such as multiply excited or multiply charge-transferred may be neglected because their contribution to the ground state will be negligible.

Making the usual weak interaction and other approximations such as the point-charge approximation for Coulombic interactions, the following terms are obtained: (1) A Coulombic term identical to E(polar) of Salem, (2) a delocalization term arising from mixing of charge-transfer configurations (A^+, B^-) with the ground state, (3) a polarization term arising from mixing of excited configurations (A^*, B) with the ground state, and (4) an exchange interaction energy resulting from interaction of filled orbitals on the two molecules, and thus present in the energy of (A, B) at separations less than infinite.

If the polarization terms are neglected, then the remaining terms are essentially those derived by Salem, except that the denominator of the terms arising from interaction of occupied orbitals with vacant orbitals (or in the CI formalism, by mixing of charge-transfer configurations with the ground configuration), are not simply differences in orbital energies. Instead, the denominator for calculation of the interaction of the (A^+, B^-) configuration with (A, B) is of the form:

$$(I_{A_a} + \Delta I_{A_a}) - (E_{B_b} + \Delta E_{B_b}) + A_{ab}$$

I_{A_a} is the negative of the SCF orbital energy—that is, the Koopmans' ionization potential of A involving removal of an electron from an orbital a—and ΔI_{A_a} is the change in this ionization potential induced by the presence of molecule B. Similarly E_{B_b} is the Koopmans' electron affinity of orbital b on B, and ΔE_{B_b} is the change caused by the presence of A. The ionization potential of A in the presence of B is less than that of isolated A, because additional electron repulsion (between an electron in ϕ_a and electrons in the orbitals of B) is relieved upon removal of ϕ_a. The electron affinity of B will be decreased by the presence of A due to increased electron repulsion. The term A_{ab} is essentially the attraction of the positive hole in ϕ_a for the electron in ϕ_b, or alternatively, but equivalently, A_{ab} is the change in electron repulsion which occurs upon electron transfer from A to B when these molecules are in close proximity, as opposed to this change when the molecules are totally separated.

The polarization energy, which arises from interaction of monoexcited configurations with the ground configuration has, in the numerator, the interaction between the half-occupied orbitals of A* and the filled orbitals of B, and, in the denominator, the "effective" excitation energy of A in the presence of B. Because the excited configurations A*B and AB* are often of higher energy than charge-transfer configurations, the polarization terms are usually much less important than the charge transfer, or delocalization, terms.

By means of electron density plots with varying interactions, Fukui has shown that the charge-transfer interactions, or occupied–vacant interactions, are primarily responsible for intermolecular bond formation.

4. Klopman

Klopman and Hudson (1967) and Klopman (1968, 1974) have also derived a generalized perturbation equation for intermolecular interactions based on polyelectronic perturbation theory. This equation consists of two terms: (1) an electrostatic term which is just the same as Salem's Coulombic term except that the effective solvent dielectric constant is included, (2) a solvation term, and (3) a charge transfer term which is identical to those discussed earlier, except that the denominator is $\epsilon_a{}^* - \epsilon_b{}^*$, where "effective" orbital energies called orbital electronegatives are included. The quantity $\epsilon_a{}^*$ is the energy required to remove an electron from molecule A in the complex, and $\epsilon_b{}^*$ is the energy required to add an electron to the molecule B in the complex. For reactions of ions with each other, the calculation of $\epsilon_a{}^*$ and $\epsilon_b{}^*$ requires a knowledge of the *IP*'s and *EA*'s of the neutral atoms, the effective ionic radii of the corresponding ions, and the dielectric constant of the solvent. Klopman applied this theory in detail to the explanation of the origin of the phenomena of "hardness" and "softness" of acids and bases. For the reactions of neutral molecules, $\epsilon_a{}^*$ and $\epsilon_b{}^*$ would be calculated in a fashion similar to that described in Fukui's approach.

B. Approximations Involved in Frontier Molecular Orbital Theory

Having discussed various approaches to the calculation of the energetic consequences of intermolecular interactions by perturbation theory, we can now more clearly delineate the approximations that are involved in frontier molecular orbital theory. The strict application of frontier orbital theory is equivalent to the adoption of the following expression for the change in energy upon interaction of two molecules

$$\Delta E = 2\frac{(\sum_i C_a{}^i C_b{}^i H_{ab}^1)^2}{\epsilon_{\text{HOMO}(a)}^* - \epsilon_{\text{LUMO}(b)}^* - Q} + 2\frac{(\sum_i C_a{}^i C_b{}^i H_{ab}^1)^2}{\epsilon_{\text{HOMO}(b)}^* - \epsilon_{\text{LUMO}(a)}^* - Q}$$

That is, only the delocalization terms (or occupied–vacant terms) involving the HOMO's and the LUMO's of the two addends are considered. Q is a factor correcting the denominator to compensate for the proximity of the molecules (*vide infra*). The frontier molecular orbital theory picks out only one pair of stabilizing interactions. This pair of interactions, or even one of these interactions taken alone, constitutes the largest stabilization, so the technique affords a good first approximation. All of the destabilizing interactions such as closed-shell repulsion, Coulombic repulsions (or attractions), and desolvation energies are neglected in the frontier orbital approximation.

C. Derivations in Terms of Charge-Transfer, or Configuration Interaction, Formalisms

Although the recent method of Fujimoto and Fukui is a configuration interaction method, earlier derivations of this type were proposed by others. Thus, Brown (1959) proposed a charge transfer model for the transition state of electrophilic substitutions, and derived a reactivity index, *Z*, based on this model. The theory of charge-transfer complexes was propounded by Mulliken

(1950, 1954), and a voluminous body of experimental and theoretical literature on the subject has developed (see, for example, Mulliken and Person, 1969). Epiotis (1974) has recently discussed all pericyclic reactions in terms of a configuration interaction model. This treatment is discussed in later sections. The perturbation equations given in the previous section can be derived from

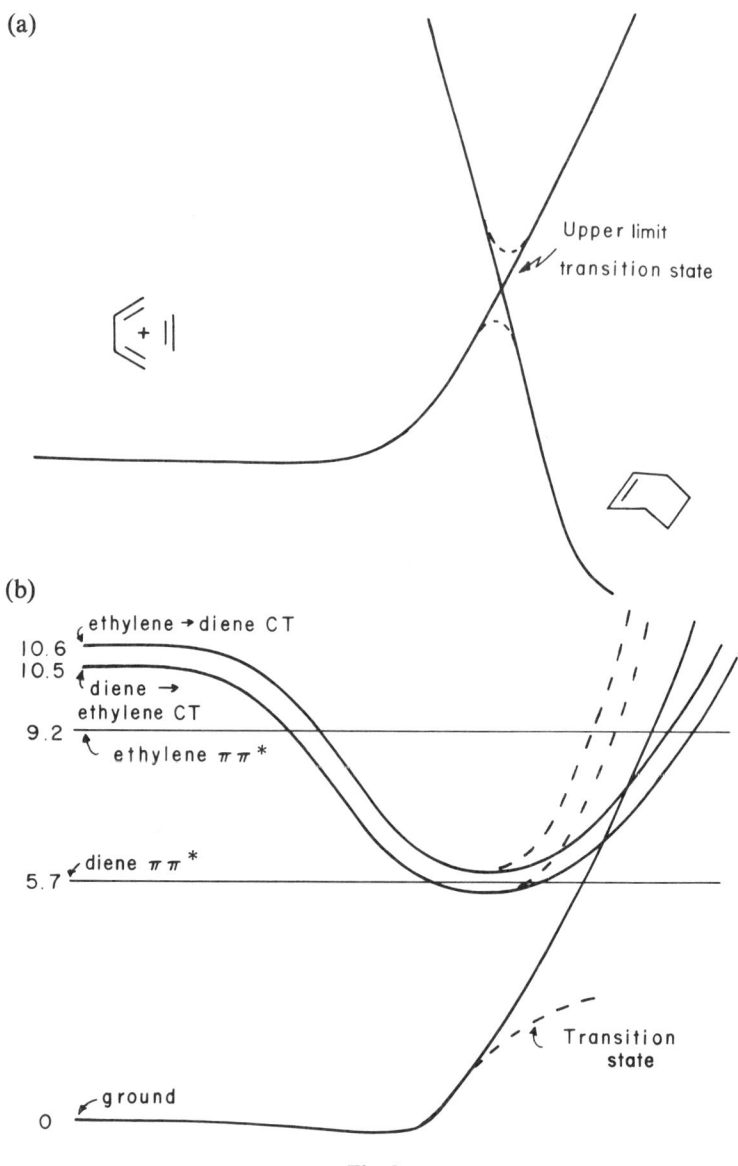

Fig. 2.

(c)

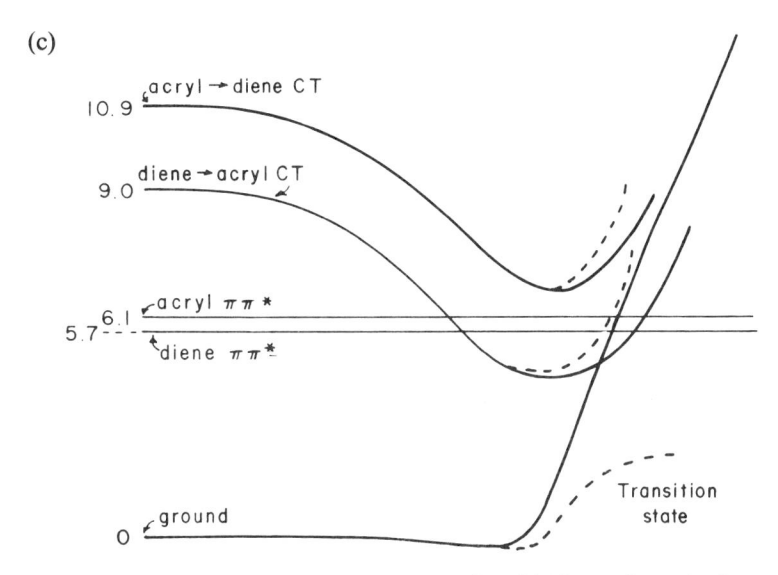

```
           acryl→diene CT
     10.9 ─────────────

           diene→acryl CT
      9.0  ─────────────

                acryl ππ*
      6.1  ─────────────
      5.7  ─ ─ ─ ─ ─ ─ ─
                diene ππ*

                                              Transition
                                              state
           ground
      0   ─────────
```

Fig. 2. (a) Evans' model for the Diels–Alder reaction. (b) The configuration interaction model for the Diels–Alder reaction of butadiene and ethylene. (c) The configuration interaction model for the Diels–Alder reaction of butadiene and acrylonitrile.

a charge-transfer complex model for pericyclic reaction transition states. A similar model has been used in our recent work on cyanoalkene cycloaddition reactivity (Houk and Munchausen, 1976).

In Fig. 2, we have shown schematically potential surfaces at various levels of approximation for an intermolecular pericyclic reaction. The particular example used here is a Diels–Alder reaction. In the spirit of Evans–Polanyi theory applied to hydrogen abstraction reactions (Evans and Polanyi, 1938), and to the Diels–Alder reaction (Evans and Warhurst, 1938; Evans, 1939), Fig. 2a shows two surfaces, the intersection of which represents the first approximation to the transition state of a Diels–Alder reaction. The surface rising from left to right represents the energy of butadiene plus ethylene as the two molecules distort, while the electronic state remains essentially that in the isolated molecules. As the molecules move near each other, there may be a shallow minimum due to van der Waals forces, but at shorter distances, there will be strong closed-shell repulsive forces. Furthermore, as the molecules are distorted from their equilibrium geometries, the energy will increase essentially as indicated by classical force constants. The curve rising from right to left is a similar potential surface for cyclohexene distorted toward the transition state geometry. As pointed out by Evans (1939), the point of intersection of these curves is an upper limit to the activation energy. If there can be interaction between the electronic configurations represented by the two intersecting curves, then the transition state will be stabilized, as represented by the dotted lines in Fig. 2a. This stabilization was noted by Evans (1939) and attributed to the formal similarity between the transition state of the Diels–Alder reaction and benzene.

This type of treatment can be extended to include the effects of higher energy configurations as follows: first, we can focus our attention on the left-hand side of the diagram only, realizing that any factor which stabilizes the ground state surface in the region of crossing of reactant and product curves will lower the transition state energy. In Fig. 2b curves are drawn for several possible electronic configurations. The lowest curve, rising toward the right of the diagram is for the ground configuration of butadiene plus ethylene as before. On the far left are the experimental energies of other electronic configurations, which, for molecules at infinite separation, are state energies. The charge transfer configuration energies are just the difference between the ionization potential of the donor and the electron affinity of the acceptor. However, as the two molecules are brought closer together, the energies of the charge-transfer configurations drop by an amount, Q, due to the increase in Coulombic attraction between the positively charged donor and the negatively charged acceptor. It is worth emphasizing again that the heavy lines in Fig. 2b are energies for hypothetical electron configurations, with no interaction between these permitted. However, these configurations may interact, and, as with orbital interactions, the lower configuration will be stabilized and the upper destabilized. However, only configurations of the same symmetry may interact, and this is where the congruence between the CI model and the orbital interaction model occurs. The energy of interaction between two configurations can be calculated by perturbation theory to be

$$E_{\text{GS}} \text{ (after CI)} = E_{\text{GS}}^0 + \frac{H_{\text{GS,CT}}^2}{E_{\text{GS}}^0 - E_{\text{CT}}^0}$$

$$E_{\text{CT}} \text{ (after CI)} = E_{\text{CT}}^0 - \frac{H_{\text{GS,CT}}^0}{E_{\text{GS}}^0 - E_{\text{CT}}^0}$$

These equations are derived by neglecting overlap. $H_{\text{GS,CT}}$, the interaction integral, is evaluated by calculating interactions between the orbitals which differ in occupation in the two configurations. This integral will vanish if the two configurations differ in the occupation of more than two orbitals. Thus, in this model, the frontier orbital approximation is equivalent to considering only the interaction of the two lowest energy charge-transfer configurations with the ground configuration. Although not included on the diagram, other charge-transfer configurations, such as one involving electron transfer from the lowest π orbital of butadiene to the ethylene LUMO are of less importance in general because of their higher energies (13 eV at infinite separation, and approximately 8 eV at a reasonable transition state distance). In the particular case of butadiene and ethylene in a C_s geometry, the diene (Ψ_1) → ethylene (Ψ_2) configuration is also of the wrong symmetry to interact with the ground configuration (Houk, 1975).

As the molecules are brought closer together, the ground configuration and charge-transfer configuration move closer in energy, and interact more strongly. Thus, the ground state surface is stabilized (dotted line), and the activation energy for reaction is decreased. The $\pi\pi^*$ configurations, which give rise to "polarization stabilization" in the Fukui model, have not been discussed as yet, since these are of wrong symmetry to interact with the ground configuration, and do not interact mutually under the influences of a one-electron perturbation, since they differ in occupation of four orbitals.

The mechanism of stabilization of the Diels–Alder transition state by frontier orbital interactions is quite easily seen in this model by a comparison of Figs. 2b and 2c. The ground configuration energy is assumed to be the same for the butadiene–ethylene reaction and the butadiene–acrylonitrile reaction represented in Fig. 2c.

Since one of the charge transfer configurations is substantially lower in energy in the case of the butadiene–acrylonitrile reaction than for the butadiene–ethylene reaction, there will be more stabilization of the ground state in the vicinity of the transition state in the former reaction.

The only difference between this discussion and the classical treatment of charge-transfer complexes is that the latter is concerned with the energy minimum in the ground state surface which can be relatively deep if there are one or more low-lying charge transfer configurations to stabilize the ground configuration, whereas the reactivity discussion centers on the transition state stabilization afforded by the same mechanism.

The previous sections have discussed the approximations involved in perturbation theory from the viewpoint of orbital interaction, or of configuration interaction, both of which, when formulated in terms of perturbation theory, give rise to the same numerical expressions for estimations of transition state stability. These models may also be applied to intramolecular pericyclic reactions if the molecule is arbitrarily broken up into fragments, and interactions between the orbitals of two fragments in the transition state geometry for the unimolecular reaction are calculated. This method will be returned to after the following discussion of a more rigorous way to treat unimolecular reactions.

D. THE BADER–SALEM–PEARSON FORMALISM FOR UNIMOLECULAR REACTIONS

Bader (1960, 1962) pointed out that second-order perturbation theory could be used to calculate the change in energy which occurs upon distortion of a molecule from its equilibrium nuclear configuration along some normal coordinate. Salem (1969) made several important contributions to this method, and the theory described below should be attributed to both Bader and Salem, while Pearson (1969, 1971, 1972) has discussed a variety of applications of this method.

Assuming that the extremes of various possible molecular vibrations are transition states for different conceivable unimolecular reactions, then this theory gives a method for calculation of the relative ease of allowed versus forbidden pericyclic reactions, or for calculation of concerted pathways versus nonconcerted pathways, or for calculations of the relative facility of different possible pericyclic reactions. In principle, this method can allow predictions of reaction rates and products, and its formulation, particularly in terms of the frontier orbital approximation, can reveal the origin of the facility of a particular reaction pathway. Many of the applications of this theory have been made by Salem and Wright (1969) and by Pearson (1969, 1971) and the following description draws heavily on these contributions.

Bader's formula for the change in energy $\Delta E(Q_i)$ of a molecule distorted along a normal coordinate Q_i is, in simplified form

$$\Delta E(Q_i) = \Delta E_0(Q_i) + \sum_j \frac{\left[\int \psi_0 \left(\frac{\partial V}{\partial Q_i}\right) \psi_j \, d\tau\right]^2}{E_0 - E_j} Q_i^2$$

ΔE_0 is the change in energy (necessarily an increase in energy) that occurs from moving the nuclei away from their equilibrium configuration, leaving the electronic functions unchanged. This is essentially the change in energy resulting from the classical force constant, and is like the energy surface for the ground configuration of two approaching molecules shown in Fig. 2, before interactions with higher configurations are allowed.

The second term allows for electronic relaxation to follow the motions of the nuclei. In this perturbation treatment, stabilizing electron "relaxation" occurs by mixing in higher energy electronic configurations with the ground electronic configuration. This is, of course, very similar to the discussion of intermolecular interactions discussed in the previous section.

The numerator of the second expression

$$\int \psi_0 \left(\frac{\partial V}{\partial Q_i}\right) \psi_j \, d\tau$$

is a measure of the extent of interaction between the ground state, ψ_0 and the excited state, ψ_j, under the influence of the perturbation, $\partial V/\partial Q$, which is the operator by which the change in nuclear–nuclear and nuclear–electronic potentials from their values at the equilibrium geometry is calculated. The only difference between this formulation and the intermolecular formulation given in the previous section is that

$$\int \psi_0 \left(\frac{\partial V}{\partial Q}\right) \psi_j d\tau$$

will be of somewhat greater complexity than the intermolecular perturbation,

$$\int \psi_a H \psi_b \, d\tau$$

discussed previously. This increased complexity will arise because both orbitals will be delocalized over the same atoms, and evaluation of interactions at each atom must be made, rather than just at two pairs of atoms as assumed for the intermolecular case.

If the nuclear motion described by Q_1 has a certain symmetry, then for the excited state to mix with, and stabilize, the ground state, the excited state and Q_1 must have the same symmetry. For molecules without symmetry, the numerator will be proportional to the overlap between the ground state and the excited state.

The denominator indicates, as before, that for two excited states of similar overlap with the ground state, the lowest energy excited state can afford more stabilization.

To bring this model into congruence with intermolecular perturbation theory, it is useful to discuss the Bader–Salem–Pearson model as a configuration interaction model. That is, assume that the MO's of a molecule in its equilibrium geometry are known, and that we wish to know whether a particular distorted geometry—our guess for a transition state—is stabilized by configuration interaction. Those excited configurations will stabilize the ground configuration which overlap with the ground configurations, and are low in energy *at the particular geometry under consideration* (Salem, 1969). The extent of overlap is calculated by comparing overlaps of the orbitals by which the ground and excited configurations differ.

The frontier orbital approximation is arrived at in this theoretical treatment by assuming that only the excited configuration involving a HOMO → LUMO excitation is sufficiently low in energy to stabilize the ground configuration to any appreciable extent. In fact, other excited configurations may stabilize the transition state, but, in general, less so than the HOMO → LUMO configuration.

An example applied to pericyclic reactions is perhaps pertinent at this stage. Figure 3 shows the three topmost occupied and lowest two unoccupied MO's of cyclobutene, taken from Jorgenson and Salem's (1973) book. On the right are shown symmetry classifications with respect to a plane of symmetry. For such a disrotatory distortion, the only low-lying excited configuration which will be of proper symmetry to stabilize the ground configuration

198   K. N. HOUK

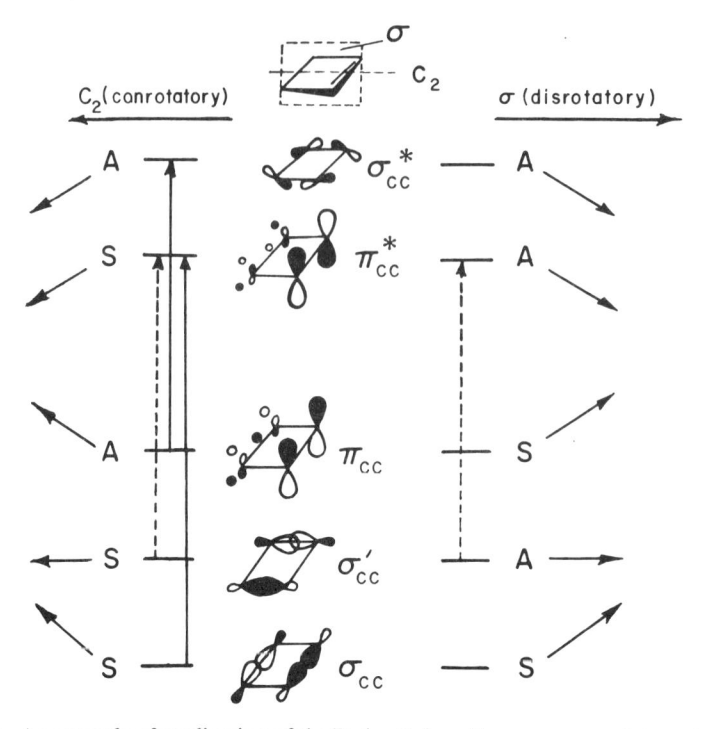

Fig. 3. An example of application of the Bader–Salem–Pearson approach to unimolecular reactions.

is the $\sigma'_{cc} \to \pi^*_{cc}$ configuration. Even for large distortions, the overlap between $\sigma'_{cc} \to \pi^*_{cc}$ will be small, and so little stabilization can be obtained. As the reaction proceeds, the remaining orbitals shown come closer in energy, so there will be low-lying excited configurations involving $\sigma \to \pi^*$ and $\pi \to \sigma^*$ transitions. These are of improper symmetry to stabilize the ground configuration for the disrotatory motion, but are of proper symmetry (as shown by full arrows on the left) to stabilize the conrotatory motion. Thus, the conrotatory motion is easier, and, as is well known, is the thermally allowed pathway.

The favorability of the photochemical pathways can be discerned by evaluation of whether the lowest excited state can be stabilized by other singly excited configurations. The $\pi\pi^*$ state of cyclobutene is lowest in energy and involves an S, A transition in either geometry so that it has A symmetry. Other excited configurations of the proper symmetry to stabilize the disrotatory motion are the $\pi_{cc} \to \sigma^*_{cc}$, $\sigma_{cc} \to \pi^*_{cc}$, and $\sigma_{cc} \to \sigma^*_{cc}$, whereas only the $\sigma_{cc} \to \sigma^*_{cc}$ and $\sigma'_{cc} \to \sigma^*_{cc}$ are of proper symmetry to stabilize the $\pi\pi^*$ configuration in the conrotatory motion. The disrotatory motion is photochemically allowed.

This method has quantitative capabilities, and provides insight into how a reluctant reaction can be facilitated, or how a reactive molecule can be stabilized. It is, however, somewhat less easily applied, in general, than similar methods based on orbital interaction, rather than configuration interaction, models. In his treatment of unimolecular reactions, Fukui (1971) divides molecules up into fragments, with the divisions corresponding to regions where new bonds will be formed in the product of the reaction. Then the frontier orbital method is applied as in intermolecular reactions. For the cyclobutene opening, the σ bond joining atoms 3 and 4 and the π bond are the localized regions of interest since these are the bonds broken in the electrocyclic opening. In the conrotatory opening, there are stabilizing interactions between the π_{HOMO} and the σ_{LUMO}. Both of these are absent in the disrotatory motion. This can be seen to be essentially identical to the Bader–Salem approach described in Fig. 3, except the latter is couched in the configuration interaction language.

Zimmerman's "MO Following" is another very useful approach to the analysis of pericyclic reactions. In his method, Zimmerman (1972) follows the energies of each orbital as the geometry is deformed from reactant, through transition state, to product. Once again, the method is based on interactions of orbitals of the same symmetry—or of high overlap, in the absence of symmetry—and is essentially the same as the method described in this section, except that the whole reaction surface is treated. Zimmerman has analyzed a large number of pericyclic reactions by this method.

IV. Practical Applications of Frontier Molecular Orbital Theory

A. Calculated and Experimental Quantities Used in Frontier Orbital Applications

In the previous sections of this chapter, the various methods of estimating transition state energies based on reactant MO's have been discussed. The frontier orbital method is a good first approximation to the calculation of transition state energies, particularly if it is kept in mind that all reactions will have repulsive barriers not considered when only occupied–vacant interactions are treated. In this section, we wish to show how the perturbation methods, in general, and the frontier orbital method, in particular, can be applied practically.

In any of the various approaches to the calculation of orbital, or configuration, interactions, the quantities needed are the differences in energies between the various orbitals, or configurations, as well as the symmetries, at the very least, or the coefficients at various centers, for each orbital. One approach to obtaining these quantities is to calculate all of them by one of the many molecular orbital methods now available. While this approach does have internal consistency and provides a purely theoretical approach to the prediction of reaction rates, it has certain shortcomings, among which the most important are the fact that many semiempirical methods are parameterized for, and thus are most successful for, a restricted class of compounds or a particular quantity of interest (e.g., CNDO/2 for SCF orbitals of diatomics, MINDO/2 for heats of formation, SPINDO for ionization potentials, etc.). The more general calculational methods (for example, *ab initio* SCF methods) are more time consuming and expensive, and calculations for all the systems of interest would be a major project. For this reason, some rational mixture of theoretical and experimental data on isolated systems seems the most useful approach to the treatment of reaction rates, and such approaches are described below.

The choice of theoretical or experimental parameters has been made in various ways by different authors, and these applications will be reviewed in the following sections. In the model that follows, the configuration interaction language is adopted, since this model makes the relationship between experimental and theoretical quantities more apparent. Figure 2 should be reconsulted in connection with this model.

1. Intermolecular Reactions

According to Koopmans' theorem, the negative of the SCF orbital energy is the ionization potential involving that orbital (Koopmans, 1934). This is because in SCF treatments, which include electron–electron repulsion, the orbital energies are:

$$\epsilon_i = H_{ii} - \sum_j^{occ} (2J_{ij} - K_{ij})$$

where H_{ii} is the energy of the electron in orbital i in the field of the nuclei (and the inner shell electrons if not explicitly included), and

$$\sum_j^{occ} (2J_{ij} - K_{ij})$$

is the total electron repulsion between electron in orbital i and the rest of the electrons in occupied orbitals, j, in the molecule. Although it is well known that Koopmans' theorem can be a poor approximation due to its neglect of electron correlation and orbital reorganization, the experimental vertical ionization potentials as measured by photoelectron spectroscopy are exact measures of the amount of energy required to remove an electron and to form the various radical cation states. In the CI model, this is one of the quantities needed to find the difference in energy between the ground configuration and various charge transfer configurations. The paramount reason for use of experimental, rather than calculated, ionization potentials in the application of perturbation theory from our group, lies in the exact nature of the experimental ionization potentials and the relatively unreliable nature of ionization potentials obtained *via* calculations based on Koopmans' theorem. The vertical ionization potential is the Franck–Condon maximum in the transition, classically thought of as the value of the ionization potential for formation of the radical cation in the geometry of the ground state. Ionization potentials from other sources, such as mass spectrometric appearance potentials, are adiabatic values, and these are less easily interpreted, since they measure the difference in energy between the ground state of the molecule and the most stable geometry of the radical cation.

Other indirect estimates of ionization potentials are available, such as energies of charge transfer maxima in molecular complexes involving acceptors of known electron affinity. Oxidation potentials are essentially solution ionization potentials, and may also be converted to ionization potentials (Briegleb, 1964).

Since SCF orbital energies contain the electron repulsion, the negatives of the vacant orbital energies are, by Koopmans' theorem, equal to the electron affinities of a molecule. However, "Koopmans' electron affinities" appear to be even more approximate than Koopmans' *IP*'s.

Unfortunately, there have been relatively few experimental measurements of electron affinities for organic molecules, and many of these have been estimated from charge transfer spectra or from reduction potentials (Briegleb, 1964).

Even for ideal cases where ionization potentials and electron affinities of molecules are known, further quantities must be calculated or measured. In the configuration interaction model (Fig. 2), the value of Q must be known for each charge-transfer state, if the calculations are to be carried out for the transition state, rather than for very weakly interacting molecules.

Fortunately, several reasonable estimates may be made for the value of Q. In the CI model, Q is the electrostatic attraction of D^+ for A^-. Although the charge distribution in the D^+ and A^- states, and their separations need to be known to calculate Q, crude estimates can be made on the basis of the point-charge model. The electrostatic attraction of a point negative and a point positive charge varies from 2.9 to 7.2 eV over distances from 5 to 2 Å. For the "reasonable" transition state separations of 2.8 to 3.6 Å, the electrostatic attraction is 4 to 5 eV. This attraction will diminish for a given separation of molecules as the charges are delocalized (to a limit of $1/n$ at zero separation, where n is the number of points over which the charge is spread on each molecule).

An alternative approach has been used by Herndon and Giles (1970) and by Epiotis (1972a,b,c). In one-electron approximations, where electron repulsion is neglected, the difference between the highest occupied orbital energy and lowest unoccupied orbital energy is the electronic transition energy. Thus, if the negatives of the IP's are used for occupied orbital energies, the vacant orbital energies can be estimated as $-(I_i - \Delta E_{ij})$, where ΔE is the energy for the i → j transition.

When electron repulsion is included, the singlet–singlet transition energy is $\Delta E_{i \to j} = IP\,(i) - EA\,(j) - (J_{ij} - 2K_{ij})$, where J_{ij} and K_{ij} are the Coulomb and exchange integrals between electrons in orbitals i and j. For $\pi\pi^*$ transition $J_{ij} - 2K_{ij}$ is on the order of 4 to 5 eV (Houk et al., 1973a), the same order of magnitude estimated earlier for Q. Thus, the one-electron approach, using ΔE to estimate LUMO energies, and the CI approach, using EA's and Q's are approximately equivalent for calculations of transition state stabilization. Both are approximations, because the molecules will be distorted to some extent in the transition state of the reaction, and the state energies will, therefore, be different from those in the isolated molecules.

The other quantities needed in the calculation of transition state stabilization are the coefficients of the MO's of the two molecules at the site of interaction, and the resonance integrals for the interactions between the orbitals of the two molecules.

The coefficients can be obtained from any of a variety of calculations, usually with the same qualitative results, but occasionally, as in the HOMO's of alkenes substituted by an electron-withdrawing group, there is disagreement as to which coefficient is largest (Houk, 1973b; Alston and Shillady, 1974). In such cases, *ab initio* methods or methods parameterized to mimic *ab initio* methods (CNDO, INDO) are used to obtain reliable coefficients.

The resonance integrals may be left as an unspecified parameter, or evaluated for an assumed distance of addend separation.

2. Intramolecular Reactions

Only Salem has reported applications of the B-P-S formalism to pericyclic reactions, although the Fukui and Zimmerman methods are approximations based on similar methods.

The possibilities of quantitative applications seem better for intramolecular reactions than for intermolecular reactions, because in the former case, electronic transition energies can be used to obtain excited state energies, oscillator strengths can provide information about the extent of interaction between ground and excited states, and infrared spectroscopy could afford insight into force constants for various distortions. The only unknown quantities required for a quantitative treatment would be the amount by which these quantities change as the molecule is distorted toward the various pericyclic transition states.

B. FRONTIER MOLECULAR ORBITALS

In quantitative applications of frontier molecular orbital theory to pericyclic reactions, the use of qualitative orbital energies and coefficients often suffices. For the purpose of facilitating discussions in later sections of this chapter, some of the frontier orbitals of representative alkenes are shown in Figs. 4–8. In each figure, the negative of the ionization potential of the alkene is given under the horizontal line for the HOMO, and the negative of the electron affinity is given under the LUMO level. The units are eV. Many of the electron affinities are estimated by methods discussed in our earlier publications (Houk et al., 1973a), for example, by the empirical estimate $EA = IP - \Delta E(\pi\pi^*) - 4.5$ eV. In the figures, the AO coefficients for the frontier MO's are given where available, as calculated by CNDO/2. As noted before, the absolute values of these coefficients should not be considered of great significance, but the trends shown here would be reproduced by more sophisticated techniques, such as *ab initio* SCF methods.

Since the qualitative derivations of the orbital shapes and energies have been discussed before (Houk, 1975) and the trends and examples have been cited previously, (Houk, 1973b; Houk et al., 1973a), only a few comments are in order here. For the electron-rich alkenes (Fig. 4), the trend of decreasing HOMO coefficients as the ionization potential decreases results from greater

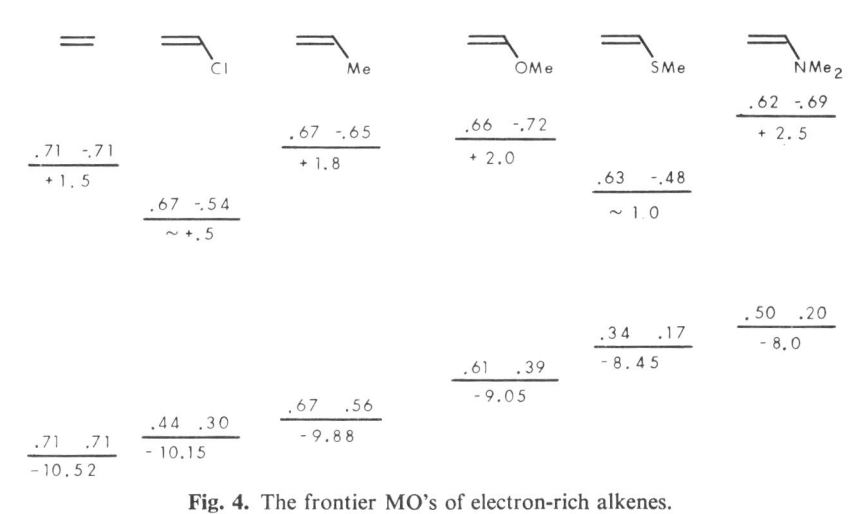

Fig. 4. The frontier MO's of electron-rich alkenes.

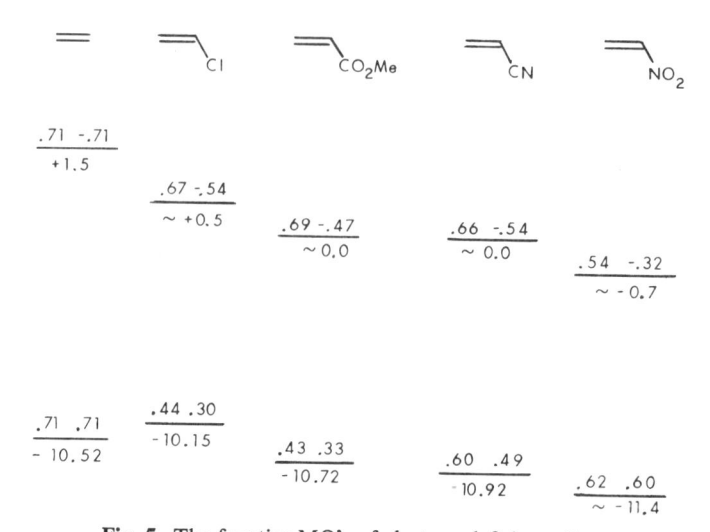

Fig. 5. The frontier MO's of electron-deficient alkenes.

admixture of substituent orbitals with the ethylene π orbital as the group becomes a better donor. This has the effect of tempering somewhat the increasing reactivity of the electron-rich species with electrophilic reagents, as discussed more in the following section of this article. Similar conclusions about the LUMO's of electron-deficient alkenes can be drawn (Fig. 5). The haloalkenes, as exemplified by vinyl chloride, deviate from the general trends somewhat. The halogen is quite electronegative, which causes a lowering of both the HOMO and LUMO, but is also a lone-pair donor, which raises the HOMO energy but has little effect on the LUMO energy.

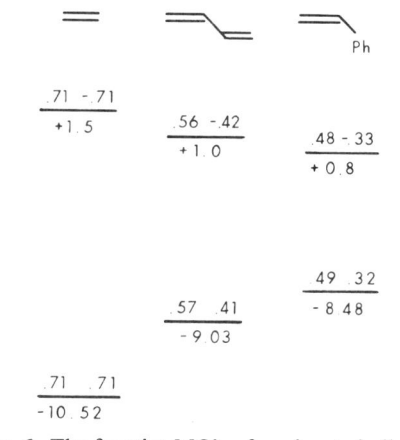

Fig. 6. The frontier MO's of conjugated alkenes.

4. FRONTIER MOLECULAR ORBITAL THEORY   205

The conjugated alkenes (Fig. 6) have raised HOMO's and lowered LUMO's as compared to ethylene, but as noted in the following section, the reactivity of these species toward electrophiles or nucleophiles will be tempered somewhat by the smaller coefficients.

The frontier MO's of cumulenes are shown and compared to those of ethylene in Fig. 7. Allene resembles propene and has degenerate HOMO's and LUMO's, perpendicular to those shown. Ketenes, isocyanates, and sulfenes have been the subject of a more detailed study (Houk et al., 1974b). Calculated coefficients and estimated energies are shown in Fig. 7. The HOMO's of these three species all resemble those of electron-rich alkenes in shape, but for the last two, the estimated energies are more like those of electron-deficient alkenes. The LUMO's of ketene and isocyanic acid are in-plane (ip) orbitals, resembling perturbed π^*_{co} orbitals.

Finally, Fig. 8 shows the frontier MO's of a few selected 1,3-dipoles. The HOMO–LUMO gap in these species is relatively small, so that their reactivities are quite high. For planar or linear 1,3-dipoles, the HOMO and LUMO coefficients follow the pattern shown in the figure, regardless of the type of calculation (Caramella et al., 1977). However, bending of either terminus can reverse these generalization, as discussed in more detail later (Caramella and Houk, 1976).

Although not shown here, these generalizations may be extended to dienes, trienes and the like. The effects on orbital energies and coefficients are essentially the same at vinylogous positions.

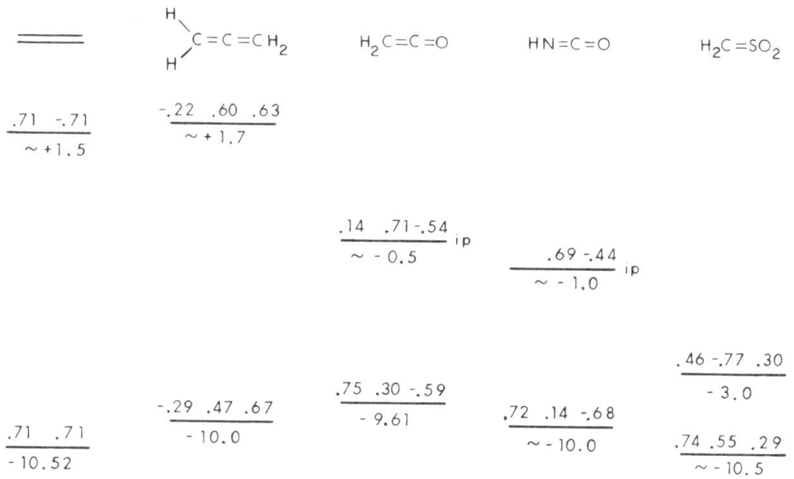

Fig. 7. The frontier MO's of cumulenes.

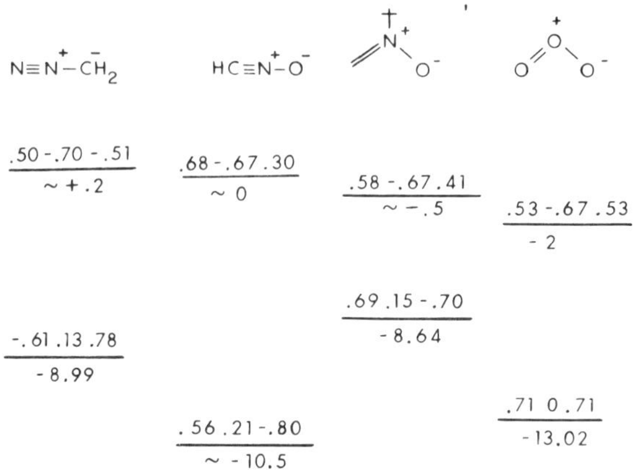

$$N\equiv\overset{+}{N}-\overset{-}{C}H_2 \qquad HC\equiv\overset{+}{N}-O^- \qquad \overset{\overset{+}{\underset{|}{N}}}{\diagup\diagdown}_{O^-} \qquad \overset{\overset{+}{O}}{\diagup\diagdown}_{O^-}$$

$$\frac{.50 \; -.70 \; -.51}{\sim +.2} \qquad \frac{.68 \; -.67 \; .30}{\sim 0} \qquad \frac{.58 \; -.67 \; .41}{\sim -.5} \qquad \frac{.53 \; -.67 \; .53}{-2}$$

$$\frac{.69 \; .15 \; -.70}{-8.64}$$

$$\frac{-.61 \; .13 \; .78}{-8.99}$$

$$\frac{.56 \; .21 \; -.80}{\sim -10.5} \qquad \frac{.71 \; 0 \; .71}{-13.02}$$

Fig. 8. The frontier MO's of 1,3-dipoles.

C. Charge-Controlled, Frontier-Energy-Controlled, and Frontier-Density-Controlled Reactions

Before discussing the applications of frontier molecular orbital theory to specific pericyclic reactions, it is of interest to examine certain general characteristics of perturbation theory. Klopman (1968) and Hudson (1973) have discussed the concept of "charge" versus "frontier" control as applied to electrophilic and nucleophilic reactions. Klopman (1968) proposed a general perturbation treatment of chemical reactivity (discussed in Section III,A,4 of this chapter), and used it to explain the phenomena of acid-base "hardness" and "softness." Hard Lewis acids have relatively high positive charges and low electron affinities, so their interaction with Lewis bases is mainly Coulombic in nature. Similarly, hard Lewis bases have relatively large negative charges, and high ionization potentials. Soft acids and bases have low charges and high electron affinities or low ionization potentials, respectively, and acid-base interactions occur mainly by mixing of frontier orbitals of the two molecules.

An example cited by Klopman for "charge-controlled" nucleophilic addition is the preferential attack of "hard" nucleophiles such as hydroxide on the 2 position of pyridinium ions. This is the carbon site with greatest positive charge. However, the pyridinium ion has the largest LUMO coefficient at position 4, so that nucleophiles with low ionization potentials attack this position; that is, in our previously described configuration interaction language, the soft nucleophiles, such as cyanide, preferentially react at position 4 because the transition state for this reaction is stabilized most by a low-lying charge transfer configuration. In frontier orbital terminology, the interaction of the cyanide HOMO and pyridinium LUMO is greatest when attack occurs at the 4 position.

A variation on this idea, applicable to pericyclic reactions, can be appreciated by reference to the *IP*'s and coefficients of the occupied MO's of ethylene shown in Fig. 9. Comparisons of reactivity of ethylene and butadiene based on frontier orbital energies alone would indicate that butadiene ($IP = 9.1$ eV) should be much more reactive than ethylene ($IP = 10.5$ eV) with all electrophilic species. However, the coefficients at any atom in the butadiene HOMO are smaller than those of ethylene, mitigating the activating influence of the lower HOMO ionization potential of butadiene. Calculations based on the numerical values in Fig. 9 for different assumed electrophile EA's indicate that for simultaneous bonding at atoms 1 and 2 in the transition state, butadiene should be slightly more reactive than ethylene for poor electrophiles, but that the difference in reactivity should increase as the *EA* of the electrophile increases. Using a Coulombic correction (Q) of 5 eV, butadiene is calculated to be 14% more reactive than ethylene for reaction with an electrophile of $EA = -2.5$ eV, and 27% more reactive for an electrophile with $EA = +2.5$ eV. In the case of the butadiene reaction with the poor electrophile, 58% of the transition state stabilization is provided by the diene HOMO, while for the reaction with the powerful electrophile, 72% of the stabilization comes from the diene HOMO. Of course, assuming Q is a constant for all electrophiles is equivalent to the assumption that the transition state has approximately the same structure for all electrophiles. The increased selectivity predicted here will be partially counteracted by Hammond postulate-like considerations. That is, as the electrophilicity of the reagent increases, transition state structure will shift toward reactants, partially or wholly overcoming the increased selectivity predicted on the basis of the higher frontier orbital energy of butadiene. In terms of the Evans–Polanyi model, for more exothermic reactions, the product dissociation curve will

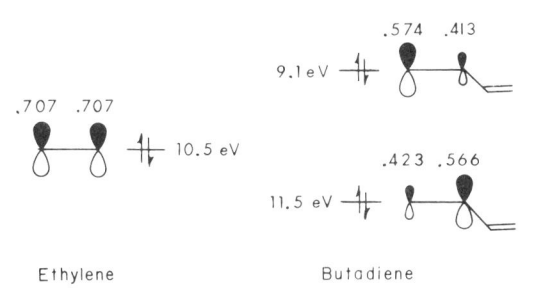

Fig. 9. The occupied π orbitals of butadiene and ethylenes used to illustrate the concepts of "frontier-energy-controlled" and "frontier-density-controlled" reactions.

cross the reactant curve closer to the reactants, or before charge-transfer stabilization becomes appreciable.

The ethylene–butadiene comparison is an example of a general phenomenon. The coefficients, as well as the energies, of the frontier orbitals must be taken into account in treatments of reactivity. For poor electrophiles, which are "hard" in the sense that they have low EA's, the total charge on the alkene carbons is of major importance, and the orbital energies are of minor importance. That is, both ethylene and butadiene have total π charges of zero, and butadiene is of slightly higher predicted reactivity because the average of its two IP's is smaller than the ethylene IP. As the electrophile EA increases, the low butadiene first IP achieves greater significance in spite of its lower coefficients. The latter is a case of "frontier-energy control"; the former is "charge-control." Here charge-control implies the same reactivity, since the total charges at relevant carbons are identically zero for both molecules. We have stressed the importance of frontier MO coefficients in a discussion of styrene reactivity (Houk et al., 1975b). Since the density in the frontier orbitals, rather than net charge is of importance, in some cases, we name this special case "frontier-density control."

V. Applications to Thermal Intermolecular Cycloadditions

The most extensive applications of perturbation theory have been in the field of cycloadditions. Rather than attempt to review all of the applications, many of which have been cited (Herndon, 1972; Houk, 1973b, 1975), a brief summary of the types of problems to which perturbation theory is applicable will be covered.

1. Allowed versus Forbidden Reactions

Fukui (1964) noted that the HOMO's and LUMO's of common Diels–Alder dienes and dienophiles, respectively, had the same nodal properties, so that both bonds could be partially formed in the transition state of the reaction. Hoffmann and Woodward (1965a) used the correlation diagram method to show in general which cycloadditions are "allowed" and which are "forbidden" to be concerted. The allowed reactions are those in which a four-center transition state (Fig. 10b) can be stabilized to a greater extent than the two-center transition state (Fig. 10a), while forbidden reactions are those in which the four-center transition state must be stabilized to a smaller extent than the two-center. The frontier MO derivation of the Woodward–Hoffmann rules can be made in the following way (Goldstein and Hoffmann, 1971).

(a)   (b)

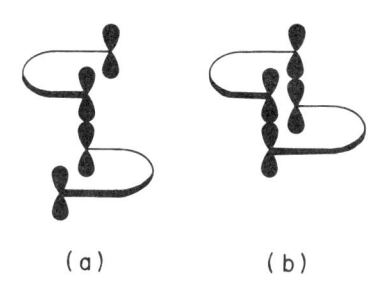

Fig. 10. (a) Two-center transition state for the cycloaddition of two polyenes. (b) Four-center transition state for the cycloaddition of two polyenes.

All planar polyenes with an even number of π electrons, whether neutral or charged, are of two types. For those with $(4n + 2)\pi$ electrons (e.g., ethylene, allyl cation, hexatriene, pentadienyl anion), the coefficients of the terminal atomic orbitals are of the same sign in the HOMO and of opposite sign in the LUMO, while for molecules with $4n\pi$ electrons (e.g., allyl anion, butadiene, pentadienyl cation, octatetraene) the terminal coefficients are of opposite sign in the HOMO, and the same sign in the LUMO. These relationships are summarized in Fig. 11. If the molecules have symmetry planes perpendicular to their molecular planes, then the HOMO's and LUMO's of $4n + 2$ systems are symmetrical (S), and antisymmetrical (A), respectively, while the HOMO's and LUMO's of $4n$ systems are A and S, respectively. In this derivation of the Woodward–Hoffmann rules, symmetry is a simplifying, but unneccessary, feature.

The two-center interaction of any two polyenes will result in stabilization of the HOMO's (Fig. 12b), but for four-center interactions (Fig. 12c) a HOMO will be stabilized only if it can interact in a bonding fashion at both termini with the LUMO of the other molecule. Allowed reactions are those in which the HOMO of each molecule has the same nodal (symmetry)

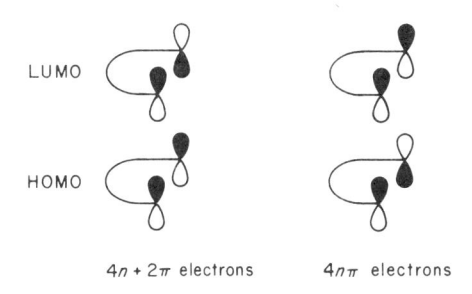

LUMO

HOMO

$4n + 2\pi$ electrons $4n\pi$ electrons

Fig. 11. Nodal properties of the two types of polyenes.

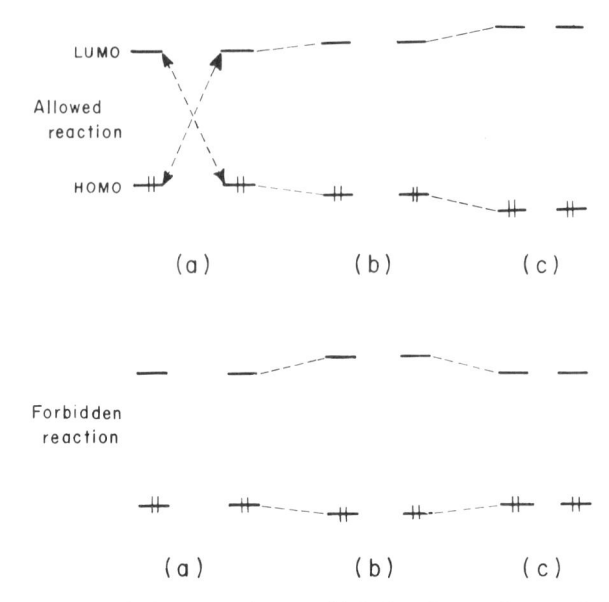

Fig. 12. Frontier orbital energy changes (a) before interaction, and upon (b) two-center, and (c) four-center, interactions for allowed and forbidden reactions.

properties as the LUMO of the second molecule, while for forbidden reactions, this pair of orbitals has opposite symmetry. Thus, for suprafacial stereochemistries, the allowed reactions involve one $4n + 2$ molecule and one $4n$ molecule, giving the requirement of a total of $(4n + 2)\pi$ electrons for the reaction to be allowed. Forbidden reactions involve two $4n$ or two $4n + 2$ systems, or a total of $4n\pi$ electrons. Generalizations to other stereochemistries and more than two addends are straightforward in the frontier MO formalism.

The alternative derivation by Dewar (1971), which uses PMO theory to determine whether a transition state is isoconjugate with an aromatic hydrocarbon (allowed) or with an antiaromatic hydrocarbon (forbidden), gives the easiest mnemonic for allowedness and forbiddenness of cycloaddition reactions.

If the reactants are highly unsymmetrical, so that the coefficients are very different at the two termini, some of the allowed–forbidden distinction breaks down. However, the $4n + 2$ cases are still clearly favored over the $4n$. For example, Fig. 13 shows the terminal coefficients (CNDO/2) for the HOMO's and LUMO's of several molecules. As far as the overlap part of the stabilization is concerned, going from butadiene–ethylene to the unsymmetrical diene–alkene pair decreases the four-center overlap ($\sum_i C_A^i C_B^i$) somewhat, because

4. FRONTIER MOLECULAR ORBITAL THEORY

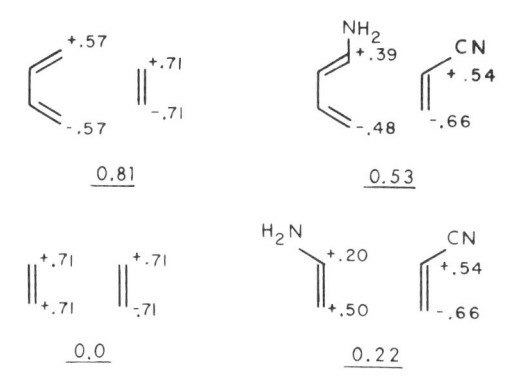

Fig. 13. HOMO (molecules on left) and LUMO (molecules on right) coefficients for symmetrical and unsymmetrical allowed and forbidden processes. The numbers under each example are $\Sigma_i C_A{}^i C_B{}^i$.

the coefficients are smaller, but increases the stabilization of the concerted forbidden [2 + 2] transition state, because the symmetry restriction against net HOMO–LUMO overlap is removed. However, the two-center transition state for [2 + 2] reactions of unsymmetrical alkenes will still be favored over the concerted four-center [2 + 2] transition state.

The question of allowed versus forbidden reactions has been complicated recently by the interesting work of Epiotis (1974), who showed that certain formally forbidden reactions between a good donor alkene and a good acceptor alkene can occur by concerted mechanisms. Epiotis, using the CI model, pointed out that if a charge-transfer configuration were lower in energy than the "ground state" configuration for a $[_{\pi}2_s + {}_{\pi}2_s]$ transition state, then a concerted mechanism would be possible.

In terms of the Evans-like model (Fig. 2a–c), the situation proposed by Epiotis is shown in Fig. 14. For a very good donor–acceptor pair, the CT configuration is not very much higher in energy than the ground configuration even at large distances, and as the addends approach each other, the CT configuration may drop below the "ground" configuration in energy. However, since the CT configuration is of much lower energy than locally excited configurations, there is little further stabilization as transition state distances are approached. That is, in our opinion, a lowest CT configuration may more easily dissociate into a radical cation and a radical anion than collapse to adduct. Furthermore, unless some geometrical or steric constraints require the molecules to approach each other in a forbidden transition state geometry, other transition states leading to intermediates may be stabilized more than forbidden ones.

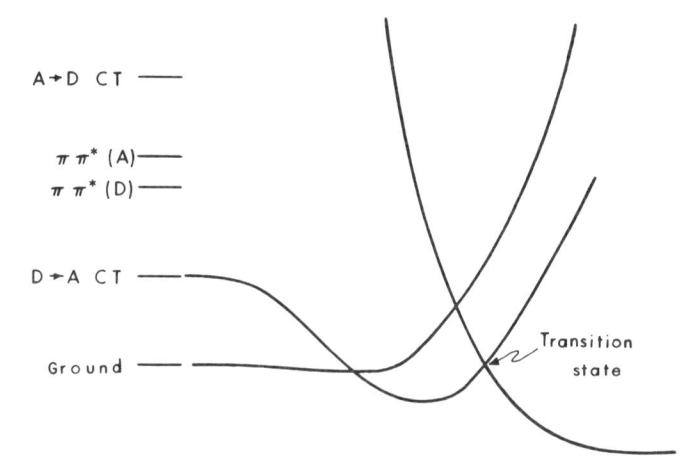

Fig. 14. Configuration interaction model for the [2$_s$ + 2$_s$] cycloaddition of a donor and acceptor alkene pair.

Fukui and co-workers have recently proposed a model for forbidden reactions similar in some ways to that of Epiotis described above (Inagaki *et al.*, 1975). The concept of "pseudoexcitation" was described as follows: for sufficiently good donor–acceptor pairs, a transition state may be stabilized by the third-order interaction between the ground, a mono-transferred, and a mono-excited configuration. Such stabilization will normally be small, but if charge transfer is sufficiently great, this third-order interaction can become important. The extent of stabilization from this source depends on HOMO–HOMO and LUMO–LUMO interactions. As a result, formally forbidden reactions can become allowed.

2. Mechanistic Types

Perturbation theory can, of course, deal not only with concerted mechanisms, but can be used to predict which type of transition state geometry will be favored in a stepwise mechanism, as well as predict the degree of charge separation in transition states.

a. [2 + 2] Cycloadditions. Figure 15 shows the limiting geometries of transition states of cycloaddition reactions involving two ethylenes. On the basis of frontier orbital arguments, the most favored geometry may be predicted as a function of the substituents present on the alkene (Houk, 1976). These geometries are proposed for the transition state, but may collapse to cyclobutane product, or to zwitterionic or diradical intermediates before collapse to product.

$2_s + 2_s$ $2_s^A + 2_a^D$ $2_a + 2_a$

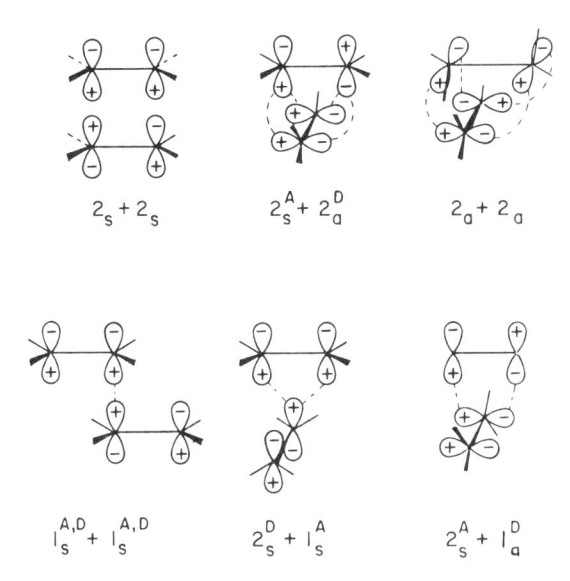

$1_s^{A,D} + 1_s^{A,D}$ $2_s^D + 1_s^A$ $2_s^A + 1_a^D$

Fig. 15. Various limiting transition state geometries for [2 + 2] cycloadditions. (Reprinted with permission from *Accounts of Chemical Research*. Copyright by the American Chemical Society.)

[$2_s + 2_s$]. This geometry is not stabilized to any extent by frontier orbital interactions, unless appreciable asymmetry caused by unsymmetrical substitution is present. For reactions between two alkenes, stabilization of this transition state can arise only from the Epiotis effect, that is, greater stabilization of the "charge-transfer" configuration than the "ground." Thus for highly complementary donor–acceptor combinations, the [$2_s + 2_s$] transition state is a viable one if a charge-transfer intermediate intervenes before formation of this transition state.

Fueno and co-workers have recently devised an interesting criterion for mechanisms which gives added insight into the feasibility of concerted [$2_s + 2_s$] cycloadditions (Yamaguchi *et al.*, 1973; Yamaguchi and Fueno, 1973; Okada *et al.*, 1974). These authors performed CNDO calculations on [2 + 2] cycloadditions of variously substituted ethylenes, and criteria for deciding the allowedness or forbiddenness, as well as the radical or nonradical nature of the transition state, were established. Those reactions which involve a change in the nodal properties of the HOMO as the reaction proceeds are "forbidden." If a HOMO–LUMO crossing occurs, then a triplet state will be the ground state in the region of this crossing. Such reactions are "forbidden radical" reactions.

For highly unsymmetrical donor–acceptor pairs (acrylonitrile and 1,1-diaminoethylene were models used in the calculation), a "forbidden nonradical" pathway was predicted. That is, the HOMO–LUMO crossing is avoided because of both the donor–acceptor character of the addends and their unsymmetrical nature. Such reactions go through species which can be considered essentially zwitterionic in spite of the cisoid geometry, so that little or no preference for the *cisoid* over the *transoid* stepwise [$1_s + 1_s$] geometry should be observed.

[$2_s^A + 2_a^D$]. This geometry is the limiting one proposed by Woodward and Hoffmann for [$_\pi 2_s + \pi 2_a$] cycloadditions. This geometry is stabilized only by interactions between the HOMO of the 2_a and the LUMO of the 2_s component. The other frontier orbital pair is noninteracting and provides stabilization only for substantial distortions from the C_{2v} geometry. Because of the stabilization by only one set of frontier orbital interactions, such a geometry is likely only for a good donor–acceptor pair.

[$2_a + 2_a$]. In addition to the unfavorable steric interactions present in such a geometry, neither pair of frontier interactions can stabilize this geometry.

[$1_s^{A,D} + 1_s^{A,D}$]. Symmetry constraints on interactions are absent in this geometry, and both sets of frontier orbital interactions are stabilizing. Such a geometry is the most likely one for nonpolar addend pairs. It can be stabilized additionally as the donor–acceptor complementarity of the addends increases. However, other geometries may be more rapidly stabilized by donor–acceptor complementarity.

[$2_s^D + 1_s^A$]. Such a geometry involve interaction of only one terminus (1_s) of the electrophilic alkene LUMO with the nucleophilic alkene (2_s) HOMO. In the absence of steric difficulties, this geometry may be stabilized as much as the [$1_s + 1_s$] for good donor–acceptor pairs, while the latter is clearly better for molecules of like polarity.

[$2_s^A + 1_a^D$]. The final geometry is a less symmetrical version of the [$2_s^A + 2_a^D$]. with good overlap of only one donor terminus with the acceptor π^* orbital.

It is also interesting that a molecule with two perpendicular π systems may simultaneously be a 1_s compenent in a [$1_s^A + 2_s^D$] interaction and a 1_a component in a [$1_a^D + 2_s^A$] interaction. Geometries of this type with these simultaneous interactions have been proposed for cyloadditions of alkenes to ketene (Woodward and Hoffmann, 1969), singlet oxygen (Kearns, 1971; Inagaki et al., 1972), benzyne (Inagaki and Fukui, 1973), and carbonyl and nitroso compounds (Inagaki et al., 1974).

These general considerations may be applied to the rationalization, or prediction, of mechanistic types in [2 + 2] cycloadditions of molecules of different polarity. In the following paragraphs, the considerations above are applied to the various classes of [2 + 2] cycloadditions.

Nonpolar [2 + 2] cycloadditions. For two molecules of similar polarity, in which neither HOMO–LUMO interaction predominates, the [1$_s$ + 1$_s$] geometry is favored over all the others. This is due both to the preferred overlap (calculated for Slater orbitals as 0, 0.0404, 0, 0.0556, 0.0474, and 0.0362, respectively, for each pair of interactions shown in Fig. 15 at an addend separation of 3 Å) and to the fact that both pairs of frontier orbital interactions stabilize the [1$_s$ + 1$_s$] geometry, but only one pair of interactions can stabilize any of the other geometries.

Polar [2 + 2] cycloadditions. If substituents which raise or lower both of the frontier orbitals simultaneously are placed on each alkene, the preference for this transition state drops somewhat, because eventually only one interaction is of any significance. Nevertheless, the [1$_s$ + 1$_s$] geometry will be favored or may compete with the [2$_s$ + 1$_s$].

Simultaneous [2$_s^D$ + 1$_s^A$] and [2$_s^A$ + 1$_a^D$] interactions may stabilize the "π-complex" geometry in special cases such as singlet oxygen and ketene referred to earlier. Such cases will occur when an alkene is not sterically hindered at one end, and when the alkene has a relatively low-lying LUMO and high-lying HOMO.

b. [4 + 2] Cycloadditions. The reactions of dienes with alkenes can proceed through any of the [2 + 2] geometries, with a vinyl group appended on one of the alkenes, and additional geometries—[4$_s^{D,A}$ + 2$_s^{A,D}$], [4$_s$ + 2$_a$], [4$_a$ + 2$_s$], and [4$_a^D$ + 2$_a^A$]—are possible. Only the first of these—the traditional parallel planes approach of diene and alkene—can be stabilized by donor diene–acceptor alkene (Alder rule) or donor alkene–acceptor diene substitution (inverse electron demand). The [4$_a$ + 2$_a$] geometry, a sterically unlikely spiroconjugated geometry, is stabilized only in the donor diene–acceptor alkene case, while neither of the (s,a) modes can be stabilized by frontier orbital interactions. Taking into account the conclusions from the previous section, the reasonable geometries are the [4$_s^{D,A}$ + 2$_s^{A,D}$], the [1$_s^{D,A}$ + 1$_s^{A,D}$], and (2$_s^D$ + 1$_s^A$). It is probable that each of these transition state geometries has some role in the reactions of some dienes and alkenes. Since the [1 + 1] and [2 + 1] geometries have been discussed before, only a comparison with the stabilization of the [4$_s$ + 2$_s$] geometry will be discussed here.

For a C_s symmetrical [4$_s$ + 2$_s$] geometry, only the frontier (HOMO–LUMO) orbital interactions are stabilizing, because the other π orbitals of the diene are of the wrong symmetry to interact with the corresponding ethylene orbitals. On the other hand, a [1$_s$ + 1$_s$] geometry can be stabilized by interactions of all filled–vacant combinations, since there are no symmetry restrictions preventing any of these interactions. Figure 16 summarizes the interactions that occur upon a symmetrical [4$_s$ + 2$_s$] approach of diene to dienophile, and upon a highly unsymmetrical [1$_s$ + 1$_s$] approach. If the former

216 K. N. HOUK

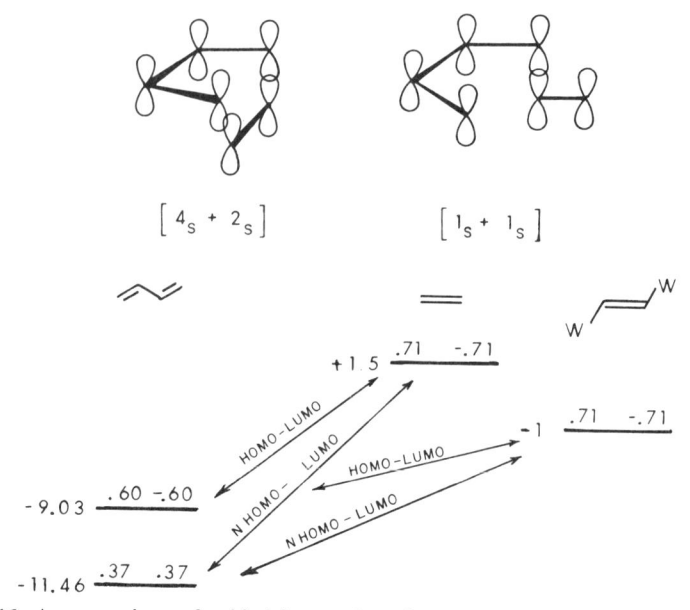

$$[4_s + 2_s] \qquad [1_s + 1_s]$$

Fig. 16. A comparison of orbital interactions for non-polar and polar Diels–Alder reactions by concerted and nonconcerted pathways. Only HOMO–LUMO interactions can occur in the symmetrical [4 + 2] transition state.

geometry is highly favored, then the traditional experimental manifestations of concert are expected—stereospecificity, formation of only [4 + 2]- and no [2 + 2]-adducts, inverse secondary isotope effects of reasonable magnitude at all four centers undergoing $sp^2 \to sp^3$ change, and large negative activation volumes (Houk, 1975). If, on the other hand, the $[1_s + 1_s]$ transition state is favored, or is nearly the same energy as the $[4_s + 2_s]$, then experimental manifestation of stepwise reactions will be observed—low (or absence of) stereospecificity, formation of both allowed [4 + 2]- and forbidden [2 + 2]-cycloadducts, inverse secondary isotope effects only at one pair of termini, and moderate negative activation volumes.

Calculations based on the interactions shown in Fig. 16 have been carried out assuming approach of addends in parallel planes separated by 3 Å. For the $[4_s + 2_s]$ transition state, in which a C_s geometry is assumed, the overlap between Slater $2p$ orbitals at each terminus in 0.042. For the $[1_s + 1_s]$ geometry, the corresponding overlap is 0.056. Using the coefficients for butadiene and ethylene obtained from CNDO calculations and a 5 eV correction for Q, stabilization of the [4 + 2] geometry is

$$\Delta E[4 + 2] = 2 \frac{[2K (0.60) (0.71) (0.042)]^2}{9.03 - (-1.5) - 5} = 4.6 \times 10^{-4} K^2$$

For simplicity, only the diene occupied-dienophile vacant interactions have been considered. For the [$1_s + 1_s$] geometry, interactions of both diene filled π orbital contributes to stabilization, and the stabilization is:

$$\Delta E[1+1] = 2\frac{[K(0.60)(0.71)(0.056)]^2}{9.03 - (-1.5) - 5} + 2\frac{[K(0.37)(0.71)(0.056)]^2}{11.46 - (-1.5) - 5}$$

$$= 2.1 \times 10^{-4}K^2 + 5.4 \times 10^{-5}K^2 = 2.6 \times 10^{-4}K^2$$

The difference between the [4 + 2] and [1 + 1] transition state stabilization is 2.0×10^{-4} K^2. In the CI model discussed earlier, these are the amounts by which the transition states are stabilized as compared to the energies they would have in the absence of electronic "relaxation." The NHOMO–LUMO interaction contributes only slightly (20%) to the transition state stabilization for the [1 + 1] transition state.

As the electron affinity of the dipolarophile increases, the stabilization of the [4 + 2] and [1 + 1] transition states increases rapidly. For simplicity, the coefficients of the LUMO are retained in the model calculation, since otherwise, all of the vacant orbitals of the dienophile would need to be taken into account to give a proper estimate of increased transition state stabilization. The corresponding stabilizations for this model, using an EA of 1 eV, are:

$$\Delta E'[4+2] = 8.4 \times 10^{-4}K^2$$
$$\Delta E'[1+1] = 3.8 \times 10^{-4}K^2 + 8.0 \times 10^{-5}K^2 = 4.6 \times 10^{-4}K^2$$

The [4 + 2] transition states is now favored by 3.8×10^{-4} K^2.

If the assumption is made that the [4 + 2] transition state is favored by 5 kcal/mole in the ethylene–butadiene system, a calculation of K indicates a preference of 9.5 kcal/mole for the [4 + 2] in the butadiene, deficient alkene system. That is, changing from ethylene to the electron-deficient alkene increases the [4 + 2] transition state stabilization by 9.5 kcal/mole and the [1 + 1] transition state stabilization by 5 kcal/mole. This is but another case of increased selectivity (here toward a concerted mechanism) as the rate of a reaction increases.

Finally, it is interesting to probe whether a stepwise reaction can intervene for a highly unsymmetrical dipolarophile in polar cycloadditions. The requirement for equal preference of stepwise and concerted reaction pathways, assuming an electron affinity of 2 eV for the dipolarophile is:

$$\Delta E[4+2] = 2\frac{[K(0.60)(C_1 + C_2)(0.042)]^2}{9.03 - 2 - 5} + 2\frac{[K(0.37)(C_1 + C_2)(0.042)]^2}{11.46 - 2 - 5}$$

$$= \Delta E[1+1] = 2\frac{[K(0.60)(C_1)(0.056)]^2}{9.03 - 2 - 5} + 2\frac{[K(0.37)(C_1)(0.056)]^2}{11.46 - 2 - 5}$$

Here C_1 is the larger LUMO coefficient in the unsymmetrical dienophile. The results are that the [4 + 2] and [1 + 1] transition states are equally stabilized for LUMO coefficients of 0.95 and 0.32, or a ratio of 3. Inspection of Figs. 4 and 5 shows that this ratio is never reached in monosubstituted cases, but a way to accomplish such drastic polarization would be unsymmetrical donor–acceptor substitution, or by increasing the electron-withdrawing group electronegativity by strong complexation with a Lewis acid catalyst.

How well do these model calculations correspond to experimental observations? Too well for comfort perhaps, given that the experimental results long preceded these "predictions." Only a few of the known examples will be discussed here.

The best known examples of stepwise Diels–Alder reactions involve either nonpolar partners or highly unsymmetrical polar partners. Among the notable examples of reactions in which [2 + 2] and [4 + 2] cycloadditions compete effectively are numerous cycloadditions of haloalkenes to dienes (Roberts and Sharts, 1962; Bartlett, 1968), the cycloaddition of α-acetoxyacrylonitrile to butadiene (Little, 1965), the dimerization of chloroprene (Stewart, 1972), and the cycloaddition of ethylene to butadiene (Bartlett and Schueller, 1968). In the last case, only 0.02% of the [2 + 2]-adduct is observed at 175°.

Recently, Mark (1974) has reported stereochemical results favoring a stepwise Diels–Alder mechanism in reactions of electron-deficient dienophiles with the electron-deficient diene, hexachlorocyclopentadiene. While reactions of *cis*-1,2-dichloroethylene and maleonitrile give extensive amounts of trans adduct, electron-rich dienophiles such as *trans*-2-butene add stereospecifically.

These results are in accord with the idea that nonpolar cycloadditions have little preference for concerted over stepwise mechanisms, while polar Diels–Alder additions do show a high preference for the concerted mechanism.

In nonpolar cases, concerted [4 + 2] reactions of dienes with alkenes will be favored electronically, if only to a small extent. The [2 + 2] reaction, or [4 + 2] reactions involving stepwise mechanisms can compete with the concerted [4 + 2] reaction if some factor other than π-orbital interactions disfavors the concerted mechanism. Two factors of this type seem to be common in the previously cited examples. The first of these is steric effects, which prevent attainment of the concerted [4 + 2] transition state geometry. Because the [4 + 2] geometry will always have more net nuclear–nuclear and electron–electron repulsions than the [1 + 1] geometry, this factor will always counteract the electronic stabilization of the [4 + 2] transition state to some extent. The second effect of this type, which may add several kcal/mole to the [4 + 2] transition state, is the preference for *s*-trans over *s*-cis conformations in many acyclic dienes. The *s*-cis or *s*-trans conformations may both enter into [1 + 1] geometries, but only the former can enter into [4 + 2] geometries.

Although a few reactions of dienes and dienophiles clearly involve diradical intermediates, and many addend pairs display little or no preference for concerted [4 + 2] over stepwise [2 + 2] cycloadditions, the majority of Diels–Alder reactions show all the earmarks of concerted cycloadditions (Wollweber, 1972). In these cases, either the diene is electron-rich and the dienophile electron-deficient ("normal electron-demand"), or vice versa ("inverse electron demand").

However, the other extreme rationalized by perturbation theory can also be realized: with highly polar diene–dienophile combinations, in which at least one addend is highly unsymmetrical, stepwise reactions involving zwitterionic intermediates take place.

Examples of this type include the reactions of tetracyanoethylene (TCNE) with 4-methyl-1,3-pentadiene (Stewart, 1962) and 1,1-diphenyl-1,3-butadiene (Eisch and Husk, 1966) in which the ratio of [2 + 2]-adduct increases with increasing solvent polarity, the reaction of the enamine and the dicyanodiene shown in Fig. 17, involving an observable zwitterionic intermediate (Gompper, 1969), and the reaction of methylvinyl ketone and an enamine (Fleming and Karger, 1967), also shown in Fig. 17. Thus, the Diels–Alder reaction can occur by essentially all possible cycloaddition mechanisms. Frontier orbital theory gives at the very least a qualitative guide to which addend pairs should react by a given mechanism.

The last examples are specially interesting in light of Epiotis' CI model for cycloaddition reactions (Epiotis, 1974). Epiotis notes that the concerted $[_\pi 4_s + _\pi 2_s]$ mechanism is allowed whether the "ground" or "charge-transfer" configuration is lowest in energy, because these two configuration extensively mix, being of the same symmetry. Since the stabilization of the [4 + 2] transition state is a maximum for the extremes of complementary polarity, Epiotis deems stepwise reactions involving zwitterions to be quite unlikely. However, as noted earlier, strong asymmetry, combined with a steric or conformational difficulty in the concerted transition state, can tip the balance of stabilization over to the [1 + 1] geometry for asymmetric polar reactants.

In the reactions of dienes with tetracyanoethylene, there is a strong acceleration of formation of cyclobutane product by polar solvents. The

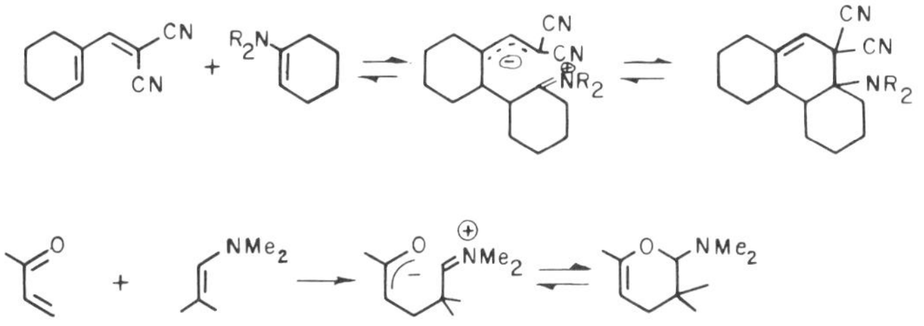

Fig. 17. Two stepwise Diels–Alder reactions involving zwitterionic intermediates.

simplest explanation for these results is that a concerted [4 + 2] and stepwise-zwitterionic [2 + 2] cycloaddition compete, with the latter favored by polar solvents. Herndon has proposed that both adducts are formed from a common "diradical" intermediate, since a perturbation treatment of the cyclization does predict that cyclobutane formation will be favored (Herndon and Feuer, 1968). On the other hand, the Epiotis model could account for the experimental observations if the [4 + 2] were concerted and occurred via a "ground-state" configuration, while the "charge-transfer" configuration could be lowest in polar solvents and for some reason favor a concerted [2 + 2] over concerted [4 + 2] mechanism.

These considerations of the Diels–Alder mechanism may be applied more generally: in cycloadditions of addends of similar polarity, the allowed concerted pathways will be only slightly favored over stepwise reactions. Any hindrance of the concerted geometry (steric, conformational, etc.) will lead to reaction via a stepwise mechanism if reaction occurs at all. As the donor–acceptor complementarity of the addends increases, the preference of concerted (if allowed) over stepwise (leading to allowed *or* forbidden products) increases. Finally, if the addends constitute a highly polar pair, and are substituted in a highly unsymmetrical fashion, stepwise mechanisms involving substantial charge separation (and possible zwitterionic intermediates) will be competitive with concerted mechanisms, and any factor (steric or conformational) which decreases the possibility of a concerted reaction pathway will result in rapid formation of the zwitterion, which may collapse to either allowed or forbidden adducts. Thus, the Woodward–Hoffmann rules are of greatest predictive values for good, but not excessive, donor–acceptor pairs.

3. *Reactivity*

The use of frontier MO coefficients and energies as reactivity indices has long been advocated by Fukui, and as noted earlier, Brown (1959) had also pointed out the importance of HOMO–LUMO interactions (IP's and EA's) upon reactivity. Konovalov and co-workers (1967) showed (Konovalov and Kiselev, 1966) that the rates of cycloadditions of a series of dienes to maleic anhydride increased as the IP of the diene decreased. Konovalov (1969) reported the use of localization energies (which, in effect, take MO coefficients into account) and IP's and EA's to more quantitatively explain Diels–Alder reactivity.

Garbisch and Sprecher (1966, 1969) and Herndon and co-workers performed perturbation calculations to explain the relative reactivities of different diene–alkene pairs (Herndon and Hall, 1967). Although all orbital interactions were considered, the major influence of the frontier interactions was noted. In his review of theories of cycloaddition reactions, Herndon (1972) outlines the simple model used in the previous section: starting from butadiene and ethylene and decreasing either HOMO–LUMO gap will accelerate the Diels–Alder reaction.

Sustmann (1971a,b) used the frontier MO model to explain reactivity in 1,3-dipolar cycloadditions and Diels–Alder reactions. He classified these allowed cycloadditions according to three types shown in Fig. 18. If A is the reference compound, then the diagram on the left shows an HOMO controlled reaction. Electron withdrawal on B will accelerate the reaction, and electron release on B will slow the reaction. The substituent effects will be large, since one interaction is strong. The middle case is one where both frontier interactions are important. Either electron release or electron withdrawal on B will accelerate reaction, but to a smaller extent than in the first case. This is because a change in energy of the MO's of B will increase a fairly weak interaction, and because one interaction will be strengthened as the second is weakened. As noted before, this case is the one likely to show little preference for allowed-concerted over stepwise pathways. The case on the right is LUMO controlled; electron-releasing substituents on B will accelerate the reaction, and electron-withdrawing ones will slow it down. This simple model is remarkably successful—even quantitative—if there are no drastic changes in A and B, other than IP's and EA's.

Eisenstein and Anh (1971) used Hückel diene and cyanoalkene HOMO and LUMO coefficients and energies to rationalize the relative rates of Diels–Alder reactions of these species.

Sustmann showed that the rates of phenyl azide cycloadditions to alkenes followed the behavior expected of a phenyl azide HOMO–LUMO controlled reaction (Sustmann and Trill, 1972). That is, a plot of the log of the cycloaddition rate constant versus the alkene ionization potential gave a rough parabola, with alkenes of IP between 9 and 11 eV reacting much slower than alkenes of lower or higher IP.

A similar plot for the log of the rate constants for Diels–Alder reactions of twelve dienes with maleic anhydride and TCNE versus the diene HOMO-

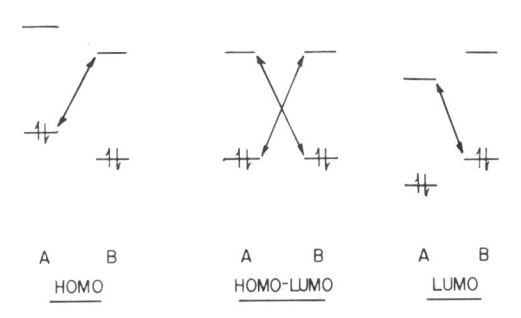

Fig. 18. The three types of cycloadditions according to Sustmann.

dienophile LUMO gap gave a rough correlation showing faster reactivity with decreasing HOMO–LUMO gap, thus conforming to a diene HOMO-controlled pattern (Sustmann and Schubert, 1972).

We have treated all common 1,3-dipoles according to this simple model and have shown that the predicted qualitative trends are adhered to very well (Houk, 1972; Houk et al., 1973a,b). The nitrile ylides (R—C≡N$^+$—C$^-$R$_2$), diazoalkanes (R$_2$C$^-$—N$^+$≡N), and azomethine ylides (R$_2$C=N$^+$(R)—C$^-$R$_2$) are HOMO-controlled 1,3-dipoles, reacting fastest with alkenes having one or more electron-withdrawing substituents. The nitrile imines (RC≡N$^+$—N$^-$R), azides (RN$^-$—N$^+$≡N), and azomethine imines (R$_2$C=N$^+$(R)—N$^-$R) are HOMO–LUMO-controlled reagents, reacting rapidly with both electron-rich, and electron-deficient dipolarophiles. The nitrile oxides (RC≡N$^+$—O$^-$) and nitrones (R$_2$C=N$^+$(R)—O$^-$) are also HOMO–LUMO-controlled 1,3-dipoles, but these species are skewed toward the LUMO controlled side. That is, moderate electron-release on the dipolarophile accelerates reaction more than moderate electron-withdrawal. Finally, species with several electronegative atoms are LUMO controlled, and the 1,3-dipoles with these characteristics include nitrous oxide, ozone, and carbonyl oxides. Since details of this treatment have been published, we wish here to point out only a few aspects of this treatment. For all of the 1,3-dipoles, a U-shaped (parabolic) correlation between the log of the reaction rate and *IP* of the dipolarophile is expected, as long as coefficients are not widely different for frontier orbitals of different dipolarophiles, and as long as the *IP* and *EA* of a molecule change in a parallel fashion. Perfect correlations would only be expected if the coefficients were constant or a linear function of *IP*'s and if the change in *EA* caused by a substituent were linearly related to the *IP* change. In spite of these limitations, useful correlations have been found for the reactions of several 1,3-dipoles with extensive series of alkenes and alkynes. Huisgen and co-workers have shown that the relative rates of cycloadditions of benzonitrile oxide, diphenylnitrilimine, and *C*-phenyl-*N*-methylnitrone to a series of alkenes are linearly related to each other (Bast et al., 1973). The minimum rate for diphenylnitrilimine cycloadditions to monosubstituted alkenes comes at an alkene *IP* of 9.5 eV as compared to the value of about 10 eV for phenyl azide. This is compatible with the somewhat more electrophilic character of phenyl azide than the nitrilimine. For benzonitrile oxide, the minimum reactivity is about 10 eV. *C*-Phenyl-*N*-methylnitrone, which has a higher EA and lower *IP* than the analog, benzonitrile oxide, is more selective as expected from the model.

In plots of rates versus *IP* of dipolarophile, Huisgen and co-workers find that acetylenic dipolarophiles are less reactive than expected on the basis of their *IP*'s (Bast et al., 1973). Since alkynes have a larger HOMO–LUMO gap than the analogous alkenes, one would expect that in reactions where interaction with the alkyne LUMO is of most importance, the alkyne will be less reactive than expected on the basis of an *IP* plot.

Several plots of cycloaddition reactivity versus *IP* of one addend were made before the frontier MO model gained wide usage. For example, Skell and Cholod (1969) showed an inverse correlation between the *IP*'s of alkenes and their reactivities with dichlorocarbene. Furthermore, a linear relationship exists between the rates of cycloaddition of dichlorocarbene and those of other carbenes and carbenoids, as well as with other electrophilic species such as bromine, peracids, and oxygen atoms. All of these electrophiles will have relatively high *EA*'s and *IP*'s, and the extent of interaction of the LUMO's with the alkene HOMO's will determine the transition state stabilization. Freeman (1975) has recently reported plots of alkene *IP*'s versus reactivity toward electrophiles for nearly all common electrophilic reagents.

For the most part, *IP* or *EA* correlations have been qualitative, with the changes in reactivity with *IP* or *EA* following the predicted trends, but little or no linearity being observed. However, we have recently reported a nearly quantitative case in the reactions of cyanoalkenes with dienes (Houk and Munchausen, 1976). The rates of Diels–Alder reactions of 9,10-dimethyl-anthracene and cyclopentadiene with all of the cyanoalkenes have been measured by Sauer *et al.* (1964), and, as shown in Fig. 19, a plot of the quantity $[IP \text{ (diene)} - EA \text{ (dienophile)}]^{-1}$ versus log k for these reactions shows excellent linearity. As we have reported (Houk and Munchausen, 1976), a plot of the full perturbation expression, taking into account only the energies and coefficients of the diene HOMO and dienophile LUMO is curved such that the less reactive species react more slowly than expected. One reason for this curvature is the frontier MO approximation itself. That is, the frontier MO approximation takes into account most of the transition state stabilization energy for the diene–tetracyanoethylene pair, because the HOMO–LUMO gap is very small; however, because the diene–HOMO, acrylonitrile–LUMO separation is much larger, calculations of only the stabilization due to this orbital pair neglects as much as half of the transition state stabilization due to other interactions.

In our work on cyanoalkene reactions, we have also performed calculations for the attack of a nucleophile at only one carbon of the cyanoalkene in the transition state. Whereas inclusion of the coefficients, as well as energies of the LUMO's make quantitative, but not qualitative, differences in calculated reactivities of the cyanoalkenes toward nucleophilic dienes ("two-bond reactivities"), the inclusion of coefficients makes a considerable difference in calculated reactivities toward "one-bond" reagents. This result arises from the fact that the sum of the cyanoalkene LUMO coefficients decreases regularly as the *EA*'s of these species increase, so that inclusion of coefficients in the calculation merely dampens the increase of reactivity as the *EA*

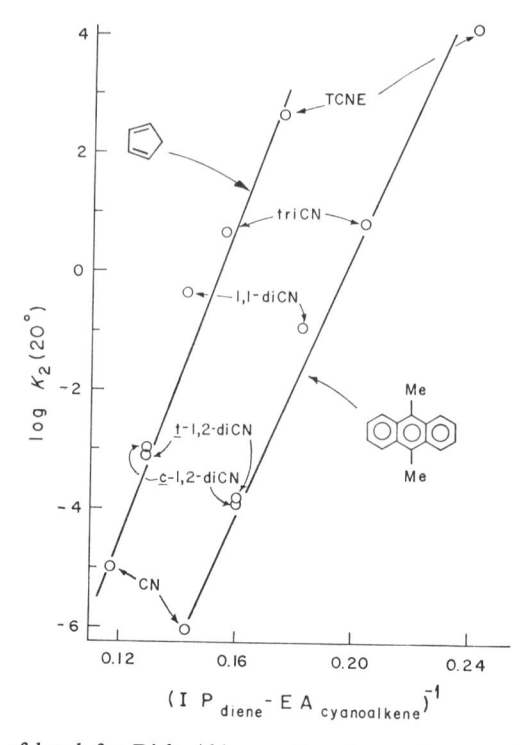

Fig. 19. Plot of log k for Diels–Alder reactions versus the quantity $[IP \text{(diene)}-EA \text{(cyanoalkene)}]^{-1}$. (Reprinted with permission from *Accounts of Chemical Research*. Copyright by the American Chemical Society.)

increases, while the larger coefficient does not decrease in this regular fashion. The EA's increase with the number of cyano groups, but the LUMO coefficients decrease in the order: $CH_2 > CHCN > C(CN)_2$. Calculations indicate that simple "one-bond" nucleophiles, or nucleophilic cycloaddends which react by stepwise mechanisms, will be most reactive with vinylidene cyanide, and acrylonitrile will be more reactive than TCNE, while concerted Diels–Alder reactions with nucleophilic dienes will give the order: TCNE > vinylidene cyanide > acrylonitrile.

Another area of interest to our group has been the study of reactivity of cumulenes (Houk et al., 1974b). This subject will be treated in more detail in later sections, but the type of reactivity expected for these species can be qualitatively discerned by inspection of Fig. 7. Whereas alkenes have relatively high-lying frontier orbitals, the heterocumulenes have strongly stabilized LUMO's, indicating that reactions will occur most readily with electron-rich alkenes. Therefore, electron release on the alkene component or electron withdrawal on the heterocumulene accelerates the reaction. Sustmann *et al.* (1972) discussed ketene cycloadditions using an SCF perturbation scheme and arrived at similar conclusions.

Classical treatments of substituent effects on reaction rates use relationships such as the Hammett or Taft equations, and discuss the sensitivity of reaction rates to changes in substituents in terms of development of partial charges in the transition state of a reaction. The relationship between such treatments and the frontier orbital treatment discussed here is qualitatively straightforward, although some quantitative aspects are less so. Qualitatively, as the electron-withdrawing strength of a substituent increases, the σ of the substituent increases, and both the IP and EA of a molecule on which this group is placed increase. On the other hand, increasingly powerful electron-releasing groups have increasingly negative Hammett σ's, and decreasing IP's and EA's. For certain substituents such as halogens, dissections into "resonance" and "inductive" effects by Hammett treatments show that these different mechanisms of electron withdrawal or release may work in opposite directions. For example, halogens are more electronegative than hydrogen, and will inductively raise both the IP and EA of a molecule. On the other hand, the relatively high-lying lone-pair orbital is electron donating, and will have a substantial effect on lowering the IP of a molecule: As shown in Fig. 5, the effect of such substituents is to lower IP's and raise EA's, increasing reactivity toward both good nucleophiles and good electrophiles.

A number of authors have discussed correlations between Hammett substituent constants and ionization potentials of substituted molecules (see, for example, Van Cauwelaert, 1971) or reduction potentials, which are indirect EA measurements (Zuman, 1967). Because of the latter correlations, Hammett plots are more or less equivalent to plots of rate versus IP or EA. In terms of the frontier orbital approximation, if the reaction of a molecule, A, with a series of para-substituted phenyl substrates, B, has a negative ρ, so that electron-releasing groups accelerate the reaction, an indication of the controlling interaction between the LUMO of A and the HOMO of B is obtained. If, on the other hand, a positive ρ is observed, control by HOMO-A, LUMO-B interaction is indicated. The magnitude of ρ is an indication of the narrowness of the HOMO–LUMO gap. That is, the narrower this gap, the more sensitive the reaction is to a change in IP of the HOMO or EA of the LUMO.

Sustmann (1974) has discussed Hammett substituent constants for cycloaddition reactions in terms of these qualitative considerations. Most of the data commented on have been measured by Huisgen and his co-workers. Para-substituted phenyl azides react with the following dipolarophiles to give the ρ's shown in parentheses: maleic anhydride (-1.1), norbornene $(+0.9)$, pyrrolidinocyclohexene $(+2.54)$. Maleic anhydride has a high EA, and the azide-HOMO, maleic anhydride-LUMO interaction is strongest. On the other hand, norbornene and the enamine have relatively low *IP*'s and the azide-LUMO, alkene-HOMO interaction is strongest. Since the enamine has a lower *IP* than norbornene, the ρ is larger. This interpretation is not unlike the classical interpretation of charge development in the transition state, in that the *IP–EA* gap controls the extent of charge transfer from donor to acceptor that occurs in the transition state. However, even large ρ's can be explained if the reaction has a symmetrical transition state, and appreciable asymmetry in bond-making processes need not be invoked.

Sustmann (1974) also discussed the interesting cases of curved Hammett plots, for example in the reactions of nitrile oxides with styrenes. In such cases, U-shaped plots of log k versus σ are obtained, where groups on styrene with both positive and negative σ's accelerate reaction. Sustmann explained these reactions by noting that in a reaction where both HOMO–LUMO gaps are comparable, decreasing either gap accelerates the reaction. However, the situation appears to be more complicated than this (R. A. Firestone, private communication to the author, 1974),* since other evidence would indicate that in reactions of styrene with benzonitrile oxides, the styrene-HOMO, nitrile oxide-LUMO gap is clearly narrowest. For example, in reactions of para-substituted benzonitrile oxides with *p*-nitrostyrene, styrene, and *p*-methoxystyrene, positive ρ's of +0.33, +0.79, and +0.90, are obtained, respectively (Dondoni and Barbaro, 1973). These results are those expected for a smallest nitrile oxide-LUMO, styrene-HOMO gap, which decreases along the series from *p*-nitrostyrene to *p*-methoxystyrene. Furthermore, all of these reactions give 5-arylisoxazolines, which we have shown to arise from dipole-LUMO, styrene-HOMO interactions (Houk *et al.*, 1973b, vide infra). Nevertheless, since both HOMO–LUMO and LUMO–HOMO interactions will contribute to transition state stabilization, donors will change the styrene *IP* most (raising reactivity with the nitrile oxide), while acceptors will change the styrene *EA* most, which also increases reactivity. A quantitative dissection of these effects is under investigation.

Catalysis of Cycloadditions. The frontier molecular orbital treatment of catalysis in cycloadditions is merely a special case of the treatment of substituent effects. Lewis acids form complexes with carbonyl groups and cyano groups by coordinating with the nonbonding orbitals on oxygen or nitrogen, respectively. A partial formal positive charge builds up on the heteroatom, and, to a first approximation, the coordinated group may be considered to have a higher group electronegativity than the uncoordinated group. For an α,β-unsaturated carbonyl compound or nitrile, coordination with a Lewis acid will lower the energies of both the HOMO and LUMO. Reactions of the coordinated species with electron-rich dienes, 1,3-dipoles, or nucleophilic

---

* I wish to thank Dr. Firestone for this and other stimulating challenges to frontier molecular orbital theory.

species in general, will be faster than reactions of the uncoordinated species, because the nucleophile HOMO-coordinated electrophile–LUMO gap will decrease, increasing the transition state stabilization energy. There may also be an effect on regioselectivity, stereoselectivity (vide infra), and even the reaction mechanism, because the coefficients of the coordinated species may be appreciably different from those of the uncoordinated.

Our treatment of catalysis in Diels–Alder reactions (Houk and Strozier, 1973) relied on calculations, as well as qualitative conclusions, to rationalize the extensive quantitative experimental data on catalyzed Diels–Alder reactions amassed by Inukai and Kojima (1971). Anh and Seyden-Penne (1973) have carried out calculations by the Hückel method, varying the Coulombic integral of X in the species, CH_2=CH—CH=X, and have come to the same conclusions. More recently, Alston and Ottenbrite (1975) have discussed the same systems treated by Houk and Strozier, but have included the effect of secondary orbital interactions to explain some regiochemical subtleties. Finally Imamura and Hirano (1975) have developed the concept of "orbital pumping," which is a perturbation method to show how the orbitals of a noncomplexed species mix under the influence of a catalyst to give the altered MO's of the complexed species.

All of these various treatments can be pictorially illustrated as in Fig. 20. Upon complexation, both frontier orbitals are lowered in energy, and the LUMO is distorted in the direction expected for conversion into an allyl cation. In reactions with nucleophilic species, in which the interaction of the

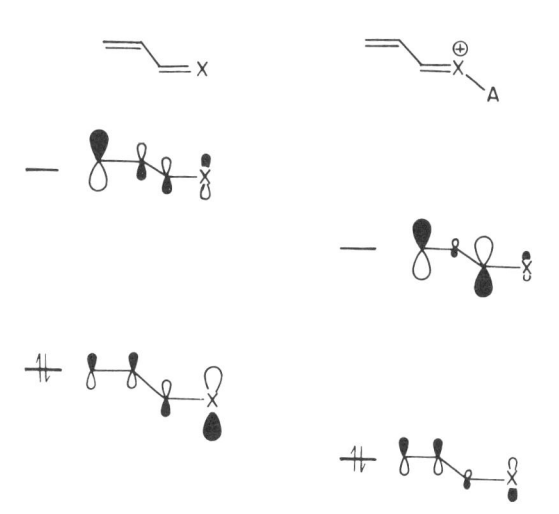

Fig. 20. Frontier MO's of a heterodiene before and after complexation by a Lewis acid (A).

LUMO of the heterodiene with the HOMO of the nucleophile is controlling, complexation lowers the LUMO energy and accelerates reaction. On the other hand, reaction with an electrophilic species will be decelerated.

4. Stereoselectivity

Hoffmann and Woodward (1965b) noted that the endo transition state of the Diels–Alder reaction could be stabilized by "secondary orbital interactions." That is, in addition to the stabilizing HOMO–LUMO interactions leading ultimately to new bonds, additional stabilizing interactions between the remainder of the diene and dienophile π systems are possible in the endo transition state, but not in the exo. The secondary orbital interactions involving one frontier orbital pair are shown for the dimerization of two dienes in Fig. 21. Woodward and Hoffmann also noted that similar stabilizing interactions would stabilize the endo transition state for the [8 + 2] cycloaddition of a triene with polyene, but would destabilize the endo [6 + 4] transition state.

The results were generalized by Houk (1970): for all allowed cycloadditions involving the reaction of a $4n$ electron polyene with an alkene with extended conjugation, the endo transition state can be stabilized by secondary orbital interaction, while for the allowed $(4n) + (4m + 2)$-cycloaddition of two polyenes, where $m \neq 0$, the endo transition state will be destabilized. This can be proven simply as shown in Figs. 22 and 23. Figure 22 shows the stabilizing secondary orbital interactions for all cycloadditions of $4n$ electron polyenes across *one* double bond of a polyene, while Fig. 23 shows the destabilizing secondary orbital interactions which ensue for interaction of the frontier orbitals of two polyenes, both utilizing more than 2 π electrons. The diagram shows a HOMO $(4n)$–LUMO $(4n + 2)$ interaction. For a HOMO $(4n + 2)$–LUMO $(4n)$ interaction, the signs of all the terminal coefficients in Fig. 23 would be reversed, also leading to destabilizing secondary orbital interaction.

Hoffmann et al. (1973) showed that secondary orbital interactions were by no means uniquely characteristic of π systems. For example, the HOMO and

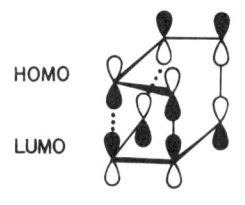

Fig. 21. Bonding (—) and secondary (···) orbital interactions stabilizing the endo Diels–Alder transition state.

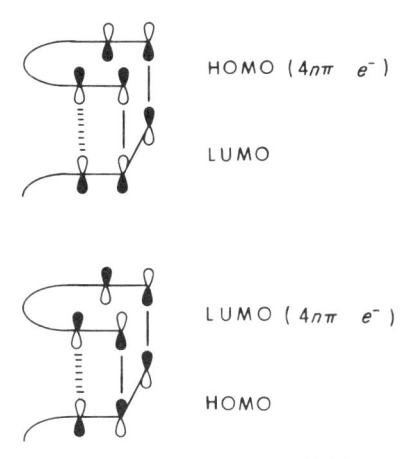

Fig. 22. Secondary orbital interactions (···) stabilizing endo transition state of [$_n4n + {_n}2$] cycloadditions.

LUMO of propene resemble the HOMO and LUMO of butadiene, where a p orbital from the methyl carbon and the antisymmetrical combination of hydrogen 1s orbitals in propene are analogous to the p orbitals of a vinyl group. Secondary orbital interactions of the propene frontier orbitals with diene orbitals can stabilize an endo transition state. This type of "steric attraction" may be responsible for the preferential formation of the sterically more crowded, and thermodynamically less stable, adducts, in a number of Diels–Alder, ketene, and carbene cycloadditions (Hoffmann et al., 1973).

5. Regioselectivity

In additions of an unsymmetrical molecule or a cycloaddend, A–B, to an unsymmetrical polyene, C–D, two different orientations of addition (regiochemistries) are possible, as shown in Fig. 24 for the case of cycloadditions. For the purposes of this discussion, reactions are said to have (z)-regioselectivity if the two termini of larger priority are united in the product, and

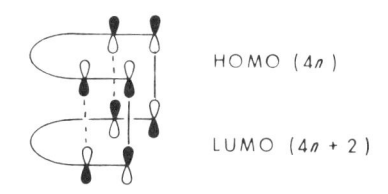

Fig. 23. Secondary orbital interactions (···) destabilizing endo transition states of [$_n4n + {_n}4m + 2$] ($m \neq o$) cycloadditions.

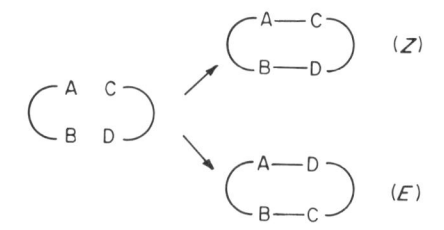

Fig. 24. Definitions of regiochemistries of cycloadditions. (If the relative Cahn–Ingold–Prelog priorities of the termini are A > B and C > D, the regiochemistries of the top and bottom reactions are (z) and (e), respectively.

(e)-regioselectivity if the terminal groups of larger priority are not united in the product.

Until recently, no plausible explanation of the high regioselectivity often observed in cycloadditions had been proposed. Because regioselectivity is quite common in radical or ionic additions to alkenes, and classical explanations of these phenomena focus on the relative stabilities of intermediates, it was quite natural to discuss cycloaddition regioselectivity in terms of zwitterionic or diradical intermediates, or to invoke large asymmetry in bond-forming processes in concerted cycloadditions. When these arguments broke down, steric effects were often invoked. None of these explanations could be applied generally, whereas frontier molecular orbital theory provides a general rationalization of, and predictive tool for, cycloaddition regioselectivity.

Herndon and co-workers were the first to propose that perturbation theory could be applied to the calculation of regioselectivity in Diels–Alder reactions (Feuer et al., 1968). They carried out perturbation calculations, using Hückel MO's, and including all π interactions, for the reactions of 1- and 2-phenylbutadienes and 1- and 2-methylbutadienes with the dienophiles, acrylonitrile, methyl acrylate, acrolein, and styrene. Assuming equal bond formation at both centers in the transition state, the correct regioisomer was calculated to have the largest transition state stabilization energy in every case. Furthermore, the HOMO–LUMO interactions contributed about 70% of the stabilization energy, and always predicted the correct regioselectivity. The two pairs of frontier orbital interactions contributed comparable amounts of stabilization energy.

Salem (1968) and Devaquet and Salem (1969) applied the more complete perturbation theory to the problem of regioselectivity in butadiene and acrolein dimerizations. They found that the predominant regioisomer in the acrolein dimerization [the (z)-Diels–Alder adduct] was not correctly predicted using Hückel MO's but was correctly predicted using SCF MO's and including Coulombic interactions.

Several other failures of the perturbation method to predict regioselectivity correctly have been noted. For example, Inukai et al. (1972) performed PMO calculations with Hückel MO's for reactions of methyl acrylate with 2-methyl-, phenyl-, chloro-, and cyanobutadiene, all of which are known to give predominantly the para isomer. The para isomer was correctly predicted in all cases except with 2-cyanobutadiene. In a similar fashion Bertrán et al. (1971) found that correct predictions were made for donor-substituted dienes and acceptor-substituted dienophiles, but when diene and dienophile were both substituted with electron-withdrawing groups, the meta isomer was predicted to be favored, contrary to experience. On the other hand, Anh and co-workers (Eisenstein et al., 1971) used frontier interactions only or all π interactions with Hückel calculations to obtain correct rationalizations for 30 reactions of all types. These authors adopted a model in which only one new bond is partially formed in the transition state. With this assumption, perfect agreement with experiment is found.

In 1973, we provided an explanation of the source of the discrepancy between calculations for a concerted model and experiment noted above, as well as an explanation of why a specific regioisomer is favored by frontier orbital interactions (Houk, 1973b). By performing extensive calculations on monosubstituted alkenes, as well as by perturbation arguments, generalizations about the shapes and energies of alkene and diene molecular orbitals and coefficients were made (Houk, 1973b, 1976). These have been discussed in the first part of the chapter. Figure 25 summarizes the frontier MO's of monosubstituted alkenes and 1- and 2-substituted dienes; it is somewhat oversimplified in that the relative coefficient magnitudes are not identical for all substituents. The situation is particularly clearcut, however, in the case of a donor or conjugatively substituted diene and acceptor alkene, or vice versa. That is, a donor or conjugative substituent at the 1-position of a diene or alkene will enlarge the remote coefficient in the HOMO, while an acceptor group at the 1-position of a diene or on an alkene will enlarge the coefficient at the most remote position in the LUMO. In the case of donor diene and acceptor alkene, the diene HOMO–dienophile LUMO interaction will be largest when the transition state involves bond formation leading to the "ortho" or (z) adduct. As shown at the bottom of Fig. 25, this is because the stabilization energy will be larger when the larger terminal coefficients and the smaller terminal coefficients of the two interacting orbitals overlap, which gives a larger net overlap in the numerator of the second-order perturbation expression, and thus a larger transition state stabilization, than if a large coefficient on one orbital interacts with a small on the second at both bond-forming centers.

Similarly, the "ortho" transition state will be favored by the diene LUMO–dienophile HOMO interaction for an acceptor diene and donor or conjugated

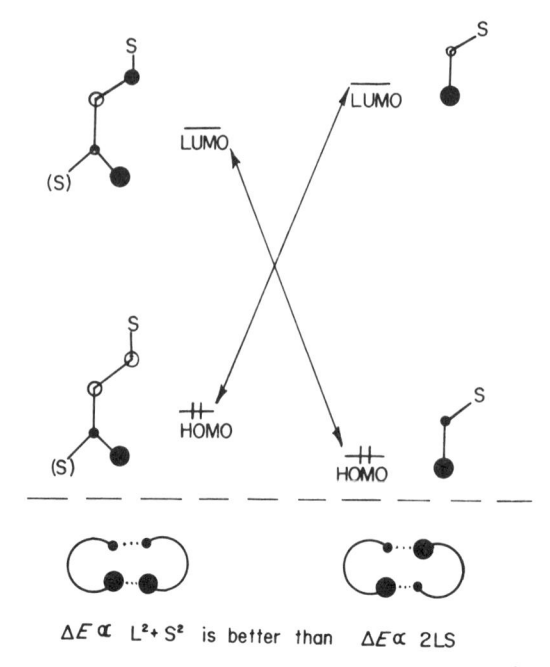

Fig. 25. Frontier orbital interactions controlling regioselectivity in Diels–Alder reactions.

$\Delta E \propto L^2 + S^2$ is better than $\Delta E \propto 2LS$

dienophile. A 2-substituted diene has the coefficient at C(1) enlarged in the HOMO for a donor or conjugative substituent and in the LUMO for an acceptor substituent. Thus the "para" (e) transition state is favored over the "meta" (z) transition state.

The "ortho" and "para" transition states are favored, but to a smaller extent, when both the diene and dienophile are substituted with electron-withdrawing groups. As noted previously, the HOMO's still have the larger terminal coefficients at the terminus shown enlarged in Fig. 25, but the difference in terminal coefficients magnitude is small. Both frontier orbital interactions will be important, but neither leads to a strong preference for the "ortho" or "para" products. Nevertheless, examples of relatively high regioselectivity are known (see Houk, 1973b for pertinent references). Alston et al. (1973) have shown that secondary orbital interactions can further strengthen the "ortho" and "para" preference. Thus, as shown in Fig. 25, the differences between the coefficients at C(2) and C(3) of substituted butadienes are substantial, and secondary orbital interaction between the position of larger coefficient and the substitutent can better stabilize the "ortho" and "para" transition states.

Finally, we turn to the case of donor substituents on both the diene and dienophile. With weak donors, such as alkyl groups, the HOMO coefficients are shown in Fig. 25, while the LUMO terminal coefficients are nearly identical. As the strength of the donor group increases, the LUMO terminal coefficients reverse in relative magnitude from those in Fig. 25, so the predicted preferred regioisomer is the "meta" in Diels–Alder reactions of donor dienes with donor alkenes. Fleming et al. (1976) have recently confirmed this prediction experimentally.

Although regioselectivity has been discussed here in terms of concerted mechanisms, exactly the same predictions are obtained regardless of the mechanism of the reaction. That is, in cases of similarly substituted dienes and alkenes, where we have previously concluded that concerted mechanisms are only slightly favored over stepwise, the larger terminal coefficients in the interacting HOMO–LUMO pair dictate the more favored bonding processes, regardless of the mechanism. The only qualification to this statement arises when interactions other than frontier interactions become comparable to the frontier interactions in magnitude. For example, with a diene and alkene both substituted by electron-releasing groups, interactions of the subjacent and superjacent (nonfrontier) π orbitals of the diene with the alkene MO's can occur in the unsymmetrical transition state, and such interactions could conceivably alter regiochemical predictions.

Epiotis (1973a) has treated the subject of concerted cycloaddition regioselectivity using generalizations similar to ours. However, in the case of acceptor-substituted dienes and acceptor-substituted alkenes, Epiotis concludes that the concerted reactions should give the "meta" products, so that the observation of "ortho" and "para" products must indicate that nonconcerted mechanisms involving the most stable diradical intermediates occur. However, as we noted in the last paragraph, the regioselectivity predictions should be independent of mechanism. The discrepancy between our regioselectivity predictions for acceptor–acceptor cycloadditions and those of Epiotis, Inukai, and Bertrán (*vide supra*) results from a disagreement among various calculations about the relative magnitudes of the HOMO coefficients in electron-deficient alkenes. As noted previously (Houk, 1973b), Hückel, extended Hückel, and in some cases, MINDO/2 calculations indicate that the HOMO has the larger terminal coefficient at the substituted carbon, while CNDO/2, INDO, and *ab initio* SCF calculations place the larger HOMO terminal coefficient on the unsubstituted carbon. Thus, frontier orbital predictions of regioselectivity in cases where the HOMO's of electron-deficient species are involved will be different for the two classes of calculations. Indeed, use of the CNDO/2 or *ab initio* coefficient magnitudes gives the correct regioselectivity predictions, and our treatments are successful for all types of molecules because of this choice.

Alston and Shillady (1974) have compared calculations by seven different techniques for acrolein, and have made conclusions similar to those described in the previous paragraph. Specifically, they have shown that correct regioselectivity rationalizations for the acrolein dimerization are made using the frontier orbital approach with INDO, CNDO/2, CNDO/S, iterative extended Hückel, MINDO/2, and two types of *ab initio* calculations, all of which give (for acrolein, but not, in our experience, with acrylonitrile) the larger alkene coefficient at the unsubstituted carbon.

Having satisfactorily rationalized the regioselectivity observed in Diels–Alder reactions, it is a simple matter to explain the increased regioselectivity often found in catalyzed Diels–Alder reactions (Houk and Strozier, 1973). As shown in Fig. 20, complexation of an electron-deficient alkene by a Lewis acid will not only lower the alkene frontier orbital energies, but will distort the coefficients of these orbitals. Both of these effects will increase regioselectivity in Diels–Alder reactions with an unsymmetrical electron-rich diene. The difference in stabilization energy for the two regioisomers is:

$$\Delta\Delta E = \frac{(C_1^D C_a^A + C_2^D C_b^A)^2 K^2}{IP_D - EA_A - Q} - \frac{(C_1^D C_b^A + C_2^D C_a^A)^2 K^2}{IP_A - EA_A - Q}$$

$$= \frac{K^2(C_1^{D2} - C_2^{D2})(C_a^{A2} - C_b^{A2})}{IP_D - EA_A - Q}$$

where C_1^D and C_2^D are the donor diene HOMO terminal coefficients and C_a^A and C_b^A are the acceptor alkene terminal coefficients. Assuming that K and Q are constant in the uncatalyzed and catalyzed reactions, the increase in EA of the alkene upon complexation will reduce the denominator of $\Delta\Delta E$, increasing the preference for formation of one isomer. Furthermore, the change in LUMO coefficients will increase the term $C_a^{A2} - C_b^{A2}$, further increasing regioselectivity.

Alston and Ottenbrite (1975) have pointed out that the secondary orbital interactions which strengthen regioselectivity in uncatalyzed reactions may also be increased upon Lewis acid complexation of an electron-deficient dienophile.

Probably the greatest success of frontier molecular orbital theory in the treatment of regioselectivity is in the area of 1,3-dipolar cycloaddition regioselectivity (Houk, 1972; Houk *et al.*, 1973a,b; Bastide *et al.*, 1972, 1973). Calculations on all of the common parent and a number of substituted 1,3-dipoles have led to the generalizations about the frontier orbitals of 1,3-dipoles shown in Fig. 26 (cf. Fig. 8 for specific examples). The HOMO's of the 1,3-dipolar systems generally have the larger terminal coefficient on the group Z, while the LUMO's have the larger coefficient at the opposite (X) terminus. The HOMO's and LUMO's of the 1,3-dipoles are qualitatively similar to those of an allyl anion but are distorted in asymmetric systems. The greatest

$$X-Y-Z$$ LUMO ⟵⟶ LUMO $\bigcirc\bigcirc_S$

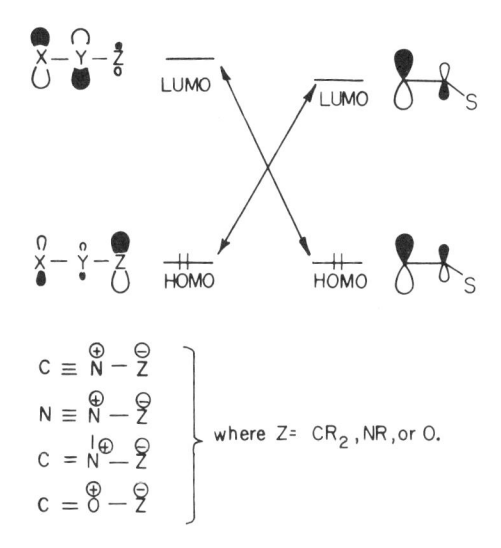

$$X-Y-Z$$ ⥮ HOMO ⥮ HOMO $\bigcirc\bigcirc_S$

$$C \equiv \overset{\oplus}{N} - \overset{\ominus}{Z}$$
$$N \equiv \overset{\oplus}{N} - \overset{\ominus}{Z}$$
$$C = \overset{|\oplus}{N} - \overset{\ominus}{Z}$$
$$C = \overset{\oplus}{O} - \overset{\ominus}{Z}$$

where Z= CR_2, NR, or O.

Fig. 26. Frontier orbital interactions leading to regioselectivity in 1,3-dipolar cycloadditions.

differences in terminal coefficients occur when the two termini differ greatly in electronegativity. The differences in calculated HOMO terminal coefficients are very small for nitrile ylides, and in the substituted cases which have been studied, an order of relative coefficient magnitudes opposite from that shown in Fig. 26 seems to obtain (Houk et al., 1973a,b). We have now determined by STO-3G optimizations that the nitrile ylide is highly bent at the HCN terminus, causing this terminus to be more nucleophilic as a result of the larger HOMO coefficient at this site (Caramella and Houk, 1976).

Once the relative coefficient magnitudes shown in Fig. 26 are realized, the origin of regioselectivity in 1,3-dipolar cycloadditions can be understood. The interaction of the dipole LUMO with the dipolarophile HOMO favors formation of the product with the substituent on carbon adjacent to Z, while the opposite frontier orbital interaction favors the opposite regioisomer. The only generalizations different from these are with the nitrile ylides, as noted above, and with very electron-rich alkenes, where the alkene LUMO coefficients are reversed in magnitude from those in Fig. 26. However, the LUMO's of electron-rich alkenes are irrelevant in virtually all cycloadditions.

In the section of this chapter on reactivity, it was noted that the ylide dipoles have high-lying frontier orbitals, so that the interaction of the HOMO's of these systems with the LUMO's of dipolarophiles will determine reactivity and regioselectivity. Thus, nitrile ylides and diazoalkanes react

Fig. 27. Preferred regioisomers in reactions of (left to right) nitrile ylides, diazoalkanes, nitrous oxide, and carbonyl ylides with monosubstituted alkenes (A = acceptor group, C = conjugative group, D = donor group).

preferentially with electron-deficient and conjugated alkenes, and give the regioisomers shown in Fig. 27. Similarly, the low-lying frontier orbitals of nitrous oxide and carbonyl oxides will facilitate dipole-LUMO, dipolarophile-HOMO interactions, and the regioisomers shown in Fig. 27 will be favored.

The other unsymmetrical species give more varied behavior, and an oversimplified summary is given in Fig. 28. The products shown are those expected for very strong donors and acceptors. Intermediate cases may prefer orientations opposite from those shown. For example, nitrile oxide and nitrones react to give mainly the 5-substituted products with weakly electron-deficient alkenes such as acrylonitrile and methyl acrylate (Christl and Huisgen, 1973). The HOMO's and LUMO's of these weakly electron-deficient alkenes both interact fairly strongly with the LUMO's and HOMO's of the nitrile oxides or nitrones, so that orientation is influenced by both interactions, with the experimental result showing that the dipole-LUMO–dipolarophile-HOMO interaction is of more influence on regioselectivity. As we predicted (Houk et al., 1973b) and later observed experimentally (Sims and Houk, 1973; Houk and Chang, 1975), strongly electron-deficient dipolarophiles and oxide dipoles give the adducts shown in Fig. 28, because only the dipole-HOMO, dipolarophile-LUMO interaction is large in these systems. A number of puzzles in 1,3-dipolar cycloadditions regioselectivity have been solved by application of these principles (Houk et al., 1973a,b: Christl and Huisgen, 1973).

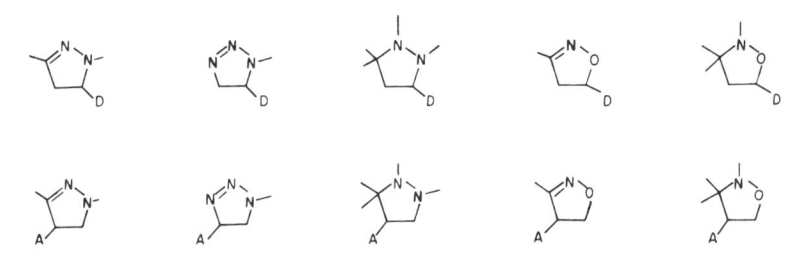

Fig. 28. Predicted regioisomers in reactions of (left to right) nitrilimines, azides, azomethine imines, nitrile oxides, and nitrones with monosubstituted alkenes.

Bastide and co-workers have independently applied perturbation theory to the problem of regioselectivity (Bastide *et al.*, 1972, 1973; Bastide and Henri-Rousseau, 1973; Henri-Rousseau, 1973). These authors carried out pertubation calculations using interactions of all dipole and alkene π orbitals, and using CNDO/2 calculated orbital energies. In all cases studied excellent correlations between calculated and observed regioselectivities were found.

Caramella and Cellerino (1974) found an excellent correlation between the electron donor or acceptor character of the allyl substituent in cyclopentenes and the regioselectivity of benzonitrile oxide cycloadditions to these species. The result was cleverly explained by frontier molecular orbital theory.

Although thermal [2 + 2] cycloadditions are, for the most part, stepwise reactions involving zwitterionic or diradical intermediates, frontier molecular orbital theory is applicable to the explanation of regioselectivity in these systems. In reactions involving cycloaddition of a nucleophilic alkene to an electrophilic alkene, the interaction of the electron-rich alkene HOMO with the electron-deficient alkene LUMO will be dominant, and since both of these have the larger terminal coefficients at the unsubstituted carbons, the head-to-head (z) regioisomer is favored. In dimerizations, or reactions of two molecules similar in donor–acceptor character, both frontier orbital interactions will be of importance. However, both the HOMO's and LUMO's of electron-deficient and conjugated alkenes have the larger terminal coefficients at the unsubstituted carbons, so head-to-head (z) regiochemistry is again favored. Epiotis (1973a,b,c,d) has come to similar conclusions, but his model generally considered these reactions to be concerted, and, since different relative coefficient magnitudes of the HOMO's of electron-deficient alkenes were assumed, incorrect predictions of head-to-tail regioselectivity were obtained for dimerizations of these species.

6. Periselectivity

Periselectivity is the selective formation of one of the possible thermally allowed pericyclic reaction products (Houk *et al.*, 1970). In the study of cycloadditions of polyunsaturated addends, a number of adducts may be thermally allowed, but the selective formation of one or more of these is quite generally observed. This is, of course, just a somewhat broader type of selectivity than regioselectivity, which refers only to orientation of addition to a π system, rather than whether 1,2-, 1,4-, or 1,6-additions are favored.

The frontier orbital treatment of periselectivity follows that of regioselectivity nearly exactly. This is, if one dominant frontier orbital interaction can be identified, that transition state will be favored in which the two largest coefficients of the two frontier MO's are united, and where the two next largest coefficients are united, as long as these unions can be simultaneously bonding—that is, that the reaction is symmetry allowed.

238 K. N. HOUK

The most thorough studies of periselectivity by frontier molecular orbital methods have been in the area of fulvene and cumulene cycloaddition periselectivity.

Fulvenes are interesting substrates for studies of periselectivity because they are polyenes with frontier molecular orbitals appreciably distorted as compared to the corresponding linear polyenes. Figure 29 shows the frontier MO's of fulvene and heptafulvene. The sizes of the π coefficients are represented by the diameters of the circles at each carbon, and are approximately the same relative magnitude in a variety of calculations (Houk et al., 1974a). The energies are either negatives of vertical ionization potentials, or are estimated if in parentheses. Whereas linear polyenes have the largest HOMO and LUMO coefficients at the terminal carbons, pentafulvene has a large coefficient at the terminal (exocyclic) carbon in the LUMO, but, because of the C_{2v} symmetry of the molecule, a node through the exocyclic carbon in the HOMO. The HOMO of fulvene is essentially like that of cyclopentadiene in both shape, and in energy, as revealed by ionization potentials. Thus fulvene acts as a normal diene in Diels–Alder reactions. However, the LUMO is lower in energy than for a simple diene, and the large density at the exocyclic carbon makes reactions different from those of cyclopentadiene possible. In reactions of dienes, D, with fulvene, F, the following symmetry-allowed, all-suprafacial, cycloadditions are possible: $[4^D + 2^F]$ (at an endocyclic or exocyclic double bond), $[4^F + 2^D]$, $[6^F + 4^D]$. In the reaction of fulvenes with simple dienes, or those with one or more electron-withdrawing groups, only the [4 + 2] modes of cycloaddition are favored, because for fulvene-HOMO–diene-LUMO interactions to stabilize the transition state, reaction must occur at an endocyclic double bond of fulvene. However, for electron-rich dienes, reaction in the [6 + 4] sense is favored, because the largest coefficient on the fulvene LUMO is at the

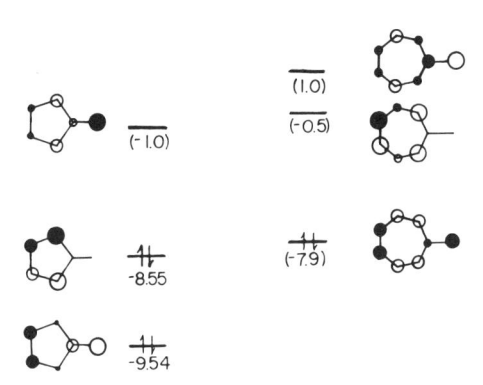

Fig. 29. The frontier MO's of fulvene and heptafulvene.

exocyclic carbon. Similarly, for 1,3-dipolar cycloadditions, only very electron-rich dipoles (e.g., diazoalkanes) will react in the [6 + 4] manner, while most 1,3-dipoles add at an endocyclic double bond in a [4 + 2] manner in order to simultaneously maximize both pairs of HOMO–LUMO interactions. These generalizations, and a discussion of substituent effects on periselectivity, are covered fully in a paper on the subject (Houk et al., 1974a). A number of experimental verifications of the rationalizations or predictions have also been published (Houk and Luskus, 1973; Sasaki et al., 1975). Extensions of these arguments may also be made for heptafulvenes. Figure 29 shows that heptafulvene has the node through the exocyclic carbons in the LUMO, and the HOMO has the largest coefficient at the exocyclic carbon. In reactions with alkenes, both [8 + 2] and [4 + 2] cycloadditions are allowed, but [8 + 2] periselectivity is expected for electron-deficient alkenes, and [4 + 2] periselectivity for electron-rich alkenes. Although qualitative agreement with these predictions can be found in the literature, the predictions have not been tested with a graded series of alkenes and a simple heptafulvene.

Paddon-Row and co-workers (1973, 1975; Paddon-Row, 1974) have independently discussed fulvene periselectivity with the aid of perturbation theory. For the most part, their predictions are based on Hückel MO calculations for reactions of fulvenes with hydrocarbon polyenes, so that many of the interesting periselectivity changes with alkene substitution are not predicted. In a recent communication, Paddon-Row et al. (1975) point out that our frontier orbital considerations may not be adequate in the case of cyclopentadiene (CP) reactions. Interaction of the CP HOMO with the fulvene NLUMO favors formation of the [$4^D + 2^F$]-adduct, as is observed, whereas consideration of only the CP-HOMO–fulvene-LUMO interaction gives a prediction of formation of a [6 + 4]-adduct. However, these considerations overlook the possible importance of the "principle of frontier orbital narrowing" (Fujimoto and Fukui, 1974). That is, frontier orbital interactions may be of more importance than is suggested by calculations on the isolated system.

Ketenes and other cumulenes can react in many thermally allowed cycloadditions. Besides the normal alkene-like reactivity of ketene, resulting from the presence of an in-plane π^*_{CO} orbital which can act like the LUMO of an electron-deficient alkene, and an out-of-plane π_{CC} orbital which is like that of an enol ether, the simultaneous presence of these two perpendicular orbitals can lead to a type of reactivity unique to ketenes and similar cumulenes. As Woodward and Hoffmann (1969) pointed out, the in-plane LUMO of ketene can interact with the HOMO of an alkene, while the ketene HOMO simultaneously interacts in antarafacial manner with the alkene LUMO. These transition state interactions can result in an overall [$_\pi 2_a + _\pi 2_s$] reaction. Even with conjugated dienes, this type of cyclobutanone-forming reaction is favored. However, as we have pointed out recently (Houk et al., 1974b; Houk, 1975), electron-rich dienes have a negligible diene-LUMO–ketene-HOMO interaction and the interaction of the diene HOMO with the ketene LUMO is strongest for the mode of reaction leading to Diels–Alder reaction across the ketene carbonyl group. For electron-deficient dienes, Diels–Alder reaction to the ketene π_{cc} orbital is favored by the diene-LUMO–ketene-HOMO interaction. These rationalizations and predictions have been summarized recently in a diagrammatic fashion (Houk, 1975).

Another aspect of the cycloadditions of polyenes can be treated by frontier molecular orbital theory. As noted in the section on mechanisms of cycloadditions, thermally allowed reactions will be highly favored over thermally forbidden only for good donor–acceptor pairs. For molecules of similar polarity, favorable competition by stepwise mechanisms, with the resulting formation of thermally forbidden products, is likely. Although very few experimental examples are known, the formation of both allowed and forbidden dimers in cycloheptatriene dimerizations (K. N. Houk, unpublished), as opposed to the formation of only one allowed [6 + 4]-adduct in the cycloaddition of tropone to cycloheptatriene (Takatsuki et al., 1970) are in accord with this generalization.

VI. Applications to Thermal Intramolecular Reactions

A. Intramolecular Reactions as Cycloadditions

There are a number of intramolecular cycloadditions in which the two π systems are joined by polymethylene chains or other bridges. These reactions can be treated in exactly the same way as intermolecular reactions, and as long as there is no strong through-bond coupling, the use of the intermolecular formalism is nearly rigorous. There is, however, a fundamental difference between intermolecular reactions and some of these intramolecular reactions when a particular transition state geometry is enforced. That is, whereas intermolecular reactions can seek the most favorable transition state geometry, intramolecular reactions may have only one or a few geometrically feasible transition state geometries available. If these geometries are forbidden ones, then the usual frontier orbital considerations say only that the reaction should not be facile. The Epiotis mechanism or control of reaction rates by interactions between orbitals other than frontier orbitals will be important in such cases.

All intramolecular pericyclic reactions can be formally discussed as cycloadditions by arbitrarily dissecting the interacting bonds into isolated units. For example, diene electrocyclizations may be considered as intramolecular [2 + 2] cycloadditions. The allowed [$_\pi 2_s + {_\pi}2_a$] geometry is quite feasible sterically, when one end of each alkene is already joined in a planar fashion, and corresponds to the Woodward–Hoffmann allowed conrotatory electrocyclization. The derivation of the Woodward–Hoffmann rules is straightforward in these terms, and this derivation will not be discussed further here. Applications to reactivity are somewhat less straightforward, as will be discussed in the following pages.

B. Electrocyclizations

1. Stereospecificity

Woodward and Hoffmann's (1965) first treatment of electrocyclic reactions pointed out that the symmetry of the HOMO of a polyene was the same as the experimentally observed mode of cyclization. That is, the stereochemistry—conrotatory or disrotatory—of all electrocyclizations occurred in such a fashion as to stabilize the HOMO. After the formulation of these reactions in terms of orbital and state correlation diagrams (Longuet-Higgins and Abrahamson, 1965; Hoffmann and Woodward, 1965a) it became clear that successful rationalizations of electrocyclization stereochemistries based on inspection of the HOMO only resulted from the fact that all orbitals of lower energy than the HOMO always correlate with bonding orbitals in the product. Thus, whether the HOMO correlates with an approximately equally bonding orbital or a more antibonding orbital in the product determines whether the state under consideration will have favorable reaction pathway (relatively low activation energy) for a particular stereochemistry. The situation is similar to Walsh–Mulliken diagrams used to predict molecular geometries. That is, the HOMO of the species apparently changes energy more than other orbitals as geometrical changes occur, so the molecule adopts that geometry which minimizes the HOMO energy (see, for example, Gimarc, 1974). Whether or not one wishes to state that the HOMO determines the reaction stereochemistry is primarily a matter of taste: while the sum of the changes in all the orbital energies, in electron–electron repulsion, and in the correlation energy will determine the reaction pathway, it is the HOMO energy change which signals whether a reaction is allowed or forbidden.

The symmetry, or nodal properties, of the HOMO's of polyunsaturated molecules depend only on the number of π electrons in the molecule: systems with $4n\pi$ electrons have antisymmetric HOMO's and those with $4n + 2\pi$ electrons have symmetric HOMO's, or, in the absence of symmetry, these molecules have terminal coefficients of opposite sign, or of the same sign, respectively, as shown in Fig. 11. Thus, the HOMO of a $4n$ system is stabilized by conrotatory motion, which joins the top of one terminal orbital with the bottom of the other, and the HOMO of a $4n + 2$ system is stabilized by a disrotatory motion which joins the tops of the terminal orbitals. The rule is applicable to neutral and charged closed-shell species, but apparently breaks down for systems of odd electrons, where confident predictions cannot be made on this basis (*vide infra*).

Tracing the changes in HOMO energy with geometry changes is essentially a first order method which succeeds, because if the HOMO is destabilized upon distortion, it will correlate with a vacant orbital in the product, and the resultant destabilization will be far greater than can be compensated for by stabilization of filled orbitals.

Epiotis (1973e) has suggested that the orbital symmetry selection rules are not invariant to substitution. One version of Epiotis' argument is based on first-order perturbation theory, and a charge-transfer model for substituent effects. As noted previously, the HOMO energy change parallels the allowed or forbidden nature of the electrocyclic reactions. An electron-withdrawing group which, in Epiotis' model, withdraws electron density from the polyene HOMO, will decrease the energy barrier for forbidden reaction and diminish the transition state stabilization of an allowed reaction. On the other hand, since the HOMO and LUMO of a polyene are of opposite symmetry, an electron-releasing group, which in Epiotis' model donates electron density into the LUMO, will also decrease the preference for allowed electrocyclizations. Epiotis concludes that the Woodward–Hoffmann rules will be reversed for strong donor or strong acceptor substituents. This interesting prediction has not yet been verified by experiment.

Epiotis (1973e) has argued for the same type of reversal of stereoselectivity on the basis of a CI argument: as substituents are placed on the polyene system, the HOMO–LUMO gap decreases, and as the molecule distorts toward a forbidden transition state, this HOMO–LUMO gap will become small enough so that CI between the ground and doubly excited configurations will be important, stabilizing the forbidden transition state. On the other hand, in a more recent treatment, thermal electrocyclizations are predicted to follow the Woodward–Hoffmann allowed route even in polar cases (Epiotis, 1974).

Radical electrocyclizations provide several experimentally verified examples of deviations from stereochemical predictions based on HOMO energy change arguments. Bauld and co-workers have reported the failure of frontier orbital arguments to rationalize the stereochemistries of several electrocyclizations of radical anions (Bauld and Cessac, 1975). In some of these cases, correlation diagrams also give incorrect stereochemical predictions, or disagree with HOMO arguments about the allowed stereochemistry of the reaction.

Woodward and Hoffmann (1965) discussed the electrocyclization of the cyclopropyl radical, and noted that the HOMO motion predicts the conrotatory mode. Longuet-Higgins and Abrahamson (1965) noted that the correlation diagrams for this reaction were equivocal, since the ground states of the reactant and product do not correlate for either stereochemistry. Fukui (1971) noted that dissection of the cyclopropyl radical into localized σ_{CC} and π regions, and use of frontier orbitals of each system, gave an interesting prediction about possible variable cyclopropyl radical electrocyclization stereochemistry. Figure 30 shows the frontier MO's of the opening σ bond and the SOMO (singly occupied MO) of the radical center. HOMO–SOMO interaction will stabilize the disrotatory opening, while LUMO–SOMO interaction will stabilize the conrotatory opening. If both of these interactions are of nearly equal importance, little preference for one stereochemistry will be observed. On the other hand, a donor substituent at the radical center—which increases the LUMO–SOMO interaction—will increase the conrotatory preference, and an electron-withdrawing substituent will increase the disrotatory preference. This prediction has not been experimentally verified.

2. Reactivity

In intermolecular cycloadditions, the relative rates of different cycloadditions may be reasonably well accounted for by frontier orbital arguments, because other energies, such as those due to closed-shell repulsions, bond stretching or compression, etc., may be more or less constant if, for example, the rates of reaction of a series of similar dienes and dienophiles are compared. That is, for such intermolecular reactions, the pre-CI Evans–Polanyi curves are approximately constant for a series of similar reactants, and differences in rates may be attributed to different extents of transition state stabilization by CI. However, for intramolecular reactions, the situation may not be as

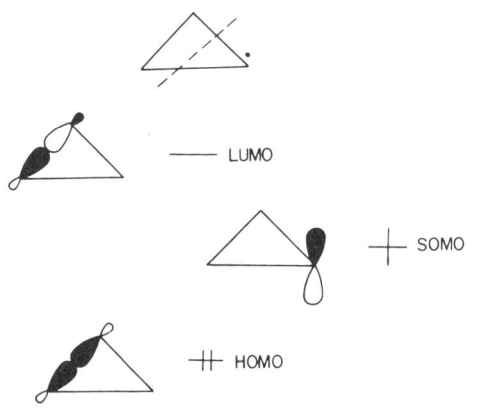

Fig. 30. Frontier molecular orbitals of the cyclopropyl radical.

clear-cut, since the interactions which stabilize the transition state may also stabilize the reactant. This complication is akin to the stabilization of charge-transfer complexes and transition states in intermolecular reactions, but a small amount of stabilization of a charge-transfer complex will not add to the activation energy of the reaction, whereas stabilization of the reactant in an intramolecular reaction will. Nevertheless, certain qualitative guides to reactivity can be gleaned from frontier orbital considerations.

The most rigorous method to treat intramolecular reaction rates would be the Bader–Salem–Pearson method, using force constants to estimate classical distortion energies, and CI between the configurations of the distorted molecules to calculate electronic relaxation. This requires considerable effort, however, and has not been applied extensively to pericyclic reactions.

The only variant of the frontier orbital method used to any significant extent in the literature to treat electrocyclizations is Fukui's method of dissection of the less unsaturated partner into polyene π and σ fragments. After this dissection, the interactions between the frontier MO's of each of these fragments are taken into account. For example, Fig. 11 would be such a treatment if localized σ_{CC}, σ_{CC}^*, π and π^* fragments were used, as shown in Fig. 31. In the ground state reaction of cyclobutene, the interaction between the σ HOMO and the π^* LUMO will be stabilizing if the ring opening is conrotatory. On the other hand, in a photochemical reaction involving the $\pi\pi^*$ excited state, the interaction of the filled σ orbital with the half-filled π orbital will be stabilizing for the disrotatory ring opening. The interaction

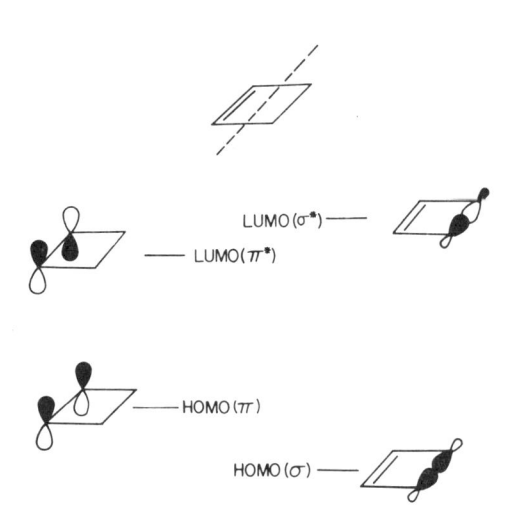

Fig. 31. Illustration of the Fukui dissection method to treat electrocyclizations by frontier molecular orbital methods.

of the vacant σ^* orbital with the half-filled π^* orbital will also be stabilizing for a disrotatory ring opening. This method of dissection also allows one to probe the effects of substituents on reaction rates. For the thermal electrocyclization of cyclobutene, the rate of reaction should be accelerated by substituents which increase either the $\sigma-\pi^*$ interaction, or the $\pi-\sigma^*$ interaction. According to the reasoning discussed earlier for intermolecular reactions, substitution of either electron donors or electron acceptors on the π orbitals will accelerate the rate of reaction. Substituents at the saturated carbons would be expected to have the same effect on reaction rates, assuming the effect on σ orbital energies is similar to their effect on π orbital energies. The greatest acceleration of such an electrocyclization will occur if the σ bond is substituted by donor groups and the π systems by acceptor groups, or vice versa.

In Epiotis' treatment of intramolecular reactions, the forbidden reaction is said to become increasingly favored as electron donors or withdrawing groups are placed on the π system of the unsaturated partners. However, the last paragraph indicates the allowed reaction should be accelerated by electron donor or acceptors more than the forbidden reaction is accelerated. The observation of acceleration of forbidden electrocyclizations by donor or acceptor substituents is most probable only for reactions in which the allowed reaction is prevented by steric effects or geometrical constraints, so that the allowed reaction cannot occur.

Few data are available to compare with this theoretical treatment, but some trends in the literature confirm the basic concepts. Schleyer *et al.* (1972) studied the solvolysis of cyclopropyl tosylates and related compounds. Substitution of β-methyls results in substantial acceleration of the reaction rate, compatible with increased transition state stabilization arising from the better mixing of the donor substituted σ orbital with the LUMO of the breaking C—OTs bond, the protocarbonium ion center. Similarly, β-methylation of *N*-chloroaziridines leads to acceleration of the solvolytic ring openings (Gassman *et al.*, 1970). The classical interpretation of these effects—stabilization of the partially formed allylic cation—is actually quite similar to the HOMO–LUMO arguments, the latter simply delving somewhat further into the physical origin of the stabilization.

In fact, ring openings which give charged products can be interpreted equally well by both classical product stability arguments and frontier MO arguments. For example, the thermal electrocyclizations of oxiranes to form carbonyl ylides, and of aziridines to form azomethine ylides are accelerated by electron-withdrawing groups on carbon (Huisgen, 1971). This is interpreted classically as arising from stabilization of the negatively charged carbons, while the frontier orbital interpretation would attribute this to facilitation of the interaction of the HOMO (lone pair) of the heteroatom subunit

246   K. N. HOUK

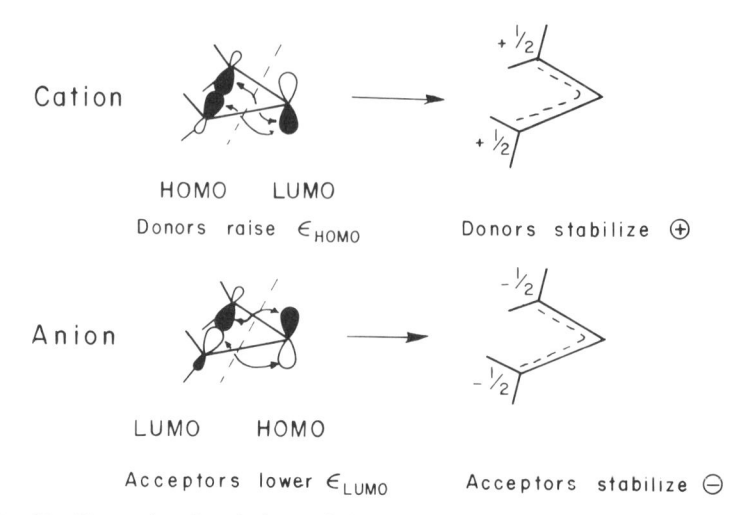

Cation

HOMO   LUMO
Donors raise ϵ_{HOMO} Donors stabilize ⊕

Anion

LUMO   HOMO
Acceptors lower ϵ_{LUMO} Acceptors stabilize ⊖

Fig. 32. Alternative descriptions of the mechanism of acceleration of cyclopropyl cation and anion electrocyclizations.

with the LUMO (σ^*) of the CC bond subunit. These alternative descriptions are compared in Fig. 32.

Similarly, frontier molecular orbital theory will make reactivity predictions similar to those deduced by "classical" considerations of "partial zwitterionic" or "partial diradical" character in electrocyclization transition states. Thus, acceleration of cyclobutene electrocyclizations by 3,4-substitution (e.g., Stephenson and Brauman, 1974) can be considered to result from increased $\sigma-\pi^*$ and $\sigma^*-\pi$ mixing, or by stabilization of the partial radical centers resulting in the transition state of σ bond breaking. More dramatic accelerations are observed by heteroatom substitution. For example, oxetenes undergo electrocyclization to α,β-unsaturated ketones 10^7 faster than the analogous cyclobutenes (Friedrich and Schuster, 1971). Although thermodynamic arguments based on product stabilities have been made, efficient LUMO (σ^*_{CO})–HOMO (π_{CC}) mixing can also be involved. Furthermore, such interactions can also be invoked to explain the much more facile electrocyclization of dienones to 2H-pyrans as compared to the corresponding triene to cyclohexadiene interconversions (Marvell et al., 1972).

The last case is one where considerations of reactivity of the more unsaturated partner in the electrocyclization need to be made. Epiotis (1974) uses the dissection method to indicate that complementary donor–acceptor substitution on the two fragments will facilitate reaction.

An alternative procedure for the treatment of the more unsaturated partner can be developed by inspection of the correlation diagram for the conrotatory electrocyclization of butadiene. Michl (1974) has proposed a similar derivation of this correlation diagram in the reverse direction. Figure 33 shows the correlation diagram built in two parts: first a rotation of orbitals without allowing any interaction between the diene termini. Both filled orbitals are destabilized because of loss of two π bonds, and both vacant orbitals are stabilized because of loss of two anti-bonds. In the second stage, interactions between the different orbitals are permitted. These arguments could, of course, be recast in a CI formalism. Since only orbitals of the same symmetry will interact, ψ_1 and ψ_3 mix and ψ_2 and ψ_4 mix to give the final cyclobutene orbitals. Although performed here in steps, both of these changes will occur simultaneously as rotation occurs. In order to determine how substituents will affect the reaction rate, it is necessary to determine how substituents will affect 1–3 and 2–4 mixing. Substituents which increase these interactions will accelerate the reaction, while those which decrease these interactions will slow down the reaction. These interactions can be facilitated in two ways: by increasing the energy of ψ_1 relative to ψ_3 and of ψ_2 relative to ψ_4, or by

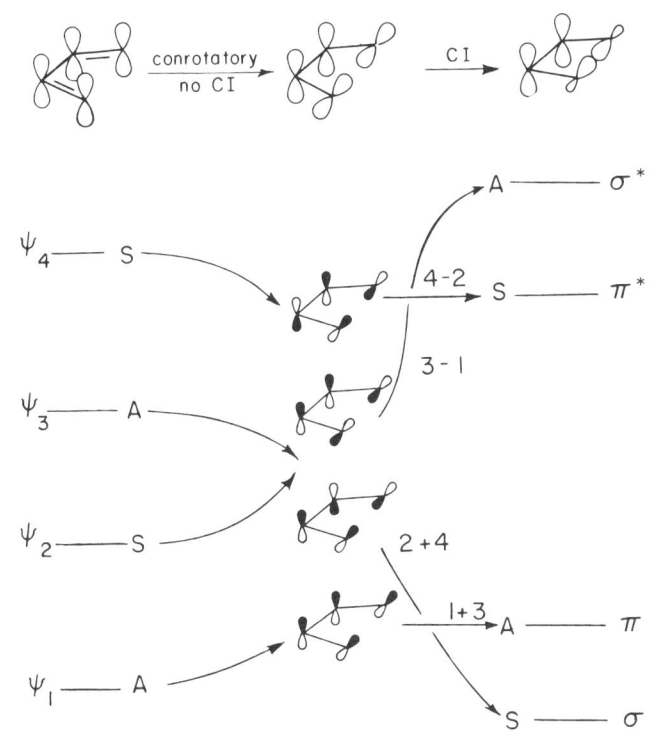

Fig. 33. Stepwise construction of the conrotatory butadiene electrocyclization correlation diagram.

increasing overlap between ψ_1 and ψ_3 or between ψ_2 and ψ_4 by concentrating these orbitals in the same region in space.

Since donors raise occupied orbital energies more than vacant orbital energies, and acceptors lower vacant orbital energies more than filled orbital energies, either type of substitution should be effective in stabilizing the electrocyclization transition state, by facilitating ψ_1–ψ_3, and ψ_2–ψ_4 interactions.

As noted earlier, Epiotis (1974) has discussed extensively the facilitation of forbidden electrocyclizations. The substitution of donor groups on one part of the polyene and an acceptor on the other can facilitate the forbidden reaction. Epiotis cites a number of examples of forbidden, but sterically enforced, disrotatory cyclobutene openings, which are facilitated by unsymmetrical donor–acceptor substitution.

3. Periselectivity

Polyenes containing more than four electrons can, in principle, undergo a number of different electrocyclic reactions. For example, a conjugated triene may undergo either the allowed four electron conrotatory electrocyclization or the allowed six-electron disrotatory electrocyclization in the ground state. Since the various orbitals involved in these two electrocyclizations are similar, the coefficient magnitudes will govern which of these reactions are favorable. As noted in the previous section the use of the fully unsaturated compound in theoretical discussion of reaction rates is far more difficult than the use of the less unsaturated partner, since use of the less saturated partner requires treatment by the full Bader–Salem–Pearson treatment. If the frontier MO energy change dictates the most favored electrocyclic reaction then the allowed one involving the largest number of atoms will be favored, because the terminal HOMO coefficients are largest. Alternatively, arbitrary dissection of the π system into smaller unsaturated fragments can be made, and the reaction treated like a cycloaddition reaction. However, this procedure has certain dangers as noted in the beginning of this section. We are aware of no published treatment of electrocyclization periselectivity by frontier molecular orbital methods, although efforts in this direction are progressing in our laboratory.

C. SIGMATROPIC SHIFTS

1. Stereospecificity

There are several variations on the frontier molecular orbital method which have been used to predict the stereospecificity of sigmatropic shifts. Woodward and Hoffmann (1969) discussed the transition state for sigmatropic

Fig. 34. The nonbonding orbitals of odd conjugated radicals.

shifts by arbitrarily dissecting the transition state into two formal radical fragments. Since all conjugated orbitals with an odd number of carbons have nodal properties as shown in Fig. 34, it is a simple matter to determine the preferred stereochemistry of union of two of these orbitals. If the two radical fragments are joined together in parallel planes with a bonding interaction at one pair of termini, simultaneous bonding at another pair of termini can occur only if the total number of atoms in the cycle is $4n + 2$, where n is an integer. If the stereochemistry at one of the radical fragments is antarafacial, simultaneous bonding at both pair of termini is possible only if the total number of atoms in the cycle is $4n$, where n is an integer. This method uses perturbation theory to directly estimate the energy of the transition state, starting from hypothetical fragments, whereas the methods discussed previously use reactant orbitals to estimate transition state energies.

The latter type of method can be used also. Woodward and Hoffmann (1969) and Fukui (1971) have applied the method of dissection of reactants into isolated systems to treat sigmatropic shifts. Figure 35 shows how both methods can be used to show that the $[1_s,5_s]$-sigmatropic shift is thermally allowed, while the $[1_a,5_s]$ is thermally forbidden. On the left side of Fig. 35, the union of pentadienyl and methyl radical nonbonding molecular orbitals

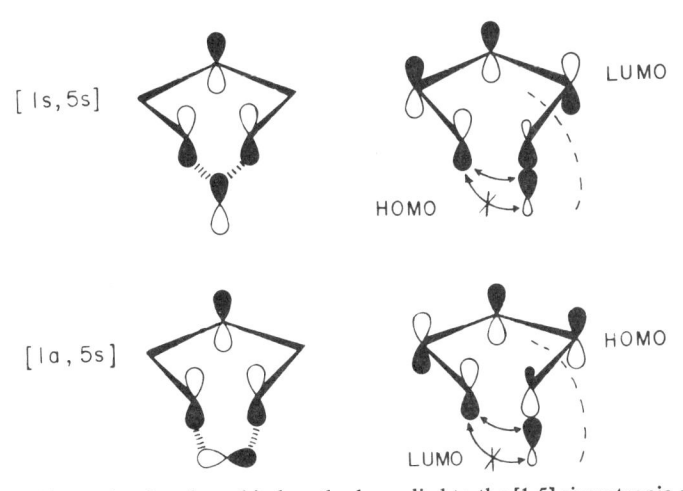

Fig. 35. Alternative frontier orbital methods applied to the [1,5] sigmatropic shifts.

(NBMO's) in [1$_s$,5$_s$] and [1$_a$,5$_s$] geometries are shown. The former leads to a bonding, stabilized transition state, while the latter is a nonbonding, unstabilized transition state. On the right-hand side of the figure, the interactions of the σ HOMO with the diene LUMO, and vice versa, are shown. Clearly, distortion in the [1$_s$,5$_s$] sense leads to favorable HOMO–LUMO mixing, while the [1$_a$,5$_s$] distortion gives no substantial HOMO–LUMO mixing. By either method, the stereochemistry of all sigmatropic shifts can be predicted.

Berson and Salem (1972) have proposed that extra-frontier orbital interactions can make formally forbidden reactions preferrable to stepwise reactions. For example, the three π MO's of the allyl radical and the π MO of a methyl radical are shown in Fig. 36. While the first-order interaction between the NBMO's can stabilize only the [1$_a$,3$_s$] transition state, interactions between the lowest allyl orbital (subjacent MO) and the methyl NBMO, or between the highest allyl orbital (superjacent MO) and the methyl NBMO, can weakly stabilize the [1$_s$,3$_s$] transition state. Several examples compatible with these have been discovered, and a more detailed theoretical treatment has been presented by Berson and co-workers (Berson, 1972).

As for other pericyclic reactions, Epiotis (1974) has suggested that formally forbidden reactions may not only be preferred to dissociation–recombination mechanisms (as Berson and Salem proposed), but may even become favored over allowed reactions if a sufficiently good donor–acceptor combination is involved. In the "union of radicals" approach, if the allyl radical fragment is made sufficiently electron-rich and the methyl radical is made sufficiently electron-deficient by substitution, then a configuration corresponding to a

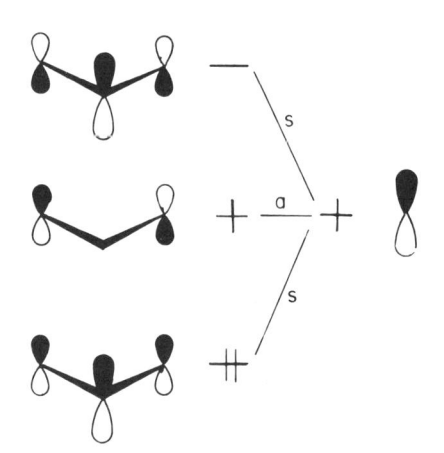

Fig. 36. π MO's of allyl and methyl radicals. The arrows marked "s" are superjacent (top) and subjacent (bottom) orbital interactions which stabilize the [1$_s$,3$_s$] transition state.

substituted allyl cation–methyl anion pair may be lowest in energy. Interactions between the anion HOMO and allyl cation NLUMO can stabilize the [$1_s,3_s$]-sigmatropic shift. However, the stronger methyl-HOMO–allyl-LUMO interaction still stabilizes only the [$1_a,3_s$]-sigmatropic shift. Since the latter will often be sterically disfavored, the concerted [$1_s,3_s$] reaction may be favored overall.

These considerations may be extended to other sigmatropic shifts. In [1,X]-sigmatropic shifts of a methyl group or related one carbon π systems, the first-order interactions between the radical fragment NBMO's always stabilize the Woodward–Hoffmann allowed transition states, and do not stabilize the forbidden transition states. However, second-order interactions between the one-carbon fragment and subjacent or superjacent orbitals will always stabilize forbidden processes. The general electronic reactivity order for sigmatropic shifts will be that proposed by Berson (1972); allowed > forbidden > stepwise.

As noted by Epiotis (1974), the allowed and forbidden reactions become more nearly equally favored for polar cases, and steric or geometrical constraints may make the forbidden reaction the most rapid.

How can these subjacent orbital effects be seen in the reactant dissection version of frontier MO theory? In the theoretical treatment of a hypothetical 1,3-methyl shift in 1-butene, the relevant MO's are the π and π^* alkene orbitals, and the σ and σ^* CH_2—CH_3 orbitals. HOMO–LUMO interactions between these systems can strongly stabilize only the allowed [$1_a,3_s$]-sigmatropic shift; however, the [$1_s,3_s$]-sigmatropic shift is not without stabilization, because the overlap at the two interacting centers are different. Thus, the dissection method would also indicate that a forbidden mechanism may be preferable to a dissociation–recombination mechanism, but the explanation of such an effect is less straightforward than by the subjacent orbital description.

2. Reactivity

Just as in treatments of stereoselectivity, there are two alternative methods by which reactivity in sigmatropic shifts may be predicted. One is to use the two odd radical fragments which make up the transition state of the reaction, and to consider the effect of substituents on the interactions of these two fragments. The other method utilizes a dissection of the reactants into σ and π fragments, and a consideration of the effect of substituents on interaction of the frontier orbitals of the two fragments. As with the treatment of electrocyclizations, there is some difficulty arising from the fact that similar interactions may lead to both ground state and transition state stabilization. However, substituent effects on transition states will probably invariably be larger than those on ground states, so no real difficulty—other than in attempted quantitative treatments—may arise.

The radical union approach has been discussed extensively by Epiotis (1974). In this method, the degenerate interaction of NBMO's leads to stabilization of allowed transition states, while subjacent interactions or superjacent stabilize forbidden transition states. Substitution of one fragment by a donor, and the other by an acceptor will decrease the extent of interaction of the original NBMO's—and decrease the stabilization of the allowed transition state. At the same time, either subjacent or superjacent interactions will be increased, leading to greater stabilization of the forbidden transition state.

Interestingly, the prediction based on the reactant dissection method is quite contrary to the radical union prediction. In the dissection method the σ(HOMO)–π*(LUMO) and π(HOMO)–σ*(LUMO) interactions facilitate reaction, and will be favored by donor substituents on the σ bond and acceptor substituents on the π bond, or vice versa.

Available experimental evidence seems most easily interpreted in terms of this latter approach. For example, the rates of [1$_a$,3$_s$]-sigmatropic shifts of bicyclo[3.1.0]pentenes increase as better donor substituents are placed on

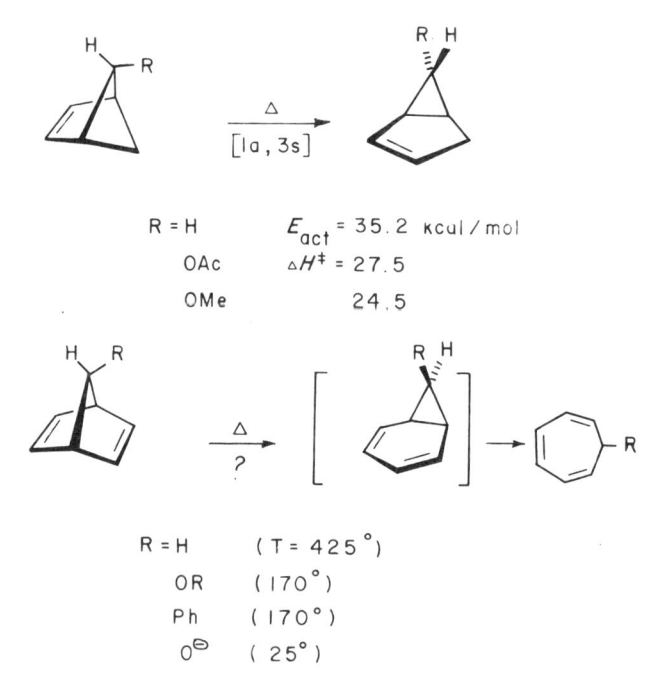

R = H E_{act} = 35.2 kcal/mol
OAc $\Delta H^{\ddagger}$ = 27.5
OMe   24.5

R = H   (T = 425°)
OR   (170°)
Ph   (170°)
O$^{\ominus}$ (25°)

Fig. 37. The effect of substituents on the rates of [1$_a$,3$_s$]-sigmatropic shifts.

one terminus of the migrating σ bond. Figure 37 shows two reactions, the first a documented [1$_a$,3$_s$]-sigmatropic shift (Scheidt and Kirmse, 1972), the second a probable [1$_a$,3$_s$]-sigmatropic shift (Franzus *et al.*, 1971). The activation energies or reaction temperatures decrease rapidly as the donor substituent on the migrating double bond increases, compatible with strengthened σ(HOMO)–π*(LUMO) interactions.

The rates of the allowed, six-electron [1$_s$,4$_s$]-sigmatropic shift reported in Fig. 38 (A) are linearly related to Hammett σ's for the benzyl substituent ($\rho = -0.3$) compatible with a dominant interaction between the nitrone HOMO and the σ*LUMO of the migrating σ bond (Schöllkopf and Hoppe, 1972). The rapidity of [1$_s$,5$_s$]-sigmatropic shifts of the silyl group in 5-tri-

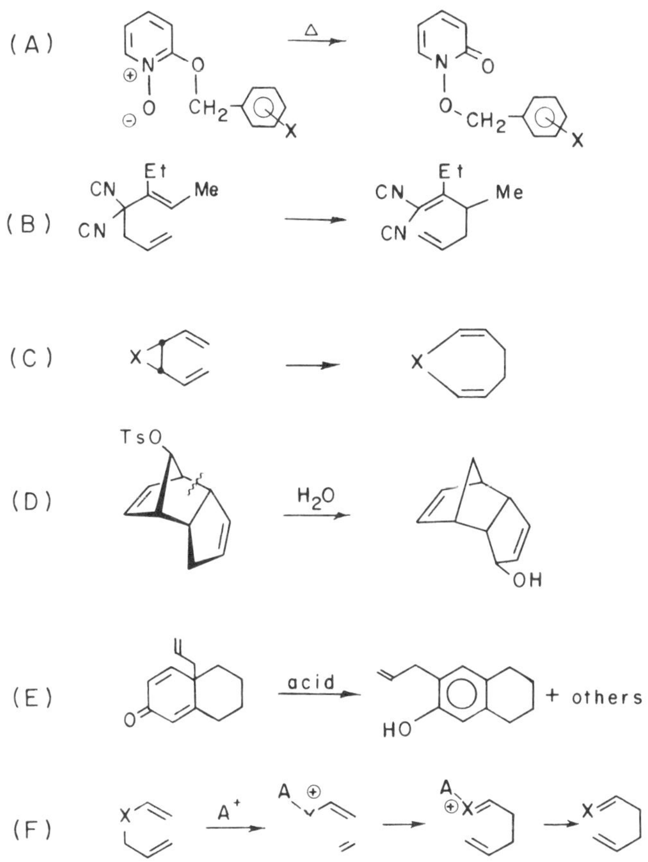

Fig. 38. Sigmatropic shifts providing evidence for frontier orbital control of reaction rate.

methylsilylcyclopentadienes (10^5 faster than hydrogen shifts; Ashe, 1970) can be attributed to the high energy of the Si–C σ orbital. There will be, however, a general difficulty in prediction of relative "migratory aptitudes" of groups in sigmatropic shifts, since a group may have an effect on both the migrating σ bond and on the π system.

Extensive data are available on rates of various [3,3]-sigmatropic shifts, although no systematic study encompassing the whole spectrum of substituents has been carried out. A few of the available examples are worthwhile to show how frontier MO theory can be applied. Sunko and co-workers have found that the Cope rearrangement shown in Fig. 38 (B) has an activation energy of 25.8 kcal/mole, much lower than the $\Delta H^\ddagger = 33$ kcal/mole measured for the Cope rearrangement of 1,1,6,6,-tetradeuterio-1,5-hexadiene (Humski et al., 1969; Malojčič et al., 1969). This indicates a favorable homodiene-HOMO–σ-LUMO interaction, whereas the radical union treatment would likely suggest little acceleration of such a substituted compound.

The homo-1,5-diene system shown in Fig. 38 (C) is interesting, since a variation in X has a profound influence on the rate of the Cope rearrangement. The activation enthalpy for the Cope rearrangement in the boat transition state (X = H,H) has been estimated as 41 kcal/mole, whereas the corresponding cyclopropyl case (X = CH_2) has an activation enthalpy of about 21 kcal/mole (Doering et al., 1968). The much lower activation energy of the latter and of the cyclobutyl analog (X = CH_2CH_2, Hammond and de Boer, 1969), can be attributed to the far greater donor ability of the cyclopropyl group. The qualitative rates of rearrangement of the heteroanalogs (X = CH_2 > NH $\gg$ O > S; Paquette and Haluska, 1970) can be attributed to the varying donor abilities of the Walsh-type occupied orbitals of these molecules (Mollere and Houk, 1977). Also compatible with this rationale is the great acceleration provided the X = CH_2 case when an oxygen replaces one of the vinyl terminal CH_2 groups (Rhoads and Cockcroft, 1969) and the increased activation energy for the X = CH_2 case as the vinyl groups are alkylated (Pickenhagen et al., 1973).

Other phenomena, such as the "solvolytic Cope" and Lewis-acid catalyzed Cope rearrangements can be treated in a similar fashion. Breslow and Hoffmann (1972) found that solvolysis of the tosylate shown in Fig. 38 (D) gives a rearranged product, and that a Cope rearrangement of the intermediate carbonium ion must occur much more rapidly than for the covalent reactant. The breaking σ bond (marked 〜〜〜) is well situated for overlap with the vacant p orbital in the intermediate cation. Greatly enhanced σ^*(LUMO)–π(HOMO) interactions are expected.

Schmid and co-workers have studied acid-catalyzed rearrangements such as that shown in Fig. 38 (E) and have reviewed Lewis acid catalysis of a variety of hetero-Cope rearrangements (Widmer et al., 1973). They introduced the concept of "charge-induced" sigmatropic shifts—in which the transition state energy is lowered because of positive charge delocalization. As we have discussed in relation to acid-catalyzed cycloadditions (Houk and Strozier, 1973), coordination of a heteroatom lone pair with a Lewis acid will have a pronounced lowering effect on the vacant π MO, and σ orbitals involving the heteroatom should be lowered also. Thus, HOMO–LUMO interactions can be greatly facilitated. In the Schmid example, the σ(HOMO)–π*(LUMO) interactions will be facilitated, while in cases such as Fig. 38 (F), the π(HOMO) –σ*(LUMO) interactions are facilitated.

Epiotis (1974) has commented on the experimental evidence for facilitation of forbidden sigmatropic shifts by donor–acceptor substitution. One of the more remarkable examples is the report by Baldwin and Andrist (1971) that the activation energies of a series of [1,3]-sigmatropic shifts are linearly related to the frequency of the lowest energy UV transitions in these molecules. Some of the examples are constrained to rearrange via a [1$_s$,3$_s$] geometry (if concerted) while others may choose either [1$_s$,3$_s$] or [1$_a$,3$_s$] pathways. The phenomenon is qualitatively compatible with the Bader–Salem–Pearson treatment of rearrangements where the facility of a particular molecular distortion will depend upon the presence of a low-lying excited state which can mix with, and stabilize, the distorted ground state.

3. Periselectivity

No detailed treatments of periselectivity in sigmatropic shifts have appeared. Although steric and conformational effects will be important factors in determining which sigmatropic shifts are geometrically possible, frontier molecular orbital theory indicates that the allowed sigmatropic shift involving the greatest number of electrons should be favored electronically in cases where two long polyene fragments are joined to a common σ bond. This prediction arises from the fact that the largest coefficients in the HOMO's and LUMO's of polyenes are at the terminal carbons, so that overlap should be best when sigmatropic shifts occur to the termini of the polyene systems. Directing effects of different types of substituents in different positions are currently under study in our laboratories.

D. CHELETROPIC REACTIONS AND ELIMINATIONS

The treatment of cheletropic reactions follows exactly that of cycloadditions, although additional subtleties arise from the fact that the fragment to which two bonds are made or broken often has a high-lying HOMO and low-lying LUMO in perpendicular orbitals. Figure 39 shows the preferred geometry of approach for such a species adding in suprafacial sense to the two classes of polyenes. In addition to these geometries, cheletropic reactions may occur in an antarafacial sense to a polyene, but the frontier orbital treatment of those cases is a simple extension of the treatment of suprafacial reactions. Inspection of Fig. 39 shows that for $4n$ polyenes, the "linear" cheletropic reactions are favored by alkene-LUMO–fragment-HOMO interactions, and to a smaller extent by alkene-HOMO–fragment-LUMO interactions. The latter may lead to some preference for an asymmetrical geometry approaching the $[1_s + 1_s]$. For $4n + 2$ polyenes, the "nonlinear" cheletropic reaction is favored by alkene-HOMO–fragment-LUMO interaction, while the alkene-LUMO–fragment-HOMO interaction will prefer the stepwise $[1_s + 1_s]$ geometry over the "nonlinear" geometry due to the poor overlap in the latter transition state (for discussions of "linear" and "nonlinear" cheletropic reactions, see Woodward and Hoffmann, 1969; Houk, 1973a).

The understanding of cheletropic reactions is hindered by the absence of experimental tests for transition state geometries. Only alkene stereochemical evidence gives a clue to the geometry at that fragment. At the current time, none of the transition states shown in Fig. 39 are established. What is established however, is that fragments expelled from molecules in cheletropic processes, or which undergo reactions with polyenes, generally are quite electrophilic, possessing a low-lying LUMO (e.g., CH_2, SO_2, N_2O, etc.), and the cheletropic reactions studied so far involve electron-rich alkene fragments. According to Fig. 39, differences in stereoselectivity will be observed with electron-deficient alkene fragments.

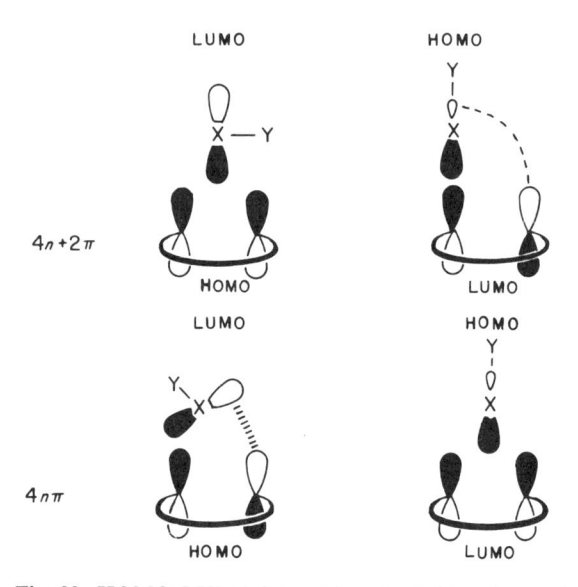

Fig. 39. HOMO–LUMO interactions in cheletropic reactions.

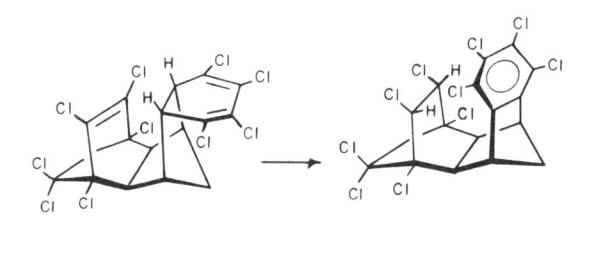

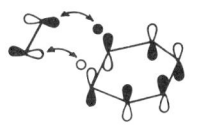

Fig. 40. A group transfer from an electron-rich system to an electron-deficient system.

Group transfers, such as dehydrogenations, generally occur from an electron-rich system to an electron-deficient one. A novel intramolecular example has been reported by MacKenzie (1969) and is shown in Fig. 40. The relevant HOMO and LUMO fragments are shown below the reaction. Dehydrogenation agents such as dichlorodicyanobenzoquinone (DDQ) may react by a similar mechanism, and the acceleration of DDQ dehydrogenations by electron releasing groups on the hydrogen donor (Stoos and Roček, 1972) is compatible with HOMO–LUMO interactions such as those shown in Fig. 40.

E. Miscellaneozations

This jocular name for reactions not classifiable in the Woodward–Hoffmann glossary encompasses a number of intramolecular reactions, usually involving more than one σ bond. Various names have been proposed for such reactions based on analogies to Woodward–Hoffmann terms—such as the 1,5-homodienyl shift for the "reverse ene" reaction. Reactions of this type are most easily treated by the dissection method. One example of this type of treatment is given here as an indication of the utility of such treatments: Gilbert and co-workers (1972) have reported that the 1,5-homodienyl shift of the compound in Fig. 41 is less favorable that the 1,7-homotrienyl shift. At 130°, the ratio of products obtained from the 1,7- and 1,5-shifts is 10:1, respectively. The authors point out the former transition state has more conjugation than the latter. A simplified frontier MO version of this is outlined in Fig. 41. The dissection into HOMO's and LUMO's in different ways leads to the rationalization of the periselectivity. Conjugation of the π orbital with the adjacent σ

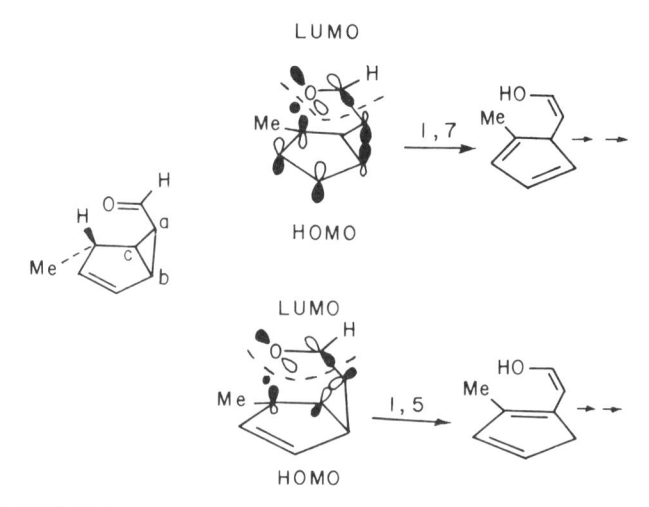

Fig. 41. HOMO–LUMO interactions involved in the competition between 1,5 and 1,7 shifts.

orbitals will result in a higher energy HOMO for the conjugated system [and thus better HOMO–LUMO (π^*_{CO}) mixing] than for the nonconjugated. In the real molecule, the orbitals will be delocalized, but more highly localized on the a–b bond then the a–c bond. HOMO–LUMO interactions will weaken the a–b bond more than the a–c bond.

Ene reactions constitute another class of reactions easily analyzed in terms of the HOMO of the allyl component and LUMO on the enophile (Hoffmann, 1969).

VII. Photochemical Pericyclic Reactions

Upon excitation of a molecule to an excited state, numerous inter- and intramolecular reactions can ensue. In general, the theoretical treatment of photochemical reactions is considerably more difficult than the treatment of thermal reactions, since, in addition to the normal problem of the determination of the favored nuclear distortion, one must be concerned with competing rates of emission or radiationless deactivation to the ground state, and internal conversion and intersystem crossing to different excited states. A complete theoretical description of a photochemical reaction must account for the relative rates of all of these processes, while at the present state of development, frontier molecular orbital theory is only capable of explaining the relative facility of various nuclear distortions which are favored beginning from a specific excited state. Nevertheless, a number of important insights into photochemical reaction rates and products have come from application of frontier molecular orbital theory. Before discussing these, a few general comments about photochemical reactions are in order.

The Woodward–Hoffmann rules state that thermally forbidden reactions are photochemically allowed (Woodward and Hoffmann, 1969). In the correlation diagram derivation of the Woodward–Hoffmann rules, this arises from the fact that the lowest excited states mutually correlate for reactions which are thermally forbidden. Dougherty (1971) has discussed various types of photochemical reactions in terms of the aromaticity (Dewar–Zimmerman) approach. Allowed pericyclic reactions have aromatic ground-state transition states and antiaromatic (destabilized) excited transition states, while ground-state forbidden reactions have antiaromatic transition states and aromatic excited transition states. Photochemical pericyclic reactions are "G-type" reactions, involving intended crossings of ground and excited surfaces, and products are formed directly in the ground state, the electronic relaxation being identical to the bond changes leading to reaction.

Michl (1972, 1974) has discussed photochemical pericyclic reactions in similar terms. Ground state forbidden reactions have large ground-state barriers caused by intended correlations between the ground states of reactant with a high-lying excited state of the product. Figure 42 shows this schematically for a $[_\pi 2_s + _\pi 2_s]$ cycloaddition of two ethylenes. The ground configurations of reactant and product each correlate with doubly excited configurations. However, in transition state regions, these configurations mix strongly, leading to a maximum in the ground-state energy curve, and a minimum in an excited surface. The $\pi\pi^*$ singlet state of the reactant correlates with the $\sigma\sigma^*$

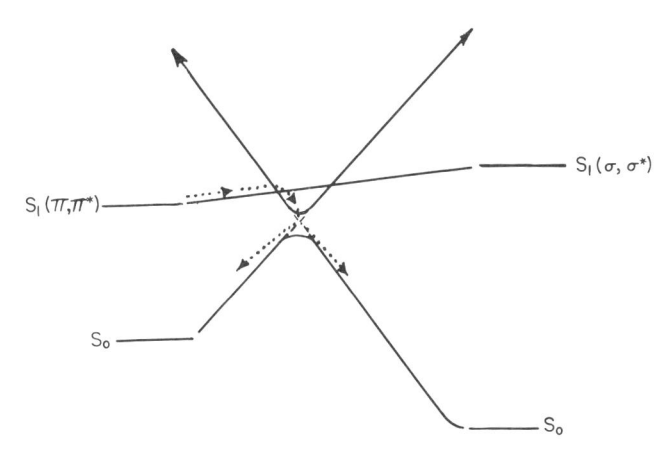

Fig. 42. State correlation diagram for the $[_\pi 2_s + _\pi 2_s]$ cycloaddition of two ethylenes.

singlet state of the product. In a dynamic sense, the lowest excited singlet of reactant is expected to follow the arrows shown in Fig. 42. The reactants remain on the lowest excited state surface until the excited and ground surfaces come very close in energy. At this point, essentially instantaneous internal conversion to the ground surface will occur. The intended crossing creates, in Michl's words, a "funnel," from which passage from the excited state to the ground state surface occurs extremely rapidly. This description is general for reactions which are thermally forbidden. Such reactions are typically, although not invariably, diabatic, beginning on the excited state surface and ending on the ground state surface. Excited state surfaces may involve barriers just like ground surfaces, and the first task of theory is to identify the lowest energy pathway to a funnel, from which deposition on the ground state surface, and subsequent nuclear reorganization to a stable product, may occur.

A. Cycloadditions

Herndon and Fukui and their co-workers pioneered the use of frontier molecular orbital theory for the treatment of photochemical cycloadditions, and many of these applications have been reviewed recently (Herndon, 1972; Herndon *et al.*, 1974). For the addition of an excited ethylene molecule to a ground-state ethylene, the frontier orbital interactions are shown in Fig. 43. The interactions of the single occupied molecular orbitals (SOMO's) of the excited ethylene with the corresponding doubly occupied and vacant orbitals of the unexcited ethylene are both first-order interactions and are symmetry

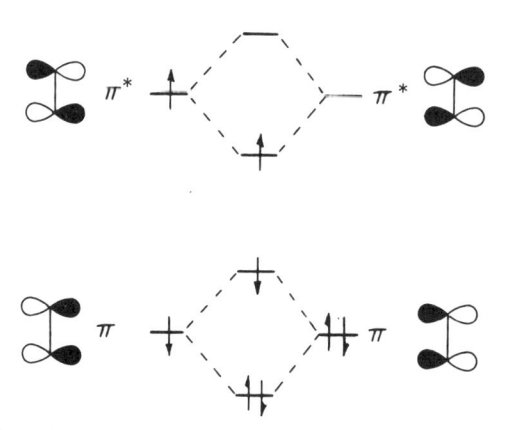

Fig. 43. Frontier orbital interactions between an excited ethylene and a ground state ethylene.

allowed. This will always be the case if the corresponding ground state cycloaddition is symmetry forbidden due to a lack of HOMO–LUMO symmetry matching. By contrast, thermally allowed reactions are those in which the symmetry of the HOMO of each addend matches the symmetry of the LUMO of the other, and these reactions will not have stabilizing first-order interactions between the SOMO's of the excited species and the corresponding orbitals of the ground state addend. Thus, ground state allowed reactions will not be favorable photochemically.

The configuration represented by the interaction diagram in Fig. 43 is basically that known spectroscopically as an exciplex. There is no net charge transfer from one molecule to the other. Since the orbital interactions between the two SOMO's of one molecule and the HOMO and LUMO of the second are nearly first-order interactions, consideration of only these frontier orbital interactions should be an excellent approximation.

Regioselectivity in photocycloadditions can be readily explained in terms of this model. In the dimerization of any substituted alkene, the head-to-head (z) regiochemistry will always be favored, since this will permit maximum overlap of the relevant orbitals. Salem (1968) has applied the more complete perturbation treatment to the dimerization of butadiene, and showed the (z) [2 + 2]-adduct and the [4 + 4]-adducts are electronically most favored.

For the [2 + 2] photocycloaddition of two different alkenes, relative reactivities as well as regioselectivities are expected to change as the substituents on the two alkenes are varied. Figure 44 shows the interaction diagram for

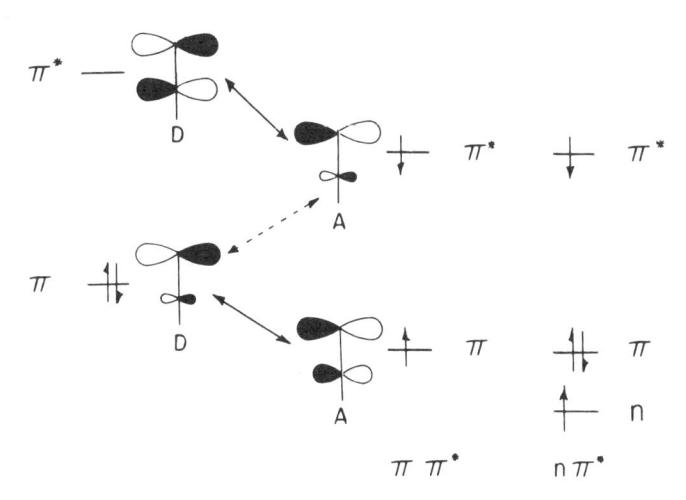

Fig. 44. Frontier orbital interactions for the photochemical reaction of an electron-rich and an electron-deficient alkene.

the photochemical reaction of a donor- and an acceptor-substituted alkene. The interactions stabilizing the [$2_s + 2_s$] transition state are diminished, and rates of cycloaddition or stabilities of intermediate exciplexes would be expected to decrease on this basis. However, a new interaction of a charge-transfer nature may become important (dashed line in Fig. 44). This interaction can best stabilize a [$2_s + 2_a$] geometry, and if this interaction becomes most important, then a change in stereochemistry of reaction can occur. A more efficient way to preferentially stabilize the [$2_s + 2_a$] geometry is if an $n\pi^*$ state of the electron-deficient alkene is involved. As pointed out by Epiotis (1974), and as shown on the right side of Fig. 44, whether the [$2_s + 2_s$] or the [$2_s + 2_a$] geometry is favored will depend upon the relative energies of the donor π and π^* orbitals, and the acceptor (e.g., α,β-unsaturated ketone) π^* orbital. The π–π interaction is absent for such a case, and the [$2_s + 2_s$] stereochemistry will be favored if the $\pi_D{}^*$–$\pi_A{}^*$ separation is smallest, while the [$2_s + 2_a$] geometry will be favored if the π_D–$\pi_A{}^*$ separation is smallest. Because of the small difference in the magnitude of the $\pi_D{}^*$ terminal coefficients, the former case is expected to give low regioselectivity, but the latter will give high regioselectivity due to the large polarization of the π_D and $\pi_A{}^*$ orbitals.

The variable stereochemistry and reactivity in Paterno–Büchi reactions (photochemical oxetane formation from an alkene and a carbonyl compound) were first discussed in terms of frontier MO theory by Herndon and Giles (1970) and have been discussed more recently by Epiotis (1972d, 1974) and Herndon et al. (1974). Typical frontier MO diagrams for an $n\pi^*$ excited state of a carbonyl compound and ground state electron-rich and electron-deficient alkenes are shown in Fig. 45. The reaction of an $n\pi^*$ excited state with an electron rich alkene will be controlled by interaction of the n SOMO and the nearly degenerate alkene HOMO. Bond formation between oxygen and the less substituted alkene carbon will lead to a biradical, which will collapse to oxetane. By contrast, electron-deficient alkenes will interact with the ketone mainly by an alkene-LUMO–carbonyl-π^*-SOMO interaction, and a [$2_s + 2_s$] geometry will be favored, with the substituted carbon attached to oxygen. Experimental support for these ideas can be found in the Herndon and Epiotis paper cited earlier.

Photochemical Diels–Alder reactions are relatively rare, but several cases of reactions which appear to be concerted are known. Herndon and Giles (1969) carried out perturbation calculations on the reaction of benzoquinone and related quinones with alkenes. They found that the concerted [$4_s + 2_s$] transition state was stabilized for both the thermal and photochemical reactions. Epiotis and Yates (1974) discussed the photochemical Diels–Alder reaction more generally, and using a correlation diagram approach, showed that concerted photochemical [$_\pi 4_s + {}_\pi 2_s$] cycloadditions were possible for

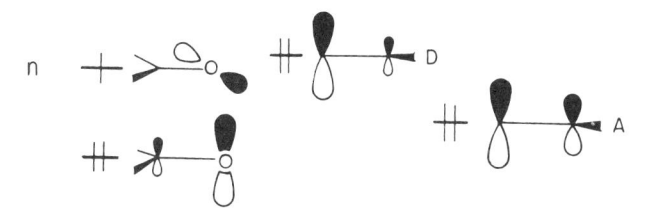

Fig. 45. Frontier MO's used in discussion of the Paterno–Büchi reaction.

good donor–acceptor pairs. A number of experimental examples were cited in support of this argument. Epiotis (1974) rephrased these arguments in terms of the CI formalism. In the frontier orbital model, the interaction between the SOMO of the excited diene (corresponding to the HOMO of the unexcited molecule) and the electron-deficient alkene LUMO can stabilize the $[4_s + 2_s]$ geometry. In Epiotis' CI language, for a very polar pair, the charge-transfer configuration is lowest in energy, so the excited state of this complex is the one corresponding in electron occupation to the ground state of the reactants. Such a configuration is stabilized as usual by HOMO–LUMO interactions.

B. Electrocyclizations

The photochemical electrocyclization of butadiene to cyclobutene is one of the most thoroughly studied reactions, from the theoretical point of view. As the prototype of photochemical electrocyclizations, the reaction provides the key to understanding of other photochemical electrocyclizations. Oosterhoff (1961) suggested that the stereospecificity of photochemical electrocyclizations observed in the vitamin D series corresponded to the symmetry of the HOMO of the π system. Woodward and Hoffmann's (1965) first paper in the field of orbital symmetry pointed out that all thermal and photochemical electrocyclizations occurred in the direction in which the HOMO of the molecule was stabilized.

It was shown later by the correlation diagram approach that photochemically allowed reactions were ones in which there was a correlation between the lowest excited states of the reactant and product. Soon thereafter, it became apparent that the correspondence between photochemical allowedness and thermal forbiddenness was crucial, and that the intended crossing of ground and excited states in thermally forbidden reactions leads to a minimum in the excited states surface which comes very close to the ground state surface. As shown in Fig. 42, the photochemically allowed reactions occur by partial rotation in the excited state, deposition near the forbidden reaction transition state and conversion to reactant or product ground state. A lucid discussion, and review of theoretical efforts, has been given by Michl (1972, 1974).

The frontier molecular orbital theory can be used to detect the presence of barriers in the ground-state surface and minima in the excited state surface. Whenever there is a HOMO–LUMO crossing along a particular reaction coordinate, there will be an intended crossing of the ground and excited states, resulting in the presence of a "funnel" on the excited state surface.

The easiest way to treat photochemical electrocyclizations by frontier orbital methods is to divide the molecule into several fragments, and to consider interaction between the SOMO's of one excited fragment and the HOMO and LUMO of the second fragment. For reactions not involving particularly polar groups, such a treatment will invariably give predictions of stereochemistry opposite to those obtained for the ground state.

While this method is straightforward when applied to the less unsaturated partner in an electrocyclization, it is not so for the more unsaturated partner. Furthermore, whereas thermal reactions can be treated in either direction because of the principle of microscopic reversibility, no such principle applies to photochemical reactions, and different "funnels" may be passed through depending on which reactant is excited.

For such cases, Fukui (1971) and Epiotis (1974) have proposed dissection into two polyene fragments, and consideration of appropriate SOMO–HOMO and SOMO–LUMO interactions. The Woodward–Hoffmann rules for photochemical electrocyclizations can be generated in this way. Epiotis (1974) has generated a prediction different from Woodward and Hoffmann's using the CI formalism. Polar photoelectrocyclizations are predicted to occur in a conrotatory fashion for $4n$ electron systems, and in a disrotatory fashion for $4n + 2$ electron systems. This results from the fact that the excited state of such a polar case is like the ground state of a nonpolar electrocyclization.

An alternative frontier orbital method to treat such reactions focuses on the stereochemistry necessary to give a "funnel" in the excited state. Since for the case of the butadiene electrocyclization, this funnel, is produced by an intended crossing of the butadiene $\psi_1{}^2\psi_3{}^2$ configuration with the ground $\psi_1{}^2\psi_2{}^2$ configuration, tracing the energy of the ψ_3 orbital will reveal for which stereochemistry the funnel exists. This method reproduces the Woodward–Hoffmann rules, and, in principle, is applicable to periselectivity in photochemical reactions. Such applications have yet to appear in the literature.

C. Sigmatropic Shifts

The principal applications of frontier orbital methods to photochemical sigmatropic shifts are Fukui's (1971) dissections and considerations of SOMO–HOMO and SOMO–LUMO interactions, as well as Epiotis' (1974) discussions. Reactivity and periselectivity have not been discussed generally.

D. Cheletropic Reactions

As for thermal reactions, photochemical cheletropic reactions can be discussed in exactly the same way as cycloadditions. Qualitative SOMO–HOMO and SOMO–LUMO diagrams have been discussed for a few cases by Fukui (1971).

E. Miscellaneozations

There are a number of photochemical reactions of the unclassifiable type, prominent among which are di-π-methane and oxa-di-π-methane rearrangements. We have described a method of tracing differences between the bond orders in the excited states and ground states of β,γ-unsaturated carbonyl compounds to explain differences in singlet and triplet reactivity (Houk et al., 1972; Houk, 1976). A simpler "dissection" version of these arguments is given in Fig. 46. The diagram on the right shows the interaction between the de-

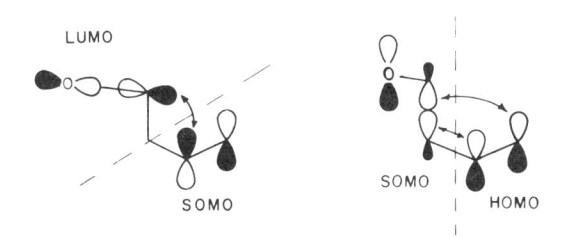

Fig. 46. Frontier orbital interactions controlling reactions of triplet ($\pi\pi^*$) and singlet ($n\pi^*$) states of β,γ-unsaturated carbonyl compounds.

localized n SOMO and the alkene π HOMO of the $n\pi^*$ singlet. These interactions favor a 1,3-shift. By contrast, the $\pi\pi^*$ triplet has dominant alkene-π^*-SOMO–carbonyl-π^*-LUMO interactions which favor carbonyl-carbon–alkene-β-bonding and lead, ultimately, to a 1,2-shift or oxa-di-π-methane rearrangement.

The regioselectivity of photochemical di-π-methane rearrangements of substituted benzonorbornadienes has recently been explained by a frontier molecular orbital model. Aryl substituents split the benzene HOMO and LUMO degeneracies and perturb the π coefficients. The resulting interactions of the singly occupied aryl MO's in the triplet state with the HOMO and LUMO of the ethylene moiety dictate the site of primary bonding (Houk and Santiago, 1976).

VIII. Conclusion

An attempt has been made in this article to outline the application of frontier molecular orbital theory to the field of pericyclic reactions. The simple theory provides a qualitative guide to stereoselectivity, regioselectivity, periselectivity, and reactivity phenomena, and, with time, it is certain that the quantitative applications that have begun to be made in the area of cycloadditions will be extended to other types of pericyclic reactions.

Acknowledgments

I am pleased to acknowledge helpful discussions and correspondence with Professor W. C. Herndon, J. Michl, and N. D. Epiotis, the pervading influence of R. Hoffmann and R. B. Woodward, and critical reading of this manuscript by Drs. P. Caramella, C. Santiago and Y.-M. Chang. Financial support of my research in this area and related areas of chemistry has been generously provided by the Research Corporation, the Donors of the Petroleum Research Fund, administered by the American Chemical Society, the National Science Foundation, the National Institutes of Health, the Camille and Henry Dreyfus Foundation, and the Alfred P. Sloan Foundation. I am indebted to the publisher of *Accounts of Chemical Research* for permission to reproduce Figures 15 and 19.

References

Alston, P. V., and Ottenbrite, R. M. (1975). *J. Org. Chem.* **40**, 111.
Alston, P. V., and Shillady, D. D. (1974). *J. Org. Chem.* **39**, 3402.
Alston, P. V., Ottenbrite, R. M., and Shillady, D. D. (1973). *J. Org. Chem.* **38**, 4075.

Anh, N. T., and Seyden-Penne, J. (1973). *Tetrahedron* **29**, 3259.
Ashe, A. J., III. (1970). *J. Am. Chem. Soc.* **92**, 1233.
Bader, R. F. W. (1960). *Mol. Phys.* **3**, 137.
Bader, R. F. W. (1962). *Can. J. Chem.* **40**, 1164.
Baldwin, J. E., and Andrist, A. H. (1971). *J. Am. Chem. Soc.* **93**, 3289.
Bartlett, P. D. (1968). *Science* **159**, 833.
Bartlett, P. D., and Schueller, K. E. (1968). *J. Am. Chem. Soc.* **90**, 6071.
Bast, K., Christl, M., Huisgen, R., and Mack, W. (1973). *Chem. Ber.* **106**, 3312.
Bastide, J., and Henri-Rousseau, O. (1973). *Bull. Soc. Chim.* p. 2294.
Bastide, J., El Ghandour, N., and Henri-Rousseau, O. (1972). *Tetrahedron Lett.* p. 4225.
Bastide, J., El Ghandour, N., and Henri-Rousseau, O. (1973). *Bull. Soc. Chim. Fr.* p. 2290.
Bauld, N. L., and Cessac, J. (1975). *J. Am. Chem. Soc.* **97**, 2284 (and references therein).
Berson, J. A. (1972). *Acc. Chem. Res.* **5**, 406.
Berson, J. A., and Salem, L. (1972). *J. Am. Chem. Soc.* **94**. 8917.
Bertrán, J., Carbó, R., and Moret, T. (1971). *An. Quim.* **67**, 489.
Breslow, R., and Hoffmann, J. M., Jr. (1972). *J. Am. Chem. Soc.* **94**, 2111.
Briegleb, G. (1964). *Angew. Chem., Int. Ed. Engl.* **3**, 617.
Brown, R. D. (1959). *J. Chem. Soc.* pp. 2224 and 2232.
Caramella, P., and Cellerino, G. (1974). *Tetrahedron Lett.* p. 229.
Caramella, P., and Houk, K. N. (1976). *J. Am. Chem. Soc.* **98**, 6397.
Caramella, P., Gandour, R. W., Hall, J. A., Deville, C. G., and Houk, K. N. (1977). *J. Am. Chem. Soc.* **99**, 385.
Christl, M., and Huisgen, R. (1973). *Chem. Ber.* **106**, 3345 (1973).
Coulson, C. A., and Longuet-Higgins, H. C. (1947a). *Proc. R. Soc. London, Ser. A* **191**, 39.
Coulson, C. A., and Longuet-Higgins, H. C. (1947b). *Proc. R. Soc. London, Ser. A* **192**, 16.
Coulson, C. A., and Longuet-Higgins, H. C. (1948a). *Proc. R. Soc. London, Ser. A* **193**, 447 and 456.
Coulson, C. A., and Longuet-Higgins, H. C. (1948b). *Proc. R. Soc. London, Ser. A* **195**, 188.
Devaquet, A. and Salem, L. (1969). *J. Am. Chem. Soc.* **91**, 3793.
Dewar, M. J. S. (1952). *J. Am. Chem. Soc.* **74**, 3341, 3345, 3350, 3353, and 3357.
Dewar, M. J. S. (1969). "The Molecular Orbital Theory of Organic Chemistry." McGraw-Hill, New York.
Dewar, M. J. S. (1971). *Angew. Chem., Int. Ed. Engl.* **10**, 761.
Dewar, M. J. S., and Dougherty, R. C. (1975). "The PMO Theory of Organic Chemistry." Plenum, New York.
Doering, W. von E., Jones, J., and Saunders, M. (1968). *Tetrahedron* **23**, 3943.
Dondoni, A., and Barbaro, G. (1973). *J. Chem. Soc., Perkin Trans.* 2 p. 1769.
Dougherty, R. C. (1971). *J. Am. Chem. Soc.* **93**, 7187.
Eisch, J. J., and Husk, G. R. (1966). *J. Org. Chem.* **31**, 589.
Eisenstein, O., and Anh, N. T. (1971). *Tetrahedron Lett.* p. 1191.
Eisenstein, O., Lefour, J.-M., and Anh, N. T. (1971). *Chem. Commun.* p. 969.
Epiotis, N. D. (1972a.). *J. Am. Chem. Soc.* **94**, 1924.
Epiotis, N. D. (1972b). *J. Am. Chem. Soc.* **94**, 1935.
Epiotis, N. D. (1972c). *J. Am. Chem. Soc.* **94**, 1941.
Epiotis, N. D. (1972d). *J. Am. Chem. Soc.* **94**, 1946.
Epiotis, N. D. (1973a). *J. Am. Chem. Soc.* **95**, 1191.

Epiotis, N. D. (1973b). *J. Am. Chem. Soc.* **95**, 1200.
Epiotis, N. D. (1973c). *J. Am. Chem. Soc.* **95**, 1206.
Epiotis, N. D. (1973d). *J. Am. Chem. Soc.* **95**, 1214.
Epiotis, N. D. (1973e). *J. Am. Chem. Soc.* **95**, 5624.
Epiotis, N. D. (1974). *Angew. Chem., Int. Ed. Engl.* **13**, 751.
Epiotis, N. D., and Yates, R. L. (1974). *J. Org. Chem.* **39**, 3150.
Evans, M. G. (1939). *Trans. Faraday Soc.* **35**, 824.
Evans, M. G., and Polanyi, M. (1938). *Trans. Faraday Soc.* **34**, 11.
Evans, M. G., and Warhurst, E. (1938). *Trans. Faraday Soc.* **34**, 614.
Feuer, J., Herndon, W. C., and Hall, L. H. (1968). *Tetrahedron* **24**, 2575.
Fleming, I., and Karger, M. H. (1967). *J. Chem. Soc. C* p. 226.
Fleming, I., Gianni, F. L., and Mah, T. (1976). *Tetrahedron Lett.* p. 881.
Franzus, B., Baird, W. C., Jr., Felty, R. E., Smith, J. C., and Scheinbaum, M. L. (1971). *Tetrahedron Lett.* p. 295.
Freeman, F. (1975). *Chem. Review* **75**, 439.
Friedrich, L. E., and Schuster, G. B. (1971). *J. Am. Chem. Soc.* **93**, 4602.
Fujimoto, H., and Fukui, K. (1974). In "Chemical Reactivity and Reaction Paths" (G. Klopman, ed.), p. 23 ff. Wiley (Interscience), New York.
Fukui, K. (1964). In "Molecular Orbitals in Chemistry, Physics, and Biology" (P.-O. Löwdin and B. Pullman, eds.), p. 513. Academic Press, New York.
Fukui, K. (1971). *Acc. Chem. Res.* **4**, 57.
Fukui, K., and Fujimoto, H. (1968). *Bull. Chem. Soc. Jpn.* **541**, 1989.
Fukui, K., Yonezawa, T., Nagata, C., and Shingu, H. (1954). *J. Chem. Phys.* **22**, 1433.
Fukui, K., Yonezawa, T., and Nagata, C. (1957). *J. Chem. Phys.* **26**, 831.
Garbisch, E. W., Jr., and Sprecher, R. F. (1966). *J. Am. Chem. Soc.* **88**, 3436.
Garbisch, E. W., Jr., and Sprecher, R. F. (1969). *J. Am. Chem. Soc.* **91**, 6785.
Gassman, P. G., Dygos, D. K., and Trent, J. E. (1970). *J. Am. Chem. Soc.* **92**, 2084.
Gilbert, J. C., Smith, K. R., Klumpp, G. W., and Schakel, M. (1972). *Tetrahedron Lett.* p. 125.
Gimarc, B. M. (1974). *Acc. Chem. Res.* **7**, 384.
Goldstein, M. J., and Hoffmann, R. (1971). *J. Am. Chem. Soc.* **93**, 6193.
Gompper, R. (1969). *Angew. Chem., Int. Ed. Engl.* **8**, 312.
Hammond, G. S., and de Boer, C. D. (1969). *J. Am. Chem. Soc.* **91**, 2815.
Havinga, E., and Schlatmann, J. L. M. A. (1961). *Tetrahedron* **16**, 151.
Henri-Rousseau, O. (1973). Dissertation, Centre Universitaire de Perpignan, France.
Herndon, W. C. (1972). *Chem. Rev.* **72**, 157.
Herndon, W. C., and Feuer, J. (1968). *J. Org. Chem.* **33**, 417.
Herndon, W. C., and Giles, W. B. (1969). *Chem. Commun.* p. 497.
Herndon, W. C., and Giles, W. B. (1970). *Mol. Photochem.* **2**, 277.
Herndon, W. C., and Hall, L. H. (1967). *Theo. Chem. Acta* **7**, 4.
Herndon, W. C., Feuer, J., Giles, W. B., Otteson, D., and Silber, E. (1974). In "Chemical Reactivity and Reaction Paths" (G. Klopman, ed.), p. 275. Wiley (Interscience), New York.
Hoffmann, H. M. R. (1969). *Angew. Chem., Int. Ed. Engl.* **8**, 556.
Hoffmann, R. (1971). *Acc. Chem. Res.* **4**, 1.
Hoffmann, R., and Woodward, R. B. (1965a). *J. Am. Chem. Soc.* **87**, 2046.
Hoffmann, R., and Woodward, R. B. (1965b). *J. Am. Chem. Soc.* **87**, 4388.
Hoffmann, R., Levin, C. C., and Moss. R. A. (1973). *J. Am. Chem. Soc.* **95**, 629.
Houk, K. N. (1970). *Tetrahedron Lett.* p. 2621.
Houk, K. N. (1972). *J. Am. Chem. Soc.* **94**, 8953.

Houk, K. N. (1973a). *Surv. Prog. Chem.* **6**, 113.
Houk, K. N. (1973b). *J. Am. Chem. Soc.* **95**, 4092.
Houk, K. N. (1976). *Chem. Rev.* **76**, 1.
Houk, K. N. (1975). *Acc. Chem. Res.* **8**, 361.
Houk, K. N., and Luskus, L. J. (1973). *J. Org. Chem.* **38**, 3836.
Houk, K. N., and Chang, Y.-M. (1975). Submitted for publication.
Houk, K. N., and Munchausen, L. L. (1976). **98**, 937.
Houk, K. N., and Santiago, C. (1976). *J. Am. Chem. Soc.* **98**, 3380.
Houk, K. N., and Strozier, R. W. (1973). *J. Am. Chem. Soc.* **95**, 4094.
Houk, K. N., Luskus, L. J., and Bhacca, N. S. (1970). *J. Am. Chem. Soc.* **92**, 6392.
Houk, K. N., Northington, D. J., and Duke, R. E., Jr. (1972). *J. Am. Chem. Soc.* **94**, 6233.
Houk, K. N., Sims, J., Duke, R. E., Jr., Strozier, R. W., and George, J. K. (1973a). *J. Amer. Chem. Soc.* **95**, 7287.
Houk, K. N., Sims, J., Watts, C. R., and Luskus, L. J. (1973b). *J. Am. Chem. Soc.* **95**, 7301.
Houk, K. N., George, J. K., and Duke, R. E., Jr. (1974a). *Tetrahedron* **30**, 523.
Houk, K. N., Strozier, R. W., and Hall, J. A. (1974b). *Tetrahedron Lett.* p. 897.
Houk, K. N., Chang, Y.-M., and Munchausen, L. (1975a). Submitted for publication.
Houk, K. N., Strozier, R. W., McElwee, D. E., and Chang, Y.-M. (1975b). Submitted for publication.
Hudson, R. F. (1973). *Angew. Chem., Int. Ed. Engl.* **12**, 36.
Huisgen, R. (1971). *Pure Appl. Chem., Suppl.* **1**, 175.
Humski, K., Streklov, T., Borčić, S., and Sunko, D. E. (1969). *Chem. Commun.* p. 693.
Imamura, A. and Hirano, T. (1975). *J. Am. Chem. Soc.* **97**, 4192.
Inagaki, S., and Fukui, K. (1973). *Bull. Chem. Soc. Jpn.* **46**, 2240.
Inagaki, S., Yamabe, S., Fujimoto, H., and Fukui, K. (1972). *Bull. Chem. Soc. Jpn.* **45**, 3510.
Inagaki, S., Minato, T., Yamabe, S., Fujimoto, H., and Fukui, K. (1974). *Tetrahedron* **30**, 2165.
Inagaki, S., Fujimoto, H., and Fukui, K. (1975). *J. Am. Chem. Soc.* **97**, 6108.
Inukai, T., and Kojima, T. (1971). *J. Org. Chem.* **36**, 924 (and references therein).
Inukai, T., Sato, H., and Kojima, T. (1972). *Bull. Chem. Soc. Jpn.* **45**, 891.
Jean, Y., and Chapuisat, X. (1974). *J. Am. Chem. Soc.* **96**, 6911.
Jorgenson, W. L., and Salem, L. (1973). "The Organic Chemist's Book of Orbitals." Academic Press, New York.
Jotham, R. W. (1975). *J. Chem. Educ.* **52**, 377.
Kearns, D. R. (1971). *Chem. Rev.* **71**, 395.
Klopman, G. (1968). *J. Am. Chem. Soc.* **90**, 223.
Klopman, G. (1974). *In* "Chemical Reactivity and Reaction Paths" (G. Klopman, ed.), p. 55ff. Wiley (Interscience), New York.
Klopman, G., and Hudson, R. F. (1967). *Theor. Chim. Acta* **8**, 165.
Konovalov, A. I. (1969). *Zh. Org. Khim.* **5**, 1713.
Konovalov, A. I., and Kiselev, V. D. (1966). *Zh. Org. Khim.* **2**, 1.
Konovalov, A. I., Kiselev, V. D., and Vigdorovich, O. A. (1967). *Zh. Org. Khim.* **3**, 2085.
Koopmans, T. A. (1934). *Physica (The Hague)* **1**, 104.
Little, J. C. (1965). *J. Am. Chem. Soc.* **87**, 4020.
Longuet-Higgins, H. C., and Abrahamson, E. W. (1965). *J. Am. Chem. Soc.* **87**, 2045.
MacKenzie, K. (1969). *J. Chem. Soc. C* p. 1784.
Malojčič, R., Humski, K., Borčić, S., and Sunko, D. E. (1969). *Tetrahedron Lett.* p. 2003.
Mark, V. (1974). *J. Org. Chem.* **39**, 3179.

Marvell, E. N., Chadwick, T., Caple, G., Gosnik, T., and Zimmer, G. (1972). *J. Org. Chem.* **37**, 2992.
Michl, J. (1972). *Mol. Photochem.* **4**, 243, 257, and 287.
Michl, J. (1974). In "Chemical Reactivity and Reaction Paths" (G. Klopman, ed.), p. 230. Wiley (Interscience), New York.
Mollere, P. D., and Houk, K. N. (1977). *J. Am. Chem. Soc.* **99**, in press.
Mulliken, R. S. (1950). *J. Am. Chem. Soc.* **72**, 601.
Mulliken, R. S. (1954). *J. Am. Chem. Soc.* **74**, 811.
Mulliken, R. S. (1955). *J. Chem. Phys.* **23**, 2343.
Mulliken, R. S., and Person, W. B. (1969). "Molecular Complexes." Wiley (Interscience), New York.
Murrell, J. N., Randle, M., and Williams, D. R. (1965). *Proc R. Soc. London, Ser. A.* **284**, 566.
Okada, T., Yamaguchi, K., and Fueno, T. (1974). *Tetrahedron* **30**, 2293.
Oosterhoff, L. J. (1961). Quoted in Havinga and Schlatmann (1961).
Paddon-Row, M. N. (1974). *Aust. J. Chem.* **27**, 299.
Paddon-Row, M. N., Watson, P. L., and Warrener, R. N. (1973). *Tetrahedron Lett.* p. 1033.
Paddon-Row, M. N., Gell, K., and Warrener, R. L. (1975). *Tetrahedron Lett.* p. 1975.
Paquette, L. A., and Haluska, R. J. (1970). *J. Org. Chem.* **35**, 132.
Pearson, R. G. (1969). *J. Am. Chem. Soc.* **92**, 1252.
Pearson, R. G. (1971). *Acc. Chem. Res.* **4**, 152.
Pearson, R. G. (1972). *J. Am. Chem. Soc.* **94**, 8287.
Pickenhagen, W., Nal, F., Ohloff, G., Muller, P., and Perlberger, J.-C. (1973). *Helv. Chim. Acta* **56**, 1868.
Rhoads, S. J., and Cockcroft, R. D. (1969). *J. Am. Chem. Soc.* **91**, 2815.
Roberts, J. D., and Sharts, C. M. (1962). *Org. React.* **12**, 1.
Salem, L. (1968). *J. Am. Chem. Soc.* **90**, 543 and 553.
Salem, L. (1969). *Chem. Phys. Lett.* **3**, 99.
Salem, L., and Wright, J. S. (1969). *J. Am. Chem. Soc.* **91**, 5947.
Saskai, T., Kanematsu, K., and Kataoka, T. (1975). *J. Org. Chem.* **40**, 1201.
Sauer, J., Wiest, H., and Mielert, A. (1964). *Chem. Ber.* **97**, 3183.
Scheidt, F., and Kirmse, W. (1972). *J. Chem. Soc., Chem. Commun.* p. 716.
Schleyer, P. V. R., Swilinski, W. F., Van Dine, G. W., Schöllkopf, U., Paust, J., and Fellenberger, K. (1972). *J. Am. Chem. Soc.* **94**, 125.
Schöllkopf, U., and Hoppe, I. (1972). *Justus Liebigs Ann. Chem.* **765**, 153.
Simmons, H. E., and Bunnett, J. F. (1974). "Orbital Symmetry Papers." Am. Chem. Soc., Washington, D.C.
Sims, J., and Houk, K. N. (1973). *J. Am. Chem. Soc.* **95**, 5798.
Skell, P. S., and Cholod, M. S. (1969). *J. Am. Chem. Soc.* **91**, 7131.
Stephenson, L. M., Jr., and Brauman, J. I. (1974). *Acc. Chem. Res.* **7**, 65.
Stewart, C. A., Jr. (1962). *J. Am. Chem. Soc.* **84**, 117.
Stewart, C. A., Jr. (1972). *J. Am. Chem. Soc.* **94**, 635.
Stoos, F., and Roček, J. (1972). *J. Am. Chem. Soc.* **94**, 2719.
Sustmann, R. (1971a). *Tetrahedron Lett.* p. 2717.
Sustmann, R. (1971b). *Tetrahedron Lett.* p. 2721.
Sustmann, R. (1974). *Tetrahedron Lett.* p. 963.
Sustmann, R., and Binsch, G. (1971). *Mol. Phys.* **20**, 1 and 9.
Sustmann, R., and Schubert, R. (1972). *Angew. Chem., Int. Ed. Engl.* **11**, 840.
Sustmann, R., and Trill, H. (1972). *Angew. Chem., Int. Ed. Engl.* **11**, 838.

Sustmann, R., Ansmann, A., and Vahrenholt, F. (1972). *J. Am. Chem. Soc.* **94**, 8099.
Takatsuki, K., Murata, I., and Kitahara, Y. (1970). *Bull. Chem. Soc. Jpn.* **43**, 966.
Van Cauwelaert, F. H. (1971). *Bull. Soc. Chim. Belg.* **80**, 181.
Wang, I. S. Y., and Karplus, M. (1973). *J. Am. Chem. Soc.* **95**, 8160.
Widmer, U., Zsindely, J., Hansen, H. J., and Schmid, H. (1973). *Helv. Chim. Acta* **56**, 75.
Wollweber, H. (1972). "Diels-Alder-Reaktion." Thieme, Stuttgart.
Woodward, R. B., and Hoffmann, R. (1965). *J. Am. Chem. Soc.* **87**, 395.
Woodward, R. B., and Hoffmann, R. (1969). *Angew. Chem., Int. Ed. Engl.* **8**, 781.
Yamaguchi, K., and Fueno, T. (1973). *Chem. Phys. Lett.* **22**, 471.
Yamaguchi, K., Fueno, T., and Fukutome, H. (1973). *Chem. Phys. Lett.* **22**, 461.
Yates, R. L., Epiotis, N. D., and Bernardi, F. (1975). *J. Am. Chem. Soc.* **97**, 6615.
Zimmerman, H. E. (1972). *Acc. Chem. Res.* **5**, 393.
Zuman, P. (1967). "Substituent Effects in Organic Polarography." Plenum, New York.

5

Thermally Forbidden Reactions

JOHN E. BALDWIN

I. Introduction	273
II. Theory	275
III. Examples	280
IV. Conclusions	297
References	297

I. Introduction

The orbital symmetry correlation rules for cycloreactions proposed by Woodward and Hoffmann (Woodward, 1967; Hoffmann and Woodward, 1968, 1970; Woodward and Hoffmann, 1969, 1971) signaled a revolutionary advance toward a deeper understanding of chemical reactions. They recognized one of the most significant ways molecules could undergo transformations while maintaining bonding character of all electrons as fully as possible, and stated the insight persuasively and economically. The tendency to maintain bonding character was grasped as a more fundamental influence on reaction rates and stereochemical preferences than the principle of least motion (Tee, 1969; Tee and Yates, 1972; Tee et al., 1974; Ehrenson, 1974).

Earlier approaches toward such correlation rules had been made. The Wigner–Witmer spin and orbital momentum correlation rules (Wigner and Witmer, 1928; Herzberg, 1950; Silver, 1974) provided part of the needed understanding. Attempts to find correlations among electronic states of reactants and either intermediates or activated complexes and products, through formal group theoretical considerations (Shuler, 1953) or intuitive contemplations of reactions shown by complex organic molecules (Evans and Polanyi, 1938; Evans and Warhurst, 1938; Evans, 1939; Oosterhoff, 1961; Dauben and Willey, 1962; Fukui, 1964) can retrospectively be recognized as important anticipations of Woodward and Hoffmann's powerful generalizations.

Later attempts toward alternative formulations of the rules strove to grace them with a more convincing theoretical basis, to provide simpler derivations, and to popularize the manifold applications they seemed to fit so convincingly

(Mulder and Oosterhoff, 1970, 1971; van der Hart et al., 1972; Trindle, 1969, 1970; Goddard, 1970, 1972; Rastelli et al., 1972; Fukui, 1971; Fujimoto and Fukui, 1972; Fujimoto et al., 1973; Wilson and Wang, 1972, 1973; Mathieu, 1973; Epiotis, 1972, 1973, 1974; Seebach, 1968; Vollmer and Servis, 1968, 1970; Caserio, 1971; Katz, 1971; Perrin, 1972; Wollenberg and Belloli, 1974; Simmons and Bunnett, 1974; Weltin, 1974).

Many restatements of the Woodward–Hoffmann theory have been introduced; the basic conclusions have been duplicated with different words, degrees of rigor, and formalisms for reactions having some appropriate symmetry element maintained throughout the conversion as well as for totally asymmetrical processes (Trindle and Collins, 1971).

Photochemical processes (van der Lugt and Oosterhoff, 1968, 1969, 1970; Dougherty, 1971; Michl, 1972, 1974; Muszkat and Sharafi-Ozeri, 1973; Salem, 1974; Devaquet, 1974), reactions of open-shell molecules (Boche and Szeimies, 1971; Merlet et al., 1974), and thermally forbidden interconversions (George and Ross, 1971) require a more detailed and careful analysis than the original treatment provided. The latter topic has been (Baldwin and Andrist, 1971a; Baldwin et al., 1972b) and is here one of our major concerns.

The first Woodward and Hoffmann communication (Woodward and Hoffmann, 1965) recognized and generalized the significant influence molecular orbital symmetry factors may exert on thermal and photochemical electrocyclic transformations. Certain stereochemical paths, according to their hypotheses, were predicted to be favored; unfavored processes were expected to occur more slowly and only under very energetic conditions.

Their second communication (Hoffmann and Woodward, 1965) extended the insight to concerted intermolecular cycloaddition reactions with the aid of molecular orbital correlation diagrams. For the $[_\pi 2_s + _\pi 2_s]$ combination of two ethylene molecules, bonding–antibonding molecular orbital level correlations were deduced and taken as indicative of a highly unfavorable ground-state process. The $[_\pi 2_s + _\pi 4_s]$ Diels–Alder addition, without such a bonding–antibonding molecular orbital correlation accompanying progress along the reaction coordinate, represented the correlation type characteristic of permitted thermal reaction. There followed rules for reactions that would be allowed thermally and allowed photochemically.

And it was asserted, "those cycloadditions which do occur in cases prohibited by our selection rules for concerted reactions must proceed through multistep mechanisms."

This assertion has gained wide currency and acceptance; it offers such a seductively appealing guide to the interpretation of experimental observations! The failure of a chemical transformation to follow the stereochemical course associated with the conservation of orbital symmetry is, according to the assertion, a sufficient condition for excluding concert. It has been applied by some to implicate nonconcerted multistep reaction mechanisms whenever an orbital symmetry forbidden product could be detected. It has been employed by others as a blindfold justifying such pronouncements in journal articles as: "There is no conclusive evidence for the occurrence of orbital symmetry forbidden reactions." "Symmetry forbidden reactions are precluded." "There are no authentic symmetry-forbidden thermal hydrocarbon reactions."

The assertion is, however, not valid. Reactions may occur according to a stereochemical path designated as unfavored or prohibited or forbidden in light of the Woodward–Hoffmann rules without the intervention of intermediates. This fact has become obvious to most theoreticians, but it has remained unassimilated by most experimentalists.

This chapter will present, in an elementary form, a general theoretical model accommodating concerted reaction profiles for forbidden thermal reactions. Some examples of specific reactions to which the model has been or might be applied will be summarized. We shall not consider forbidden reactions that are promoted by an additional reactant, such as a metallic (Mango, 1972, 1973, 1974) or enzymic (Ferreira, 1973) catalyst.

II. Theory

The modifier "forbidden" has long been used by scientists to describe an event which, in terms of some appropriate theory, cannot occur. When such a forbidden process is observed, a more complete or more refined theoretical model is sought to account for the apparent anomaly. Thus, the lowest energy $\pi \to \pi^*$ transition ($^1B_{2u} \leftarrow {}^1A_{1g}$) in D_{6h} benzene is forbidden by symmetry. Yet, it may be observed as a weak band near 260 nm; individual molecules may be in vibrational states distorted from D_{6h} symmetry and the transition is not forbidden, according to theory taking cognizance of such vibrational interaction. Or, intersystem crossing between, say, the lowest excited anthracene singlet and triplet states, S_1 and T_1, is spin forbidden; yet, it occurs and can be explained by including a spin-orbit interaction in an improved theoretical analysis.

In the same way, the occurrence of thermal reactions forbidden by orbital symmetry theory must be treated through a more complete theory, one which recognizes the limitations of the simple orbital symmetry approach; it cannot give a realistic account of molecules or reactions in which single electronic configurations are not accurate approximations to electronic states. The Woodward–Hoffmann approach will fail when a single-product wave function, or configuration, is not representative of the actual wave function; configuration interaction or some alternative, such as single-determinant wave functions based on complex molecular orbitals (Pople, 1971) or spin-density wave functions (Jordan and Silbey, 1973) is required for reasonable accuracy in a molecular orbital treatment.

This limitation was explicitly noticed first by Longuet-Higgens and Abrahamson (1965) in their careful assessment of the distinctions between molecular orbital, electronic configuration, and state correlation diagrams. The figures they included showed, in dotted lines, "what would happen to a given electron configuration if there were no configuration interaction; the full lines (represented) a more realistic view of the correlations between states, since electron repulsion prevents states of the same symmetry from crossing."

An electronic configuration in molecular orbital theory is a specification of orbital occupancy; it is an assignment of electrons to orbitals. The ground-state π-electronic configuration ϕ_0 of butadiene, for instance, is represented by the notation $\chi_1{}^2\chi_2{}^2$; χ_1 and χ_2, the two lowest-energy π-molecular orbitals of the molecule, contain two electrons apiece. The configuration $\phi_* = \chi_1{}^2\chi_3{}^2$ specifies a higher-energy apportionment of electrons among the four π-molecular orbitals.

An electronic state wave function in a molecular orbital representation restricted to a single electronic configuration is the normalized Slater determinant based on that configuration, which assures proper conformity to the Pauli exclusion principle. For an n-orbital, $2n$-electron ground-state molecule, $\psi_0 = \det(\chi_1\bar{\chi}_1 \ldots \chi_n\bar{\chi}_n)$, with normalization assumed.

An electronic state in a molecular orbital theory which may include two or more electronic configurations has, as a wave function, a variationally optimized linear combination of Slater determinants, one from each of two or more electronic configurations of the same symmetry: $\psi = c_1\psi_1 + c_2\psi_2$. When two electronic configurations are well separated energetically, the state will be nearly as well described by a single determinant as by two, and $c_1 \gg c_2$. But, when the two configurations are of comparable energy, either one determinantal wave function alone would give a poor description of the state, and $c_1 \approx c_2$.

In the butadiene case, the configurations ϕ_0 and ϕ_* have the same state symmetry and may interact configurationally. The highest occupied molecular orbitals in each configuration are of different symmetry (for instance, a and s with respect to the mirror plane), and these molecular orbitals cross in the course of a forbidden disrotatory isomerization to form cyclobutene; but the configurations have the same symmetry and they do not cross. The linear combination of the two configurations varies the relative values of c_1 and c_2 along the reaction coordinate to minimize the energy of the lowest state.

Molecular events are predicated on the properties of electronic states, not on the denouements of orbital dramas. The theoretically calculable crossing of single determinantal wave functions having the same spatial symmetry and spin will never correspond with reality. Single determinantal wave functions corresponding to different electronic configurations and having the same symmetry will always mix to a greater or lesser degree and will never cross.

The characteristic molecular orbital correlation diagram for a forbidden reaction (Pople, 1971) exhibits a set of lower orbital levels correlated between reactant and product, a highest-occupied reactant orbital correlated with the lowest unoccupied product orbital and vice versa, and sets of higher orbitals unoccupied both before and after reaction. The highest-occupied molecular orbitals of starting material and product are of different spatial symmetry with respect to the appropriate symmetry element of the reaction. The electronic states which may arise from the distribution of two electrons to these highest occupied and lowest unoccupied orbitals during the reaction are four: one singlet and one (three component) triplet from the configurations ... $\chi_n \bar{\chi}_{n+1}$ and ... $\chi_n \chi_{n+1}$ and two symmetric singlet states from the configurations ... $\chi_n^2 \chi_{n+1}^0$ and ... $\chi_n^0 \chi_{n+1}^2$. The configuration interaction series suitable for describing the forbidden reaction will be dominated by the substrate ground-state configuration, ... $\chi_n^2 \chi_{n+1}^0$, which becomes the product doubly excited configuration, and the substrate doubly excited configuration, ... $\chi_n^0 \chi_{n+1}^2$, obtained from the ground state configuration by transferring a pair of electrons from the highest occupied bonding orbital to the lowest unoccupied or virtual antibonding orbital, which becomes the ground state configuration of the product as the nuclear motions characteristic of the reaction occur. The two determinant configuration interaction wave function may describe a smooth energy curve for the reaction: emphasis changes from the ground state configuration of substrate to ground state configuration of product as the reaction takes place in a continuous manner. Two configurations will be sufficient for a qualitative description of such reactions. The use of a two-term expression for the wave function, $\psi = c_1 \psi_1 + c_2 \psi_2$, provides the flexibility needed to model forbidden reactions which maintain an appropriate symmetry element. All electrons in the molecular system may remain completely correlated in the lowest-energy singlet state throughout the forbidden reaction.

This flexibility is the basis of the perturbative configuration interaction treatment using localized orbitals (Langlet and Malrieu, 1972; Daudey et al., 1974): diexcited and ground state substrate determinants provide the needed corrective to the Woodward–Hoffmann theory, which assumes that interaction terms between determinants are always small and negligible. They are small for allowed processes; they are large and essential in theoretical molecular orbital treatments of forbidden reactions.

A single configuration molecular orbital treatment applied to a forbidden reaction is simply inadequate (Lowe, 1974). Molecular orbital correlation diagrams for a forbidden process postulating violation of the noncrossing rule, as in the butadiene–cyclobutene disrotatory electrocyclic interconversion (Nohira, 1974), are as artificial as those honoring the noncrossing principle: the issue becomes moot within the context of a configuration interaction treatment.

Detailed expositions of LCAO MO theory, including configuration interaction applied to the hydrogen molecule, have been provided (Slater, 1963; Dewar and Kelemen, 1972). While the fragmentation of diatomics and triatomics to closed-shell fragments (e.g., $H_2F^+ \rightarrow H^+ + HF$) may be described reasonably through single determinant wave functions, dissociations to radicals ($HF \rightarrow H\cdot + F\cdot$) require configuration interaction or equivalent computational techniques (Jakubetz et al., 1971).

The dissociation of a closed-shell organic molecule into two identical radicals, such as the fragmentation of ethane into two methyl radicals, may be treated by configuration interaction of the ground state with the lowest doubly excited configuration; at infinite separation, the results obtained by this approach and by the "half-electron method," based on the assumption that the electronic structure of a radical may be approximated by a closed shell structure in which the highest occupied molecular orbital contains two half-electrons with opposite spins, are identical (Kollmar, 1971; Čársky and Zahradník, 1972). At lesser separations, for instance, when bond cleavage in a cyclopropane or cyclobutane leads to a molecular rearrangement by way of a presumed diradical intermediate or transition state, the half-electron method is *not* equivalent to a configuration interaction calculation. A reasonably faithful description of the reactions of organic radicals, such as the isomerization of cyclopropyl radical to allyl radical, also requires configuration interaction (Peyerimhoff and Buenker, 1974).

When a forbidden reaction is described through a single determinant wave function, there is a discontinuity in the energy derivative for the potential curve corresponding to the lowest state; but this discontinuity is just an artifact due to the limited form of the wave function. Similarly, many artifacts arising from simple molecular orbital single configuration depictions of diradicals are resolved through calculations including configuration interaction (Michl, 1972; Salem and Rowland, 1972; Salem, 1972, 1973; Stephenson et al., 1973; Bergman, 1973; Martin, 1974a). Experimental distinctions between singlet and triplet states, and between activated complexes and intermediates, will continue to be important in processes that some describe with diradical formalisms. Whether the historically sanctioned term can be preserved through new definitions and understandings of its meanings is an open question; perhaps, it isn't necessary, scientifically speaking, in spite of the clear mellifluous superiority of "diradical" over "configurational composite."

One redefinition of the role of diradicals has already been advanced, following a clear experimental indication that the normal diradical hypothesis was mechanistically inadequate to the systems being studied (Doering and Sachdev, 1974).

Forbidden reactions are simply processes in which there is a change in predominant electronic configuration (Stohrer and Hoffmann, 1972b), a change which can be conceptually and qualitatively accommodated through the formal methods of configuration interaction. In any specific case, however, the availability of a theory suitable for rationalizing an energetically concerted reaction profile is no assurance that such a mechanistic course is followed. Orbital symmetry rules offer no reliable mechanistic truths for forbidden reactions and a more comprehensive configuration interaction theory able to treat concerted forbidden conversions sheds no light on the mechanism of any particular forbidden reaction. Some reactions which, at first glance, appear to go by a forbidden route may, in fact, be accomplished indirectly, by a series of allowed steps. Others will occur indirectly, utilizing forbidden, as well as allowed, steps. Others may be completely concerted. And, finally, some may involve interstate crossing and triplet intermediates.

Most of the reactions used here as illustrations of these possibilities involve hydrocarbons, for they seem to offer experimentalists and theoreticians the best current prospects for a meaningful dialog on the nature of prototype thermally induced molecular transformations. And they are less subject to idiosyncratic substituent effects than molecules containing heteroatoms (Tanny et al., 1972; Libit and Hoffmann, 1974). Hydrocarbons may, of course, give reaction products in dependence both on structure and reaction conditions. For example, 5,5-dimethylcyclohexa-1,3-diene (Spangler and Boles, 1972) and the pentamethylhomofulvene derivative shown (Criegee et al., 1973) give products in decalin solution or as neat liquids in sealed tubes that may be rationalized through radical-chain mechanisms.

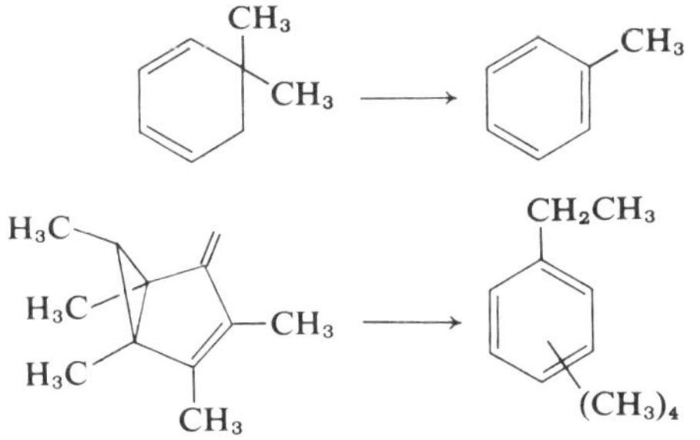

In the gas phase, the major isomeric products obtained are quite different; the dimethylcyclohexadiene gives initially 1,5-dimethyl-1,3-cyclohexadiene; in time, the other two dihydro-*m*-xylenes and *m*-xylene may be found in the product mixture. The *exo*-pentamethylhomofulvene system affords the endo isomer, hexamethylbenzene, and 5-vinylpentamethylcyclopentadiene.

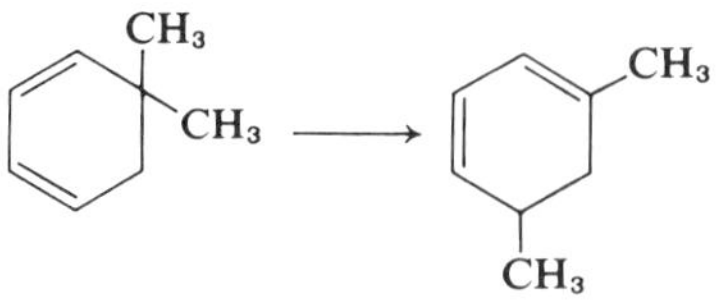

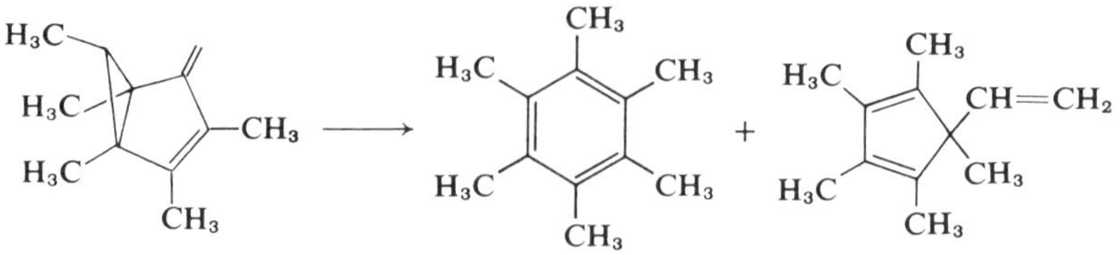

III. Examples

Molecules will do the most astounding things to follow orbital symmetry allowed paths (Hoffmann and Woodward, 1972; Hoffmann, 1973)! The few examples formulated give vivid reminders of this significant truth as they stretch our conceptions of how very flexible reacting molecules may be (Goldstein and Benzon, 1972; Grimme and Doering, 1973; Meinwald and Kapecki, 1972; Nelsen and Gillespie, 1972; Berson *et al.*, 1971; Hoffmann *et al.*, 1968; Meinwald and Gruber, 1971; Meinwald *et al.*, 1975; Baldwin and Fleming, 1972, 1973; Bertrand *et al.*, 1974; J. E. Baldwin and G. D. Andrews, unpublished results, 1975).

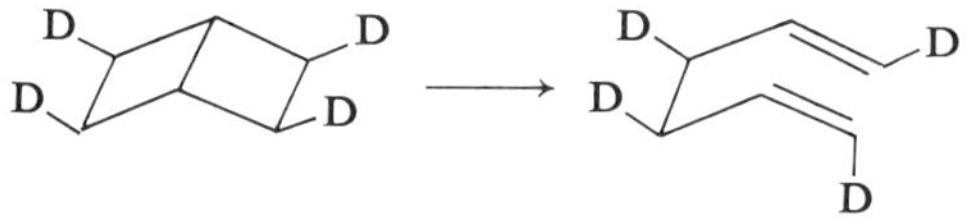

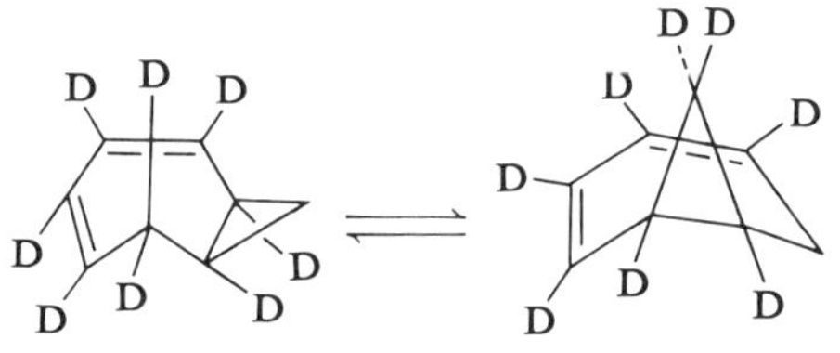

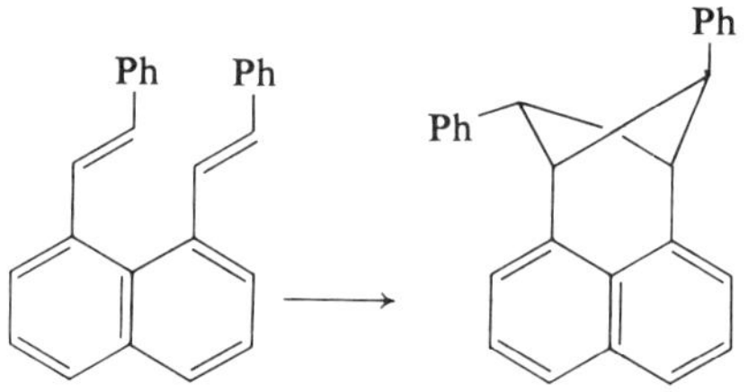

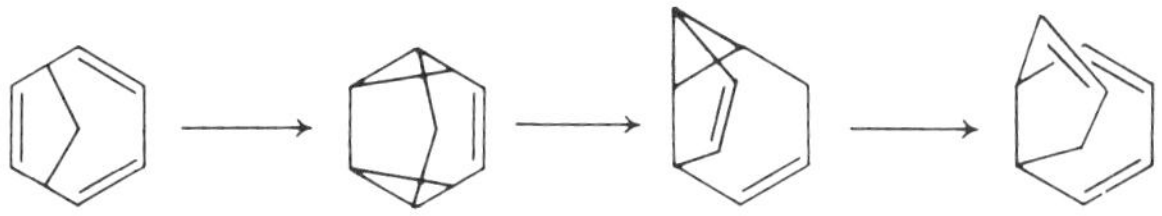

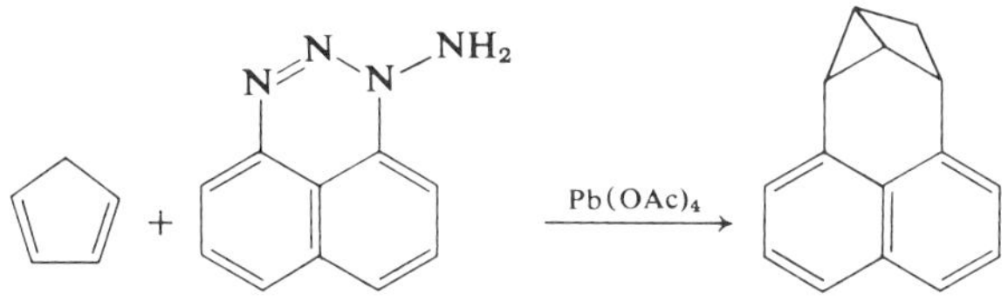

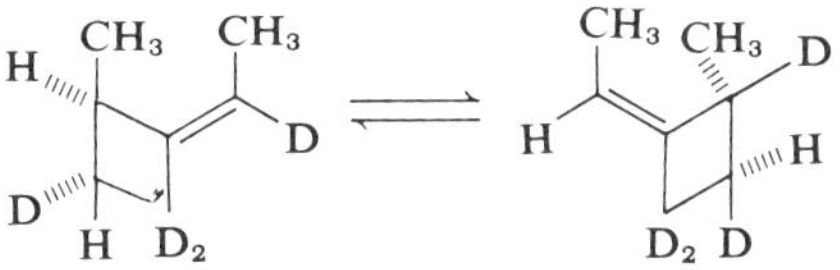

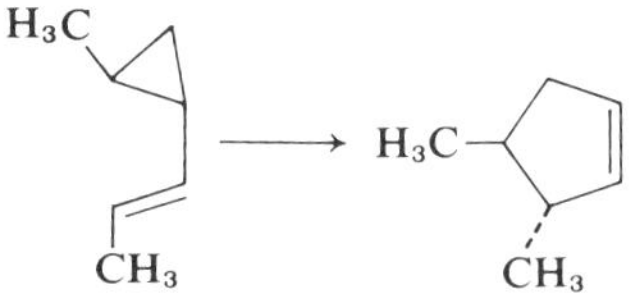

As the last two examples demonstrate, the [1,3]-sigmatropic shift of carbon can occur with the allowed but seemingly awkward antarafacial, retention stereochemistry. This long-sought stereochemical path eluded detection in investigations on a system designed to force retention of configuration on the migrating group (Berson et al., 1974).

Aware of this tendency, one anticipates that some reactions which appear to be forbidden as formulated may follow some devious but more efficient and fully allowed course.

If one were to observe a conrotatory ring closure of a hexatriene, for instance, it would appear that a forbidden reaction had occurred.

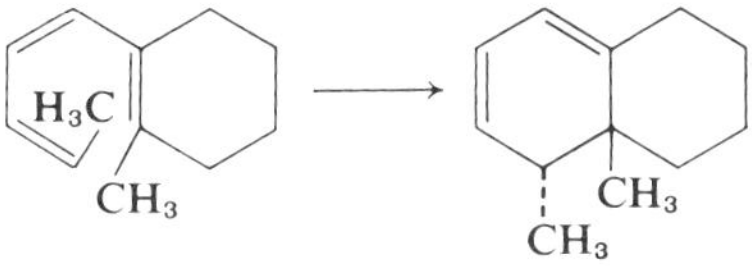

But another possibility, isomerization of the triene through two allowed 1,7-hydrogen shifts, followed by an allowed disrotatory cyclization, would be a reasonable mechanistic alternative (Ramage and Sattar, 1970; Heimgartner et al., 1972).

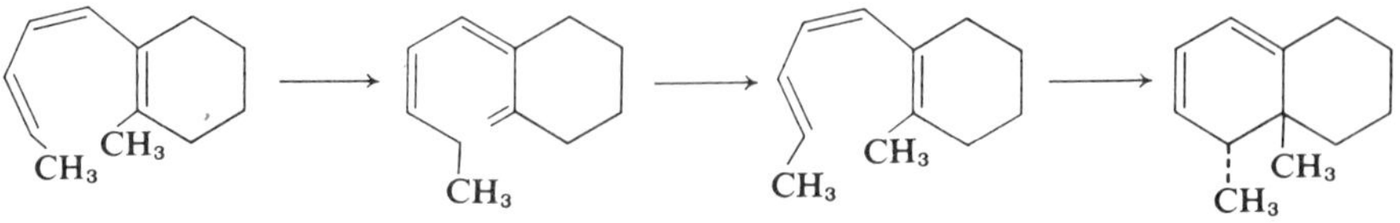

Appropriate experimentation would, of course, resolve the ambiguity in this and related instances.

The thermal conversion of basketene to Nenitzescu's hydrocarbon directly would be a forbidden fragmentation; indirectly, through a retro-Diels–Alder process followed by a Cope rearrangement, the isomerization may be accomplished in two symmetry-allowed steps (Dauben and Whalen, 1966; Westburg et al., 1969; Vedejs, 1971; Paquette et al., 1972; Allred and Beck, 1973).

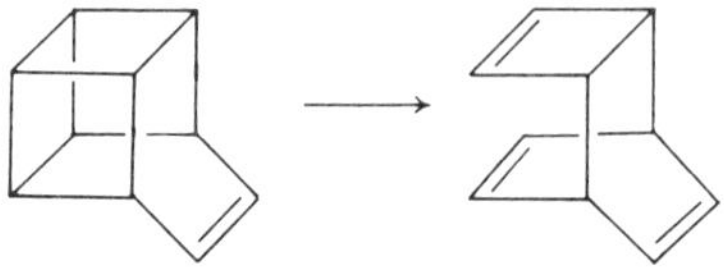

That two or more steps may be involved to achieve a particular net transformation while a seemingly more direct forbidden (or allowed) but energetically more demanding reaction course is shunned may, in a particular case, be occasion for surprise (Baldwin and Pinschmidt, 1971; Baldwin and Kaplan, 1971, 1972; Baldwin and Bryan, 1974; Boche et al., 1974). But this possibility must always be taken seriously, for chemical rates are governed by activation energies and entropies, not by a dominant tendency to minimize discrete steps.

Some reactions which seem to occur, even though they are forbidden, have been observed for such potentially complex molecular systems that hypotheses regarding forbidden steps may be very difficult to support experimentally. The thermolysis of cyclohexa-1,3-diene is a case in point. 5,6-cis-Dideuteriocyclohexa-1,3-diene at high temperatures gives benzene-d_i, and the distribution of deuterium in the product is not consistent with a concerted elimination of molecular deuterium or hydrogen (Fleming and Wildsmith, 1970). But at the temperatures required for the dehydrogenation, 1,5-hydrogen migrations might be expected to cause extensive scrambling of label in the starting material prior to the formation of benzene; second-order processes could be involved; and bicyclo[3.1.0]hex-2-ene might conceivably be an intermediate leading from the conjugated to the unconjugated cyclohexadiene (Schiess and Fünfschilling, 1972; De Maré et al., 1972; Orchard and Thrush, 1973; Rose, 1973). Whether any unimolecular forbidden 1,2-elimination of hydrogen from cyclohexa-1,3-diene reaction plays a role in the overall process remains very much in doubt.

5. THERMALLY FORBIDDEN REACTIONS

When molecular complexity is sufficiently great, and many possible concerted pathways are available, a complex product mixture may arise even without the participation of forbidden paths. Both racemic spirocyclic hydrocarbons shown rearrange thermally to all four racemic forms of 6,7-dimethylbicyclo[3.2.0]hept-1-ene, the 6,7-*cis*,*exo*, 6,7-*cis*,*endo*, 6-*exo*,7-*endo*, and 6-*endo*,7-*exo* isomers (Roth and Enderer, 1970; Gajewsky and Burka, 1972). Yet, each product is formally derivable using an allowed [$_\sigma 2_s + {_\sigma}2_a$] path!

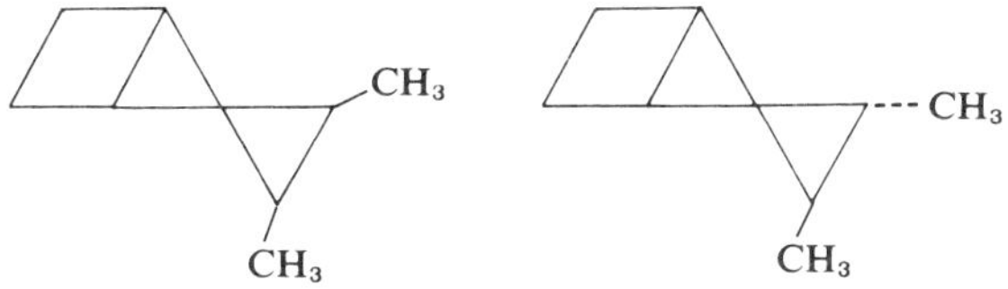

Whether the products are produced through competitive allowed pathways or through the participation of forbidden reactions is not, of course, answered by recognition of both possibilities. To sort out the mechanistic situation would demand some difficult experimentation with related optically active substrates. But the mere presence of four products does not demonstrate the intrusion of forbidden paths, since the *s* and *a* roles can be assumed by different bonds in different ways.

High pericyclic selectivity (Goe, 1972; Paddon-Row *et al.*, 1973), it is sometimes asserted, is a characteristic of reactions controlled by the conservation of orbital symmetry; it has even been supposed that, since a concerted pericyclic reaction must lead from one particular substrate to only one product, mixtures of reaction products, even when each one might correspond to an allowed path, signal the intrusion of nonconcerted reaction mechanisms! Both propositions are completely unjustified. The mechanism of a reaction consists of the sum of many individual paths, each of which contributes in proportion to its critical energy and entropy (Doering and Sachdev, 1974).

These paths need not be minor variants or slight modifications of one activated complex; they could involve widely separated transition states.

Another reaction for which questions regarding concertedness and stereochemistry await possibly demanding work is the conversion of dispiro[2.0.2.4]-deca-7,9-diene to *o*-ethylstyrene and tetralin (de Meijere, 1974).

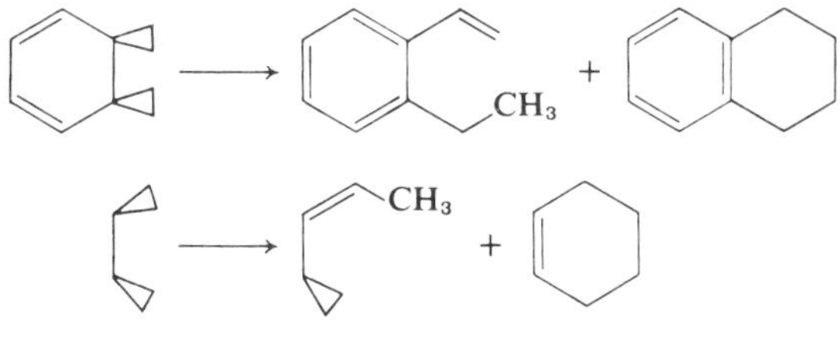

By analogy with the thermal chemistry of bicyclopropyl, where the major products are cyclopropylpropenes and cyclohexene (Flowers and Frey, 1962), one might wonder whether the dispirodiene may give in parallel paths 8-(Z)-ethylidenespiro[2.5]octa-4,6-diene and tetralin in normal [$_\sigma 2 + {}_\sigma 2$] cycloreactions, with subsequent rapid isomerization of the triene intermediate.

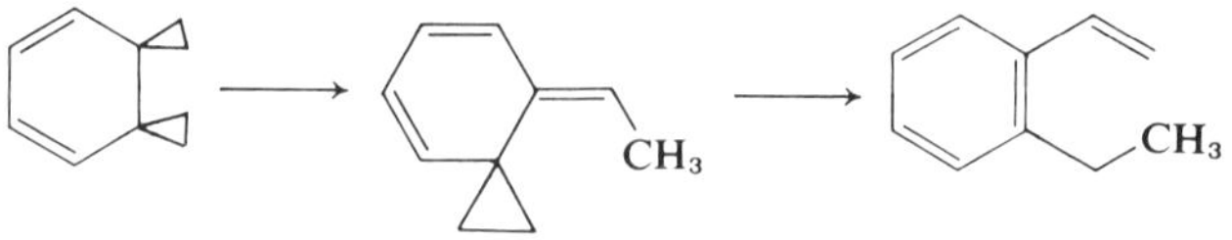

Other formulations are possible; to understand the reactions, both kinetic and stereochemical data would be required.

The same need for combined kinetic and stereochemical experimentation is evident in reports on the isomerization of allylenecyclopropanes to 1-methylenecyclopent-2-enes.

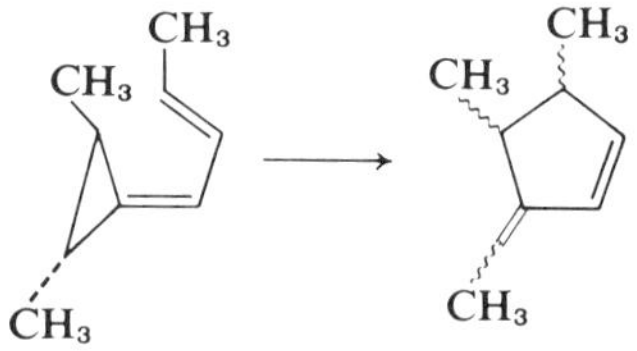

The knowledge that a stereochemically well-defined allylenecyclopropane substrate produces several isomeric 1-methylenecyclopent-2-ene products (Roth and Schmidt, 1971) is ambiguous as long as the possible involvement of 2-vinyl-1-methylenecyclopropanes as reactive intermediates and possible isomerizations among 2-vinyl-1-methylenecyclopropanes is unclarified (Kende and Riecke, 1972; Gilbert and Higley, 1973). Optically active substrates and combined kinetic and stereochemical studies could reveal whether competitive allowed or allowed and forbidden pathways are responsible for the overall stereochemical result.

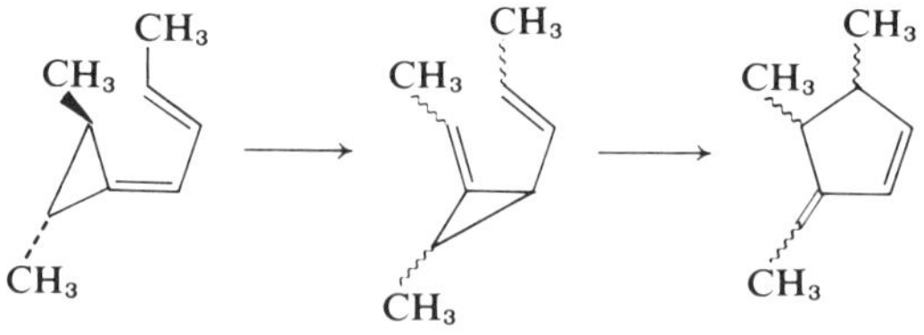

The thermal isomerization of bicyclo[2.1.0]pent-2-enes offers another example where an experimental distinction between allowed and forbidden mechanistic schemes is still awaited (McLean *et al.*, 1972). It is known that 2-methylbicyclopentene thermally rearranges to give both 2-methyl- and 1-methylcyclopentadiene as primary products: the product ratio is extremely sensitive to the reaction conditions, being 1.3:1 in the gas phase and 12:1 in hexane solution (Baldwin and Andrews, 1972; Andrews *et al.*, 1973; Brauman *et al.*, 1973).

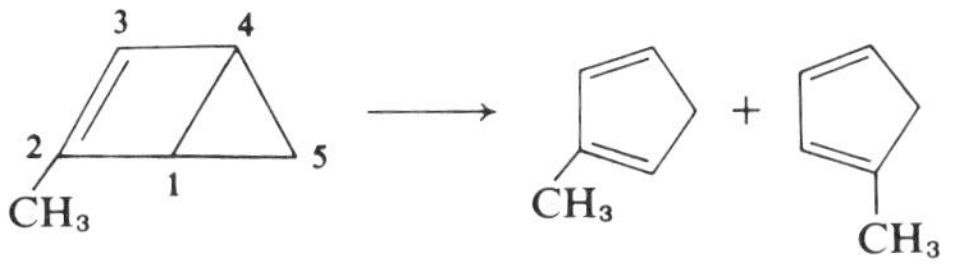

One explanation for these results postulates cleavage of the C(1)–C(4) bond to form vibrationally excited 2-methylcyclopentadiene, which then isomerizes in part to the 1-methyl isomer (Flowers and Frey, 1972; Stephenson and Brauman, 1974; Dewar, 1975; cf. Andrist, 1973). About 16% of the vibrationally excited 2-methylcyclopentadiene molecules undergo 1,5-hydrogen shifts faster than they are collisionally deactivated in hexane! Another hypothesis is that the products may be derived in part from C(1)–C(2)/C(4)–C(5) and C(3)–C(4)/C(5)–C(1) reactions. Fortunately, an experimental resolution of the issue may be obtained; it is now being sought through a study of the rearrangement of some bicyclopentene-$^{13}C_2$ molecules (J. E. Baldwin and G. D. Andrews, unpublished results, 1975).

The thermal isomerizations of bicyclo[5.2.0]nona-1,3,5,8-tetraenes to indenes present some of the same mechanistic ambiguities as the bicyclo-[2.1.0]pent-2-ene to cyclopentadiene process (Lombardo and Wege, 1972; Waali and Jones, 1973a,b).

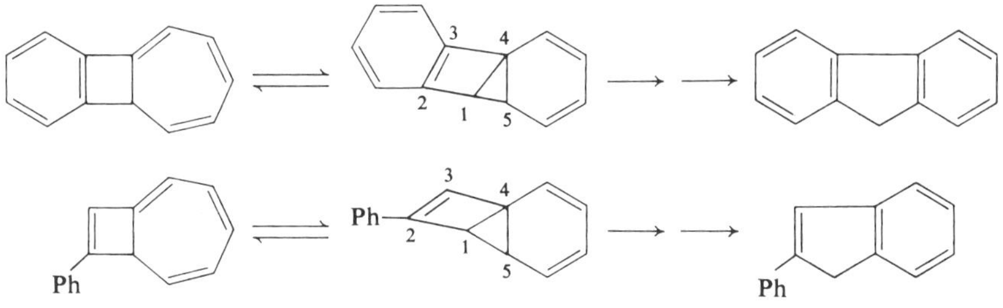

Either a forbidden disrotatory rupture of the C(1)–C(4) bond in the bridged bicyclopentene valence isomer, or a [C(3)–C(4)/C(5)–C(1)] process, followed in each case by allowed hydrogen shifts, might be used to rationalize the observed products and to plan mechanistic work designed to discriminate between the two formulations.

The ambiguity appears again in the recent demonstration that optically active 3,7-dimethyl-7-methoxycarbonylcyclohepta-1,3,5-triene racemizes (Klärner, 1974).

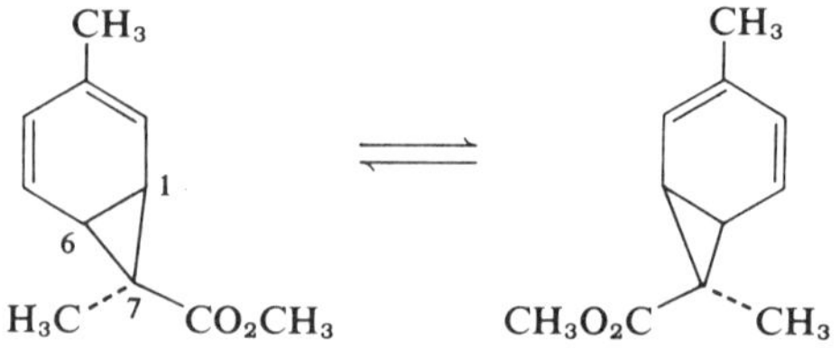

If the reaction involves breaking C(1)–C(7) (Berson, 1968) and a 1,5-carbon migration with inversion, it is a symmetry forbidden least-motion isomerization. But, if it is a C(1)–C(7)/C(6)–C(5) process, it may be allowed. The distinction between these two possibilities, here and for other isomerizations involving tropilidene–norcaradiene valence isomers, is a subtle one, inaccessible to labeling experiments: it would require kinetic resolution between Schemes 1 and 2, illustrated in abbreviated form for three tropilidene (T) and norcaradiene (C) isomer sets.

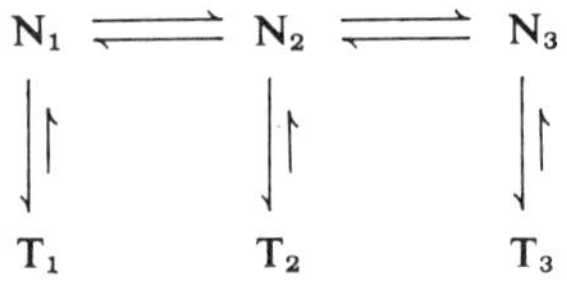

$$N_1 \rightleftharpoons N_2 \rightleftharpoons N_3$$
$$\updownarrow \quad \updownarrow \quad \updownarrow$$
$$T_1 \quad T_2 \quad T_3$$

Scheme 1

$$N_1 \quad N_2 \quad N_3$$
$$T_1 \quad T_2 \quad T_3$$

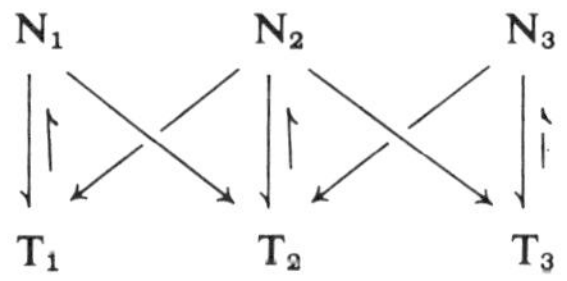

Scheme 2

Another possible mechanism for the observed racemization of 3,7,-dimethyl-7-methoxycarbonylcyclohepta-1,3,5-triene which, fortunately, is much easier to probe experimentally, would involve 1,4-dimethyl-3-methoxy-2-oxabicyclo[3.2.2]nona-3,6,8-triene as a key intermediate; all steps to and from this optically inactive molecule could be well-precedented six-electron allowed reactions.

In some reactions, it seems almost certain that a forbidden step is involved, but the precise reaction course may still be elusive. For instance, the dideuterio[4.4.2]propella-2,4-diene shown gives tetralin and both *cis*- and *trans*-1,2-dideuterioethylene when it is pyrolyzed (Paquette and Thompson, 1972).

Either a direct cycloelimination occurs with both an allowed and a forbidden stereochemical component, or perhaps a 1,3-carbon migration transpires in both *is* and *rs* modes to afford a short-lived intermediate, an ethylene–tetralin Diels–Alder adduct.

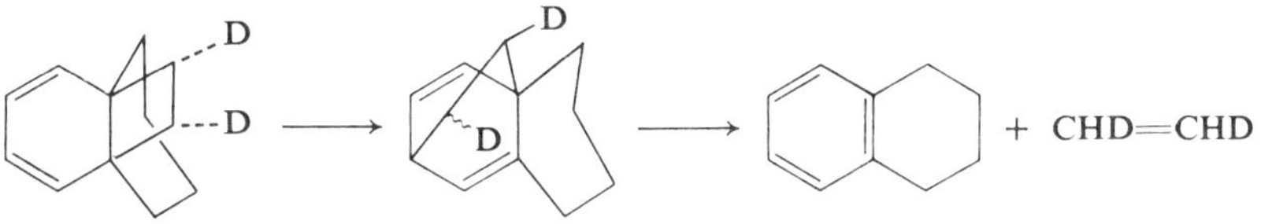

Most investigators concerned with the bicyclo[6.1.0]nona-2,4,6-triene isomerization to *cis*-dihydroindene seem agreed that a forbidden step leading to the all-cis isomer of cyclononatetraene is involved (Staley, 1971; Baldwin and Andrist, 1971b; Baldwin *et al.*, 1972a; Anastassiou and Griffith, 1973; Grimme, 1973; Boche and Schneider, 1974; Lewis and Brookhart, 1975) but whether the [5.2.0] or the [6.1.0] form of the folded conformer is the direct precursor to all-*cis*-cyclononatetraene remains unresolved.

Other cycloreactions seem to be clear examples of forbidden processes having no obvious indirect path based on either allowed or a combination of allowed and forbidden steps. Electrocyclic isomerizations, cycloadditions, and sigmatropic rearrangements may all occur with forbidden stereochemistry in apparently single stage processes (Landheer *et al.*, 1974; Vedejs *et al.*, 1974; Meinwald and Schmidt, 1969; Meinwald and Tsuruta, 1969; Zimmerman *et al.*, 1969).

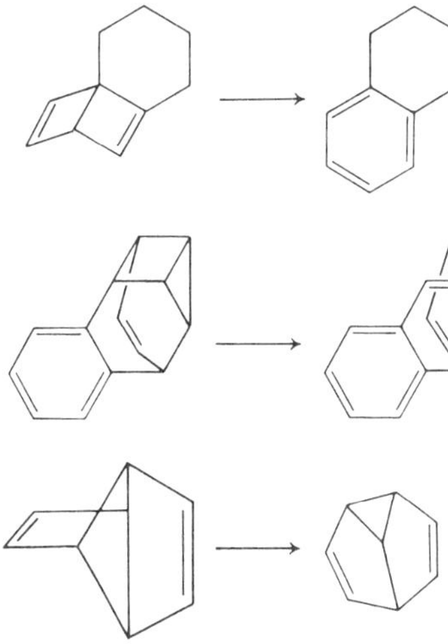

The rotation of ethylene about the C–C bond, which passes through planar and D_{2d} perpendicular forms every 180°, has been the subject of a configuration interaction molecular orbital calculation (Buenker, 1968; Buenker and Peyerimhoff, 1974). The calculated transition state is predicted to be some 43 kcal mole^{-1} lower than the hypothetical crossing of the single SCF configurations.

Geometrical distortions away from the most symmetrical of possible transition states may lower activation energies for both allowed (McIver, 1972, 1974; McIver and Stanton, 1972) and forbidden reactions (Salem *et al.*, 1975). An interesting example is provided by the addition of singlet methylene to ethylene: a one-electron molecular orbital analysis of the process suggests that the low-energy nondynamic reaction path begins as a π approach and terminates in the σ orientation (Hoffmann, 1968). Additional calculations, allowing for configuration interaction and optimal geometrical orientations of reactants, would provide some insight as to the relative importance of the two mutually dependent factors. Trajectory calculations would provide still additional appreciation of the complexities of the reaction dynamics (Wang and Karplus, 1973).

The thermal electrocyclic interconversion between the *trans*-dimethyldihydropyrene and the dimethyl[2.2]metacyclophane diene formulated is forbidden according to orbital symmetry theory; yet, it occurs quite readily ($\Delta H^{\ddagger} = 22.4$ kcal mole^{-1}). Configuration interaction between ground and doubly excited configurations provides a convincing and natural rationale for the isomerization (Schmidt, 1971).

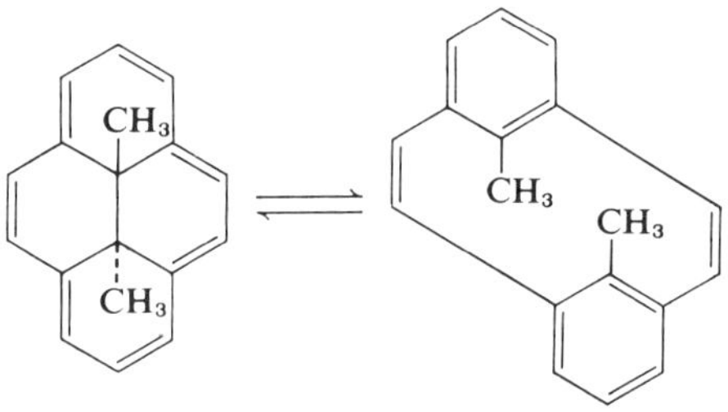

Quite generally, configuration interaction is to be expected at and near the crossing of substrate and product electronic configurations. The majority of observed forbidden sigmatropic processes may be energetically concerted, and the same inference may be valid for cycloeliminations and cycloadditions, as well, especially when substituents introducing low-lying excited singlet states are present (Schmidt, 1972).

State correlation in the thermal isomerization of *trans*-4a,4b-dihydrophenanthrene to *cis*-stilbene, a conrotatory and forbidden reaction which occurs readily at room temperature, is also achieved through configuration interaction. Here, as in the dihydropyrene–metacyclophane isomerization, the configuration interaction matrix element, according to a privately communicated estimate, was 5 kcal mole^{-1} (Muszkat and Schmidt, 1971).

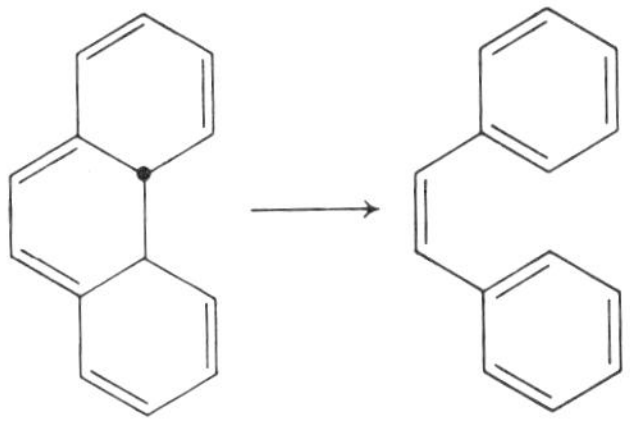

Calculations on the hypothetical interconversion of "bond" and "stretch" isomers of [2.2.2]propellane, using a two-configuration treatment, show that the calculated energy barrier is substantially lowered and the calculated transition state occurs more toward the bond isomer: the energy minima come at C(1)–C(4) bond lengths of 1.54 and 2.51 Å and the maximum at 1.95 Å, rather than at 2.35 Å, the C(1)–C(4) bond length at which the single-configuration curves cross (Newton and Schulman, 1972). Here, interestingly, the least-energy path is clearly on the singlet energy surface; the triplet state is greater than 1 eV above the singlet at the transition state near $r = 1.95$ Å. Other calculations on this novel type of isomerism in [2.2.2]propellane have been performed (Stohrer and Hoffmann, 1972a; Dannenberg and Prociv, 1973).

The reaction modeled is not likely to be observed: with the configurational mixing, the C(1)–C(4) bond becomes sufficiently antisymmetric to permit smooth reaction with the C(2)–C(3) bond and the formation of 1,4-dimethylenecyclohexane (Baldwin and Grayston, 1974).

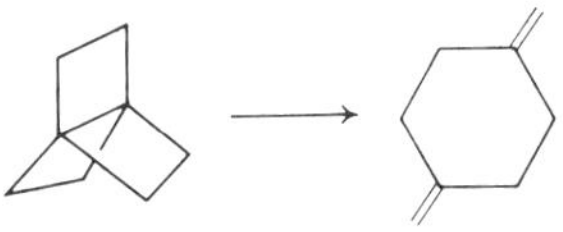

Known intramolecular cycloadditions between one single and one double bond in the *ss* stereochemical sense may be understood in terms of the comparatively facile configuration interaction possibilities available to cyclopropyl rings; either lengthening a cyclopropyl C–C bond or rotating methylene units at the ends of a cyclopropyl C–C bond leads to extensive configurational mixing and the acquisition of substantial antisymmetric character by the C–C bond (Hoffmann, 1968; Buenker and Peyerimhoff, 1969; Siu *et al.*, 1970; Hay *et al.*, 1972; Gregory, 1974). Cycloaddition with a proximate double bond, as in the tricyclo[3.2.1.0^{2,4}]oct-6-ene to tetracylo-[3.3.0.0^{2,8}.0^{4,6}]octane isomerizations formulated (Longone and Stehouwer, 1970; Prinzbach and Martin, 1968; Kinnel and Freeman, 1973; Block *et al.*, 1973; Martin, 1974b), then becomes a possible concerted reaction.

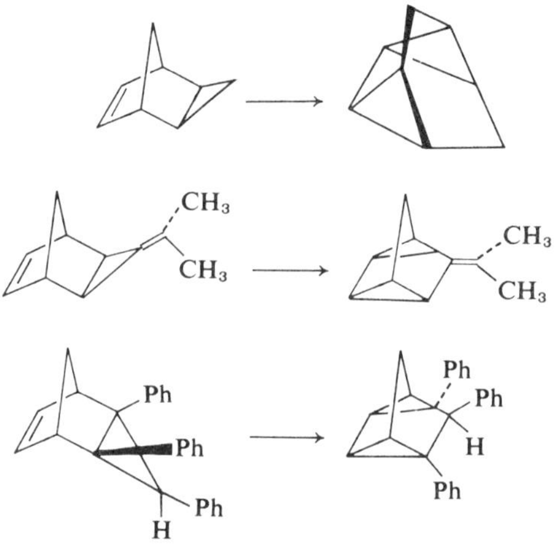

In the last case, it will be interesting to determine whether the endo isomer gives the triphenyltetracyclooctane directly, or by way of the *exo*-tricyclooctene.

The same basic rationale also explains the special propensity of cyclopropyl C–C bonds to show intramolecular cycloadditions to H–C and C–C single bonds, even in systems prohibited by stereochemical constraints from adopting the geometry requisite for a π-cyclopropane intermediate (Baldwin and Ollerenshaw, 1972; Flowers *et al.*, 1972; Baldwin and Grayston, 1974; J. E. Baldwin and F. Anderson, unpublished results, 1974).

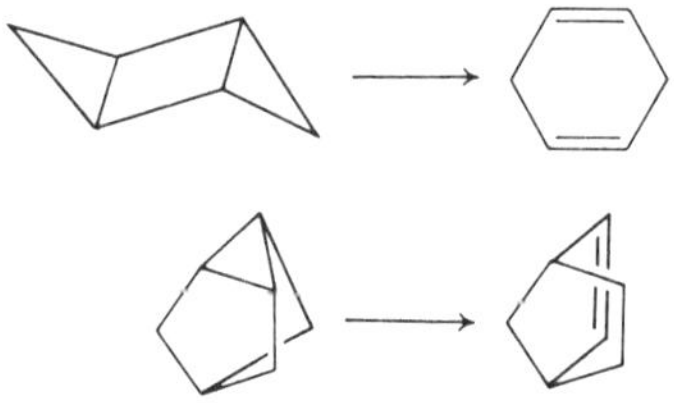

In some cases, the structure of the product may indicate which [$_\sigma 2 + _\sigma 2$] pathway is dominant. The tricyclohexane to cyclohexa-1,4-diene conversion is a C–C/C–C rearrangement, while the nortricyclene molecule prefers a C–C/H–C path: the C–C/C–C product, bicyclo[3.1.1]hept-2-ene (norpinene), is not observed (J. E. Baldwin and F. Anderson, unpublished results, 1974). In other isomerizations, the product may stem from several distinct processes, and isotopic labeling studies are required to make distinctions. In the bicyclopentane and bicyclo[3.1.0]hex-2-ene cases formulated (Baldwin and Andrews, 1973; Cooke and Andrews, 1973), C–C/H–C modes take place to the apparent exclusion of C–C/C–C rearrangements. In 2,4-dehydrohomoadamantane, both are of comparable importance according to carbon-13 labeling results (Baldwin and Grayston, 1974).

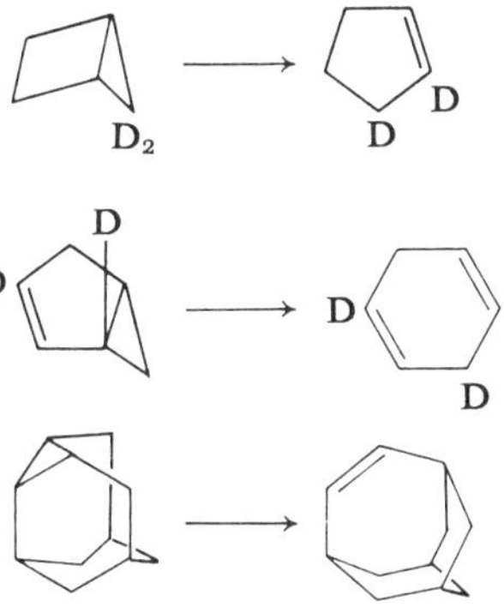

An *ab initio* minimum basis set calculation for the rectangular decomposition of cyclobutane to two molecules of ethylene has been done including configuration interaction between the two wave functions built from the dominant configuration of cyclobutane[...$(3b_{3u})^2(4b_{2u})^0$] and the dominant configuration used to describe two separate face-to-face ethylene molecules [...$(3b_{3u})^0(4b_{2u})^2$] (Wright and Salem, 1972). Near the transition state, both configurations are of importance: the interaction or variational mixing of the two gives a smooth reaction profile and a calculated energy stabilization of 56 kcal mole^{-1} relative to the energy of the crossing of single-configuration wave functions. Whether the actual transition state for the fragmentation would generally correspond to a much lower energy two-step pathway, or whether configuration interaction calculations on nonrectangular, nonplanar cyclobutane fragmentation paths would bring the calculated activation energy down much further, so that the concerted reaction of forbidden stereochemistry might seem accessible theoretically, remains to be seen.

The demonstration (Baldwin and Ford, 1969) that a thermal cyclobutane fragmentation could give both the allowed cis,trans (57%) and the forbidden cis,cis (43%) pairs of ethylenic products, with more of the allowed stereochemical outcome than could be accommodated by any diradical mechanistic scheme, has been interpreted most often as evidence favoring an allowed process superimposed on a stereochemically randomizing reaction (Goddard, 1972). An alternative hypothesis, competitive concerted allowed and forbidden pathways, while inherently simpler, and equally congruent with experimental fact, has not been widely espoused.

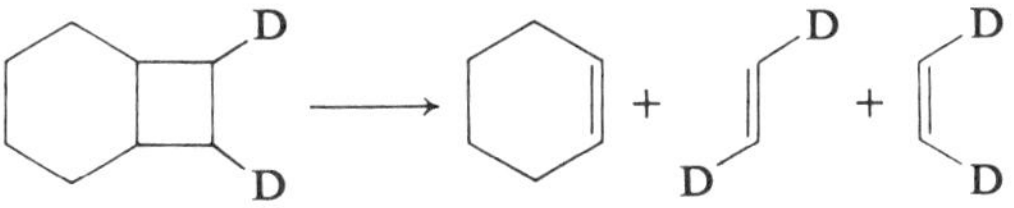

Using the MINDO/2 approximate molecular orbital method augmented through inclusion of configuration interaction with the first doubly excited configuration, the crucial additional configuration, calculated activation energies for forbidden hydrogen migrations in propene, the 1,3-suprafacial shift, and in *cis*-piperylene, the 1,5-antarafacial shift, are found to be 49 and 37 kcal mole^{-1}. These MINDO/CI results suggest both reactions may conceivably be observed in suitably prepared molecular environments (Bingham and Dewar, 1972).

The skeletal inversion of bicyclo[2.1.0]pentane has been calculated using a limited configuration interaction treatment (Collins *et al.*, 1972). The planar species halfway between the two folded versions of the hydrocarbon has almost equal weights of each of two configurations and is indicated to be a true intermediate. Both the electronic stabilization through configuration interaction and the stabilization at the planar form shown by the "core" energy, attributable to a relief of strain, contribute toward the overall effect.

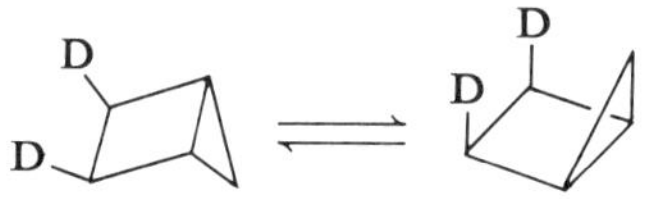

Kinetic studies on the skeletal inversions of deuterium labeled analogs of bicyclopentane and 1-methylbicyclopentane have shown that methyl substitution on the bridgehead carbon has very little effect on the activation parameters for the two-center inversion reaction (J. E. Baldwin and J. Ollerenshaw, unpublished results, 1974).

An *ab initio* study of the degenerate methylenecyclopropane rearrangement carried through with a linear combination of three singlet configurations has revealed the C_{2v}-symmetric midpoint along the reaction coordinate to be a shallow energy minimum, not a transition state maximum (Hehre *et al.*, 1974). Here, as in the bicyclopentane skeletal inversion, there is an indication of a relatively unstable singlet-state intermediate species.

An ambitious attempt to resolve the structure of the transition state for a simple yet significant reaction, the geometrical isomerization of cyclopropane, has been made utilizing SCF minimal basis set calculations and including configuration interaction between ground, a singly excited, and a doubly excited configuration in the region of the trimethylene diradical (Horsley *et al.*, 1972).

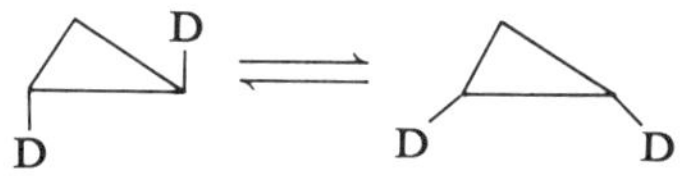

Geometrical variations over a 21-dimensional hypersurface led to "total" resolution of the structure of the transition state: the one-center epimerization reaction occurred with complicated coordination of relative motions at both ends of a trimethylene species; no secondary minimum indicative of a reaction intermediate was found. A two-step reaction mechanistic scheme with an intermediate diradical was not indicated. In accord with dominant opinion and the most exercised kinetic model, these calculations were based on the trimethylene single-center inversion formalism, yet they gave no support for involvement of an intermediate.

Whether these calculations give a faithful representation of the reaction is, however, open to question. The uncertainties in errors associated with minimum basis set calculations are explicitly recognized concerns; and the isomerization may not be a one-center epimerization! Recent data for the geometrical and optical isomerizations of optically active *trans*-1,2-dideuteriocyclopropane (Berson and Pedersen, 1975; Berson *et al.*, 1975) are consistent with all epimerization events being two-center processes!

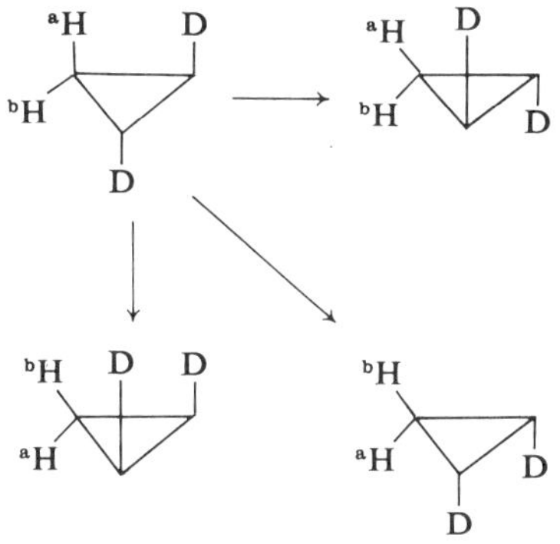

By symmetry, the three two-center epimerizations would have equal rates, and the rate constants for approach to optical and geometrical equilibrium would be equal—which is precisely the observed finding. Both diradical and one-center epimerization schemes predict other k_α/k_{geom} ratios.

This new discovery puts to confusion much conventional and conventionalized wisdom on the trimethylene species (Berson and Balquist, 1968; Carter and Bergman, 1968; Bergman and Carter, 1969; Bergman, 1973), for the diradical model applied to molecular rearrangements, fragmentations, and cycloadditions has, as an essential feature, kinetic competition between rotation about a single bond and formation (or reformation) of another. Such a formalism predicts one-center epimerizations occur faster than two-center epimerizations.

There are but few examples suggesting that one-center epimerizations of cyclopropanes may occur: bridged tricyclo[3.1.0]hexane interconversions have been formulated in these terms (Dauben and Kellogg, 1972; W. G. Dauben, private communication, 1975).

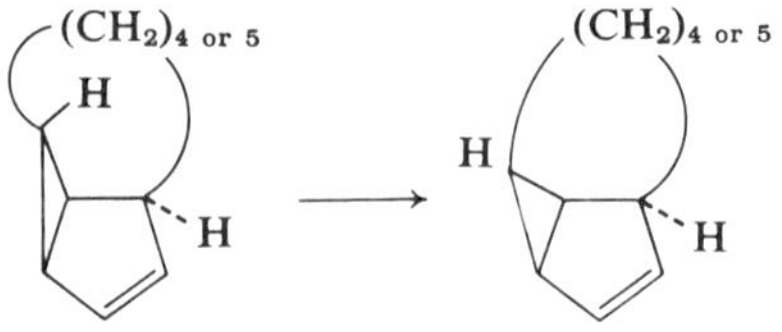

The thermal conversion of *trans*-1,2-divinylcyclopropanes to cyclohepta-1,4-dienes, which occur by way of *cis*-1,2-divinylcyclopropanes (Ullenius *et al.*, 1972; Baldwin and Ullenius, 1974), may involve one-center epimerizations or, conceivably, two-center epimerizations, at C(1) and C(3), or C(2) and C(3). The point has not yet been subjected to an experimental test. The same issue should be raised for the geometrical isomerizations of cyclobutanes, such as the cis to trans reaction of 1,2-diphenylcyclobutane (Jones and Chow, 1974). Is there epimerization at C(1), or at C(1) and C(4) simultaneously?

Whether two-center epimerizations of cyclopropanes are conrotatory or disrotatory is a hotly debated issue among some theoreticians. We note that many examples of the disrotatory reaction stereochemistry are known in bicyclopentanes, bicyclo[3.1.0]hex-2-enes, and similar fused hydrocarbons, and that the disrotatory reaction is intrinsically allowed. Counter examples of conrotatory cyclopropane two-center epimerizations are conspicuously lacking. They may well be possible, but they have not yet been demonstrated.

One of the most commonly encountered and extensively investigated type of forbidden reactions is 1,3-migration of carbon with retention, suprafacial stereochemistry, as in the isomerization of *trans*-1,2-dipropenylcyclobutanes (Berson and Dervan, 1973).

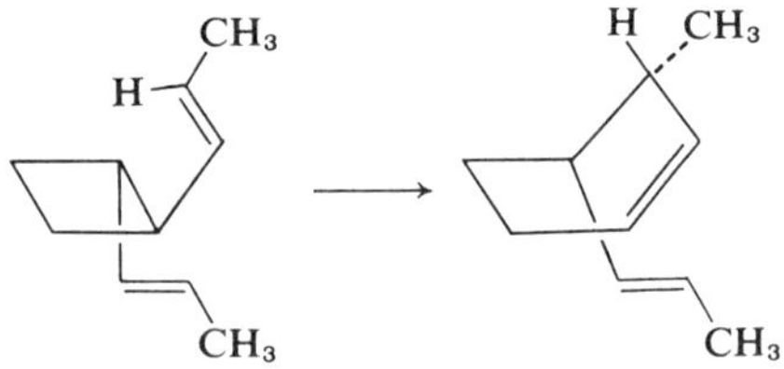

The simplest one-electron molecular orbital analysis for [1,3]-sigmatropic shifts with the forbidden retention, suprafacial stereochemistry predicts such reactions may occur without involvement of diradical intermediates (Berson, 1972; Berson and Salem, 1972). Mixing between a nonbonding orbital and other orbitals of the same symmetry leads to a net stabilization, even though the lower energy orbital stabilized through the interaction is "subjacent," that is, lower in energy, than some nonbonding orbital of another symmetry. The conclusion drawn: stereoselective forbidden reactions may be expected to the extent that the subjacent orbital factor controls the reaction. Approximate inclusion of two-electron effects in a more refined version of the same model process gives the same result (Borden and Salem, 1973).

Whether this model provides a reliable understanding of 1,3-retention, suprafacial shifts remains to be seen. It postulates a symmetrical transition state geometry; it is independent of the type of sigma bond cleaved in the rearrangement; it ignores possible through-bond interactions (Hoffmann, 1971; Gleiter, 1974) between the migrating atom and the allylic unit.

Even one example of this reaction stereochemistry for a system in which the bond broken was not disposed originally in a three- or four-membered ring, or flanked by unsaturation as in the degenerate tricyclic triene process shown (Paquette and Kukla, 1972), would give significant support for the transition state model proposed.

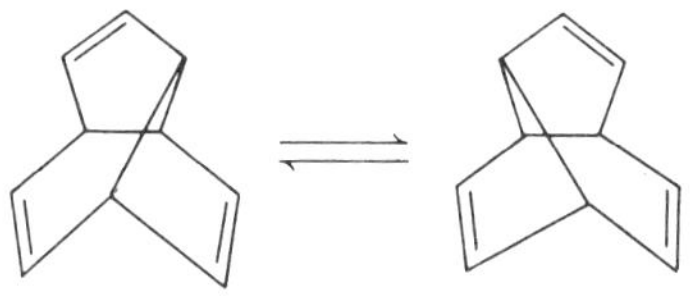

Without such an instance, one may suppose that these rearrangements occur thanks to the antisymmetric character cyclopropyl or cyclobutyl or highly conjugated single bonds assume as they are elongated, and to the opportunities they thereby gain for energetically concerted reactions with an adjacent double bond used suprafacially.

Those few 1,3-carbon retention, suprafacial shifts which are degenerate, such as those shown to contribute to the [1,3]-sigmatropic rearrangements of bicyclo[3.1.0]hex-2-enes (Cooke and Andrews, 1974; Doering and Schmidt, 1971; Swenton and Wexler, 1971) and methylenecyclobutanes (Baldwin and Fleming, 1972, 1973) may well involve intermediates at the C_s-symmetrical geometry (Baldwin and Fleming, 1970) and such extensive through-bond coupling between the migrating atom and the allylic moiety that the subjacent orbital model would be a misleading guide to structural or mechanistic characteristics of the isomerizations.

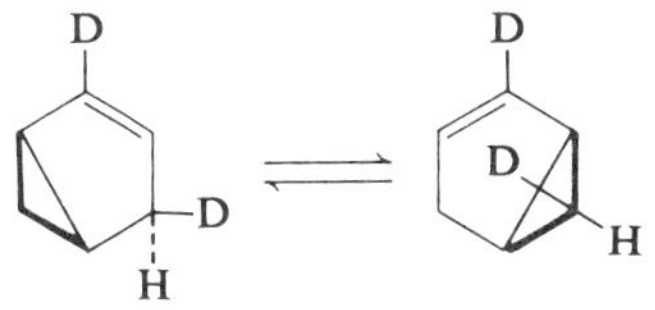

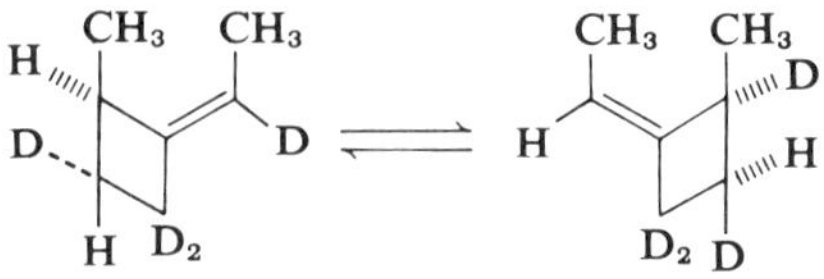

Spin-forbidden crossing from a substrate ground state to a triplet state of a product may occur under appropriate circumstances. Bicyclo[2.2.0]hexa-2,5-diene is rearranged thermally to benzene with production of an indirectly detectable yield of benzene triplets: it is estimated that only one rearranging molecule out of 10^4 gives triplet benzene, which suggests that state crossing is a participating but not a dominant mechanistic path for the reaction (Breslow et al., 1972; Lechtken et al., 1973). Similar studies on the thermal conversions of benzvalene and naphthvalene to benzene and naphthalene (and other products) have been inconclusive (Turro and Lechtken, 1973).

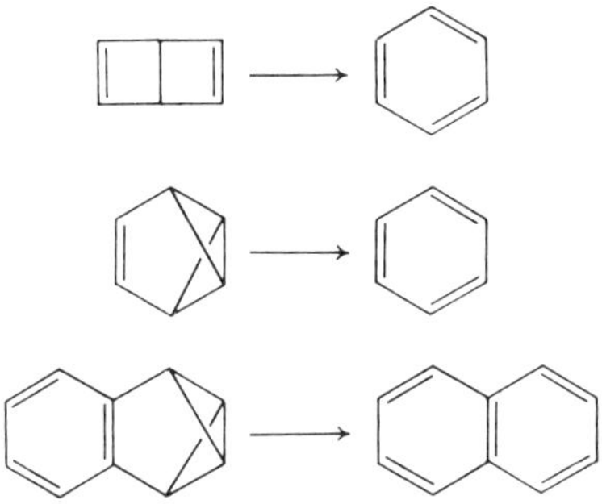

Naphtho[b-2,3]bicyclo[2.2.0]hexa-2,5-diene gives anthracene without chemiluminescence; although the energetics of the reaction would permit population of either S_1 or T_1 of the product, neither excited state seems to be produced (Yang et al., 1974).

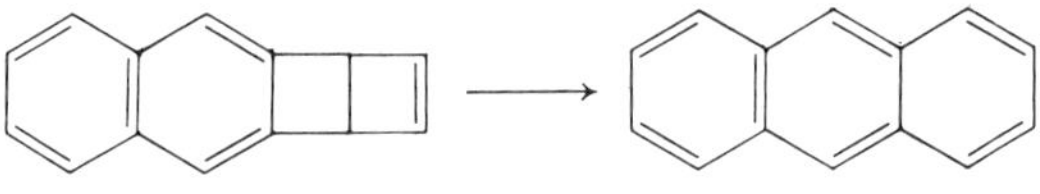

The suggestion that the syn and anti dimers of cyclobutadiene are converted thermally to cyclooctatetraene, through intersystem crossing and triplet state species (Case et al., 1974) has been disputed (Frey et al., 1975).

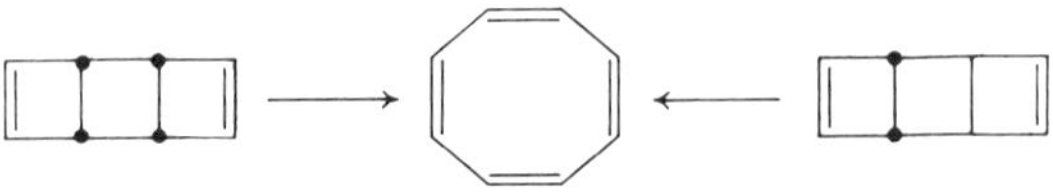

Additional work directed toward detecting chemiluminescence and other consequences of intersystem crossing accompanying thermal reactions may be anticipated.

IV. Conclusions

One electron molecular orbital methods cannot give a plausible account of thermally forbidden reactions. Through a more versatile molecular orbital theoretical model, one including configuration interaction, it becomes evident that orbital symmetry forbidden reactions may be concerted.

The extent to which configuration interaction may be included, through presently available computation methods, is severely limited; when two or three configurations are considered, the calculated potential surface for a forbidden reaction will not have the artifacts and discontinuities inherent in single-determinental molecular orbital treatments, but it will not necessarily be a quantitatively reliable approximation to the real surface or to the calculated surface one would deduce if several hundred configurations could be included. Attempts to include more configurations in calculations on the molecular transformations of small organic hydrocarbons may be anticipated in the near future (Alston *et al.*, 1975; G. Binsch, private communication, 1975).

Forbidden reactions may be concerted and they may, also, be indirect, requiring more than one elementary step. The experimental challenge, then, is to determine which alternative obtains in given cases, and to attempt to correlate such determinations with structural idiosyncrasies and theoretical predictions. This challenge is not rendered inoperative by a Law of Nature limiting orbital symmetry forbidden reactions to nonconcerted reaction profiles.

Acknowledgments

Support for our research from the National Science Foundation and Hoffmann-La Roche, Inc. is gratefully acknowledged. Most of this chapter was written in 1974–1975 at the Institut für Organische Chemie der Universität Hamburg; the hospitality of Professor Dr. K. Heyns and his colleagues, and an award from the Alexander von Humboldt Stiftung, proved of great assistance to the project.

References

Allred, E. L., and Beck, B. R. (1973). *J. Am. Chem. Soc.* **95**, 2393.
Alston, P. V., Shillady, D. D., and Trindle, C. (1975). *J. Am. Chem. Soc.* **97**, 469.

Anastassiou, A. G., and Griffith, R. C. (1973). *J. Am. Chem. Soc.* **95**, 2379.
Andrews, G. D., Davalt, M., and Baldwin, J. E. (1973). *J. Am. Chem. Soc.* **95**, 5044.
Andrist, A. H. (1973). *J. Org. Chem.* **38**, 1772.
Baldwin, J. E., and Andrews, G. D. (1972). *J. Am. Chem. Soc.* **94**, 1775.
Baldwin, J. E., and Andrews, G. D. (1973). *J. Org. Chem.* **38**, 1063.
Baldwin, J. E., and Andrist, A. H. (1971a). *J. Am. Chem. Soc.* **93**, 3289.
Baldwin, J. E., and Andrist, A. H. (1971b). *J. Am. Chem. Soc.* **93**, 4055.
Baldwin, J. E., and Bryan, D. B. (1974). *J. Am. Chem. Soc.* **96**, 319.
Baldwin, J. E., and Fleming, R. H. (1970). *Fortschr. Chem. Forsch.* **15**, 281.
Baldwin, J. E., and Fleming, R. H. (1972). *J. Am. Chem. Soc.* **94**, 2140.
Baldwin, J. E., and Fleming, R. H. (1973). *J. Am. Chem. Soc.* **95**, 5249, 5256, and 5261.
Baldwin, J. E., and Ford, P. W. (1969). *J. Am. Chem. Soc.* **91**, 7192.
Baldwin, J. E., and Grayston, M. W. (1974). *J. Am. Chem. Soc.* **96**, 1629 and 1630.
Baldwin, J. E., and Kaplan, M. S. (1971). *J. Am. Chem. Soc.* **93**, 3969.
Baldwin, J. E., and Kaplan, M. S. (1972). *J. Am. Chem. Soc.* **94**, 668 and 4696.
Baldwin, J. E., and Ollerenshaw, J. (1972), *Tetrahedron Lett.* p. 3757.
Baldwin, J. E., and Pinschmidt, R. K., Jr. (1971). *Chem. Commun.* p. 820.
Baldwin, J. E., and Ullenius, C. (1974). *J. Am. Chem. Soc.* **95**, 1542.
Baldwin, J. E., Andrist, A. H., and Pinschmidt, R. K., Jr. (1972a). *J. Am. Chem. Soc.* **94**, 5845.
Baldwin, J. E., Andrist, A. H., and Pinschmidt, R. K., Jr. (1972b). *Acc. Chem. Res.* **5**, 402.
Bergman, R. G. (1973). *In* "Free Radicals" (J. K. Kochi, ed.), p. 191. Wiley (Interscience), New York.
Bergman, R. G., and Carter, W. L. (1969). *J. Am. Chem. Soc.* **91**, 7411.
Berson, J. A. (1968). *Acc. Chem. Res.* **1**, 152.
Berson, J. A. (1972). *Acc. Chem. Res.* **5**, 406.
Berson, J. A., and Balquist, J. M. (1968). *J. Am. Chem. Soc.* **90**, 7343.
Berson, J. A., and Dervan, P. B. (1973). *J. Am. Chem. Soc.* **95**, 269.
Berson, J. A., and Pedersen, L. D. (1975). *J. Am. Chem. Soc.* **97**, 238.
Berson, J. A., and Salem, L. (1972). *J. Am. Chem. Soc.* **94**, 8917.
Berson, J. A., Boettcher, R. R., and Vollmer, J. J. (1971). *J. Am. Chem. Soc.* **93**, 1540.
Berson, J. A., Miyashi, T., and Jones, G., II. (1974). *J. Am. Chem. Soc.* **96**, 3468.
Berson, J. A., Pedersen, L. D., and Carpenter, B. K. (1975). *J. Am. Chem. Soc.* **97**, 240.
Bertrand, M., Gras, J.-L., and Gil, G. (1974). *Tetrahedron Lett.* p. 37.
Bingham, R. C., and Dewar, M. J. S. (1972). *J. Am. Chem. Soc.* **94**, 9107.
Block, R., Leyendecker, F., and Toshima, N. (1973). *Tetrahedron Lett.* p. 1025.
Boche, G., and Schneider, G. (1974). *Tetrahedron Lett.* p. 2499.
Boche, G., and Szeimies, G. (1971). *Angew. Chem., Int. Ed. Engl.* **10**, 911 and 912.
Boche, G., Weber, H., and Benz, J. (1974). *Angew. Chem., Int. Ed. Engl.* **13**, 207.
Borden, W. T., and Salem, L. (1973). *J. Am. Chem. Soc.* **95**, 932.
Brauman, J. I., Farneth, W. E., and D'Amore, M. B. (1973). *J. Am. Chem. Soc.* **95**, 5043.
Breslow, R., Napierski, J., and Schmidt, A. H. (1972). *J. Am. Chem. Soc.* **94**, 5906.
Buenker, R. J. (1968). *J. Chem. Phys.* **48**, 1368.
Buenker, R. J., and Peyerimhoff, S. D. (1969). *J. Phys. Chem.* **73**, 1299.
Buenker, R. J., and Peyerimhoff, S. D. (1974). *Chem. Rev.* **74**, 127.
Čársky, P., and Zahradník, R. (1972). *Theor. Chim. Acta* **26**, 171.
Carter, W. L., and Bergman, R. G. (1968). *J. Am. Chem. Soc.* **90**, 7344.
Case, R. S., Dewar, M. J. S., Kirschner, S., Pettit, R., and Slegir, W. (1974). *J. Am. Chem. Soc.* **96**, 7581.

Caserio, M. C. (1971). *J. Chem. Educ.* **48**, 782.
Collins, F. S., George, J. K., and Trindle, C. (1972). *J. Am. Chem. Soc.* **94**, 3732.
Cooke, R. S., and Andrews, U. H. (1973). *J. Org. Chem.* **38**, 2725.
Cooke, R. S., and Andrews, U. H. (1974). *J. Am. Chem. Soc.* **96**, 2974.
Criegee, R., Schönleber, D., Huber, R., Schweickhardt, C., Wolf, R., and Ramirez, R. (1973). *Chem. Ber.* **106**, 857.
Dannenberg, J. J., and Prociv, T. M. (1973). *Chem. Commun.* p. 291.
Dauben, W. G., and Kellogg, M. S. (1972). *J. Am. Chem. Soc.* **94**, 8951.
Dauben, W. G., and Whalen, W. L. (1966). *Tetrahedron Lett.* p. 3743.
Dauben, W. G., and Willey, F. G. (1962). *Tetrahedron Lett.* p. 893.
Daudey, J. P., Langlet, J., and Malrieu, J. P. (1974). *J. Am. Chem. Soc.* **96**, 3393.
De Maré, G. R., Huybrechts, G., and Toth, M. (1972). *J. Chem. Soc., Perkin Trans. 2* p. 1256.
de Meijere, A. (1974). *Chem. Ber.* **107**, 1702.
Devaquet, A. (1974). *Plenary Lect. IUPAC Photochem. Symp., Enschede, Holland.*
Dewar, M. J. S. (1975). *Chem. Br.* **11**, 97.
Dewar, M. J. S., and Kelemen, J. (1972). *J. Chem. Educ.* **48**, 494.
Doering, W. von E., and Sachdev, K. (1974). *J. Am. Chem. Soc.* **96**, 1168.
Doering, W. von E., and Schmidt, E. K. G. (1971). *Tetrahedron* **27**, 2005.
Dougherty, R. C. (1971). *J. Am. Chem. Soc.* **93**, 7187.
Ehrenson, S. (1974). *J. Am. Chem. Soc.* **96**, 3778 and 3784.
Epiotis, N. D. (1972). *J. Am. Chem. Soc.* **94**, 1924, 1935, 1941, and 1946.
Epiotis, N. D. (1973). *J. Am. Chem. Soc.* **95**, 1191, 1200, 1206, and 1214.
Epiotis, N. D. (1974). *Angew. Chem., Int. Ed. Engl.* **13**, 751.
Evans, M. G. (1939). *Trans. Faraday Soc.* **35**, 824.
Evans, M. G., and Polanyi, M. (1938). *Trans. Faraday Soc.* **34**, 11.
Evans, M. G., and Warhurst, E. (1938). *Trans. Faraday Soc.* **34**, 614.
Ferreira, R. (1973). *J. Theor. Biol.* **39**, 665.
Fleming, I., and Wildsmith, E. (1970). *Chem. Commun.* p. 223.
Flowers, M. C., and Frey, H. M. (1962). *J. Chem. Soc.* p. 1689.
Flowers, M. C., and Frey, H. M. (1972). *J. Am. Chem. Soc.* **94**, 8636.
Flowers, M. C., Frey, H. M., and Hopf, H. (1972). *Chem. Commun.* p. 1284.
Frey, H. M., Martin, H.-D., and Hekman, M. (1975). *Chem. Commun.* p. 204.
Fujimoto, H., and Fukui, K. (1972). *Adv. Quantum Chem.* **6**, 177.
Fujimoto, H., Miyagi, M., Yamabe, S., and Fukui, K. (1973). *Bull. Chem. Soc. Jpn.* **46**, 1357.
Fukui, K. (1964). *In* "Molecular Orbitals in Chemistry, Physics, and Biology" (P.-O. Löwdin and B. Pullman, eds.), p. 513. Academic Press, New York.
Fukui, K. (1971). *Acc. Chem. Res.* **4**, 57.
Gajewsky, J. J., and Burka, L. T. (1972), *J. Am. Chem. Soc.* **94**, 8860 and 8865.
George, T. F., and Ross, J. (1971). *J. Chem. Phys.* **55**, 3851.
Gilbert, J. C., and Higley, D. P. (1973). *Tetrahedron Lett,* p. 2075.
Gleiter, R. (1974). *Angew. Chem., Int. Ed. Engl.* **13**, 696.
Goddard, W. A., III. (1970). *J. Am. Chem. Soc.* **92**, 7520.
Goddard, W. A., III. (1972). *J. Am. Chem. Soc.* **94**, 793.
Goe, G. L. (1972). *J. Org. Chem.* **27**, 2434.
Goldstein, M. J., and Benzon, M. S. (1972). *J. Am. Chem. Soc.* **94**, 7147.
Gregory, A. R. (1974). *J. Chem. Phys.* **60**, 1680.
Grimme, W. (1973). *J. Am. Chem. Soc.* **95**, 2381.
Grimme, W., and Doering, W. von E. (1973). *Chem. Ber.* **106**. 1765.

Havinga, E., and Schlatmann, J. L. M. A. (1961). *Tetrahedron* **16**, 146.
Hay, P. J., Hunt, W. J., and Goddard, W. A. III. (1972). *J. Am. Chem. Soc.* **94**, 638.
Hehre, W. J., Salem, L., and Willcott, M. R. (1974). *J. Am. Chem. Soc.* **96**, 4328.
Heimgartner, H., Hansen, H.-J., and Schmid, H. (1972). *Helv. Chim. Acta* **55**, 1385.
Herzberg, G. (1950). "Molecular Spectra and Molecular Structure," 2nd ed., Vol. 1, p. 315f. Van Nostrand-Reinhold, Princeton, New Jersey.
Hoffmann, R. (1968). *J. Am. Chem. Soc.* **90**, 1475.
Hoffmann, R. (1971). *Acc. Chem. Res.* **4**, 1.
Hoffmann, R. (1973). *Kem. Tidskr.* **85**, No. 3, 28.
Hoffmann, R., and Woodward, R. B. (1965). *J. Am. Chem. Soc.* **87**, 2046.
Hoffmann, R., and Woodward, R. B. (1968). *Acc. Chem. Res.* **1**, 17.
Hoffmann, R., and Woodward, R. B. (1970). *Science* **167**, 825.
Hoffmann, R., and Woodward, R. B. (1972). *Chem. Unserer Zeit* **6**, 167.
Hoffmann, R., Imamura, A., and Hehre, W. J. (1968). *J. Am. Chem. Soc.* **90**, 1499.
Horsley, J. A., Jean, Y., Moser, C., Salem, L., Stevens, R. M., and Wright, J. J. (1972). *J. Am. Chem. Soc.* **94**, 279.
Jakubetz, W., Lischka, H., Rosmus P., and Schuster, P. (1971). *Chem. Phys. Lett.* **11**, 38.
Jones, G., II, and Chow, V. L. (1974). *J. Org. Chem.* **39**, 1447.
Jordan, K. D., and Silbey, R. (1973). *Chem. Phys. Lett.* **18**, 27.
Katz, H. (1971). *J. Chem. Educ.* **18**, 84.
Kende, A. S., and Riecke, E. E. (1972). *J. Amer. Chem. Soc.* **94**, 1397.
Kinnel, R. B., and Freeman, P. K. (1973). *Tetrahedron Lett.* p. 4803.
Klärner, F.-G. (1974). *Angew. Chem., Int. Ed. Engl.* **13**, 268.
Kollmar, H. (1971). *Chem. Phys. Lett.* **8**, 533.
Landheer, I. J., de Wolf, W. H., and Bickelhaupt, F. (1974). *Tetrahedron Lett.* p. 2813.
Langlet, J., and Malrieu, J.-P. (1972). *J. Am. Chem. Soc.* **94**, 7254.
Lechtken, P., Breslow, R., Schmidt, A. H., and Turro, N. J. (1973). *J. Am. Chem. Soc.* **95**, 3025.
Lewis, C. P., and Brookhart, M. (1975). *J. Am. Chem. Soc.* **97**, 651.
Libit, L., and Hoffmann, R. (1974). *J. Am. Chem. Soc.* **96**, 1370.
Lombardo, L., and Wege, D. (1972). *Tetrahedron Lett.* p. 4859.
Longone, D. T., and Stehouwer, D. M. (1970). *Tetrahedron Lett.* p. 1017.
Longuet-Higgins, H. C., and Abrahamson, E. W. (1965). *J. Am. Chem. Soc.* **87**, 2045.
Lowe, J. P. (1974). *J. Am. Chem. Soc.* **96**, 3759.
McIver, J. W., Jr. (1972). *J. Am. Chem. Soc.* **94**, 4782.
McIver, J. W., Jr. (1974). *Acc. Chem. Res.* **7**, 72.
McIver, J. W., Jr., and Stanton, R. E. (1972). *J. Am. Chem. Soc.* **94**, 8618.
McLean, S., Findlay, D. M., and Dmitrienko, G. I. (1972). *J. Am. Chem. Soc.* **94**, 1380.
Mango, F. D. (1972). *Intra-Sci. Chem. Rep.* **6**, 171.
Mango, F. D. (1973). *Tetrahedron Lett.* p. 1509.
Mango, F. D. (1974). *Fortsch. Chem. Forsch.* **45**, 39.
Martin, H.-D. (1974a), *Nachr. Chem. Tech.* **22**, 412.
Martin, H.-D. (1974b). *Chem. Ber.* **107**, 477.
Mathieu, J. (1973). *Bull. Soc. Chim. Fr.* p. 807.
Meinwald, J., and Gruber, G. W. (1971). *J. Am. Chem. Soc.* **93**, 3802.
Meinwald, J., and Kapecki, J. A. (1972). *J. Am. Chem. Soc.* **94**, 6235.
Meinwald, J., and Schmidt, D. (1969). *J. Am. Chem. Soc.* **91**, 5877.
Meinwald, J., and Tsuruta, H. (1969). *J. Am. Chem. Soc.* **91**, 5877.
Meinwald, J., Dunkerton, L. V., and Gruber, G. W. (1975). *J. Am. Chem. Soc.* **97**, 681.
Merlet, P., Peyerimhoff, S. D., Buenker, R. J., and Shih, S. (1974). *J. Am. Chem. Soc.* **96**, 959.

Michl, J. (1972). *Mol. Photochem.* **4**, 243, 257, and 287.
Michl, J. (1974). *Top. Curr. Chem.* **46**, 1.
Mulder, J. J. C., and Oosterhoff, L. J. (1970). *Chem. Commun.* pp. 305 and 307.
Mulder, J. J. C., and Oosterhoff, L. J. (1971). In "Aromaticity, Pseudo-Aromaticity, Anti-Aromaticity" (E. D. Bergmann and B. Pullman, eds.), p. 34. Isr. Acad. Sci. Hum., Jerusalem.
Muszkat, K. A., and Schmidt, W. (1971). *Helv. Chim. Acta* **54**, 1195.
Muszkat, K. A., and Sharafi-Ozeri, S. (1973). *Chem. Phys. Lett.* **20**, 397.
Nelsen, S. F., and Gillespie, J. P. (1972). *J. Am. Chem. Soc.* **94**, 6237 and 6238.
Newton, M. D., and Schulman, J. M. (1972). *J. Am. Chem. Soc.* **94**, 4391.
Nohira, H. (1974). *Tetrahedron Lett.* p. 2573.
Oosterhoff, L. J. (1961). Quoted in Havinga and Schlatmann (1961), p. 151.
Orchard, S. W., and Thrush, B. A. (1973). *Chem. Commun.* p. 14.
Paddon-Row, M. N., Watson, P. L., and Warrener, R. N. (1973). *Tetrahedron Lett.* p. 1033.
Paquette, L. A., and Kukla, M. J. (1972). *J. Am. Chem. Soc.* **94**, 6874.
Paquette, L. A., and Thompson, G. L. (1972). *J. Am. Chem. Soc.* **94**, 7127.
Paquette, L. A., Kukla, M. J., and Stowell, J. C. (1972). *J. Am. Chem. Soc.* **94**, 4920.
Perrin, C. L. (1972). *Chem. Br.* **8**, 163.
Peyerimhoff, S. D., and Buenker, R. J. (1974). *Ber. Bunsenges. Phys. Chem.* **78**, 119.
Pople, J. A. (1971). *Int. J. Quantum Chem., Symp.* **5**, 175.
Prinzbach, H., and Martin, H.-D. (1968). *Helv. Chim. Acta* **51**, 438.
Ramage, R., and Sattar, A. (1970). *Chem. Commun.* p. 173.
Rastelli, A., Pozzoli, A. P., and Del Re, G. (1972). *J. Chem. Soc., Perkin Trans. 2* p. 1571.
Rose, T. L. (1973). *J. Am. Chem. Soc.* **95**, 3500.
Roth, W. R., and Enderer, K. (1970). *Justus Liebigs Ann. Chem.* **733**, 44.
Roth, W. R., and Schmidt, T. (1971). *Tetrahedron Lett.* p. 3639.
Salem, L. (1972). In "Computational Methods for Large Molecules and Localized States in Solids" (F. Herman, A. D. McLean, and R. K. Nesbet, eds.), p. 23. Plenum, New York.
Salem, L. (1973). *Pure Appl. Chem.* **33**, 317.
Salem, L. (1974). *J. Am. Chem. Soc.* **96**, 3486.
Salem, L., and Rowland, C. (1972). *Angew. Chem., Int. Ed. Engl.* **11**, 92.
Salem, L., Leforestier, C., Segal, G., and Wetmore, R. (1975). *J. Am. Chem. Soc.* **97**, 479.
Schiess, P., and Fünfschilling, P. (1972). *Tetrahedron Lett.* p. 5195.
Schmidt, W. (1971). *Helv. Chim. Acta* **54**, 862.
Schmidt, W. (1972). *Tetrahedron Lett.* p. 581.
Seebach, D. (1968). *Fortschr. Chem. Forsch.* **11**, 177.
Shuler, K. E. (1953). *J. Chem. Phys.* **21**, 624.
Silver, D. M. (1974). *J. Am. Chem. Soc.* **96**, 5959.
Simmons, H. E., and Bunnett, J. F. (1974). "Orbital Symmetry Papers" (ACS Reprint Collection). Am. Chem. Soc., Washington, D.C.
Siu, A. K. Q., St. John, W. M., and Hayes, E. F. (1970). *J. Am. Chem. Soc.* **92**, 7249.
Slater, J. C. (1963). "Quantum Theory of Molecules and Solids. Vol. I. Electronic Structure of Molecules," p. 62f. McGraw-Hill, New York.
Spangler, C. W., and Boles, D. L. (1972). *J. Org. Chem.* **37**, 1020.
Staley, S. W. (1971). *Intra-Sci. Chem. Rep.* **5**, 149.
Stephenson, L. M., Jr., and Brauman, J. I. (1974). *Acc. Chem. Res.* **7**, 65.
Stephenson, L. M., Jr., Gibson, T. A., and Brauman, J. I. (1973). *J. Am. Chem. Soc.* **95**, 2849.

Stohrer, W. D., and Hoffmann R. (1972a). *J. Am. Chem. Soc.* **94**, 779.
Stohrer, W. D., and Hoffmann, R. (1972b). *J. Am. Chem. Soc.* **94**, 1661.
Swenton, J. S., and Wexler, A. (1971). *J. Am. Chem. Soc.* **93**, 3066.
Tanny, S. R., Grossman, J., and Fowler, F. W. (1972). *J. Am. Chem. Soc.* **94**, 6495.
Tee, O. S. (1969). *J. Am. Chem. Soc.* **91**, 7144.
Tee, O. S., and Yates, K. (1972). *J. Am. Chem. Soc.* **94**, 3074.
Tee, O. S., Altmann, J. A., and Yates, K. (1974). *J. Am. Chem. Soc.* **96**, 3141.
Trindle, C. (1969). *J. Am. Chem. Soc.* **91**, 4936.
Trindle, C. (1970). *J. Am. Chem. Soc.* **92**, 3251 and 3255.
Trindle, C., and Collins, F. S. (1971). *Int. J. Quantum Chem., Symp.* **4**, 195.
Turro, N. J., and Lechtken, P. (1973). *Pure Appl. Chem.* **33**, 363.
Ullenius, C., Ford, P. W., and Baldwin, J. E. (1972). *J. Am. Chem. Soc.* **94**, 5910.
van der Hart, W. J., Mulder, J. J. C., and Oosterhoff, L. J. (1972). *J. Am. Chem. Soc.* **94**, 5724.
van der Lugt, W. T. A. M., and Oosterhoff, L. J. (1968). *Chem. Commun.* p. 1235.
van der Lugt, W. T. A. M., and Oosterhoff, L. J. (1969). *J. Am. Chem. Soc.* **91**, 6042.
van der Lugt, W. T. A. M., and Oosterhoff, L. J. (1970). *Mol. Phys.* **18**, 177.
Vedejs, E. (1971). *Chem. Commun.* p. 536.
Vedejs, E., Steiner, R. P., and Wu, E. S. C. (1974). *J. Am. Chem. Soc.* **96**, 4040.
Vollmer, J. J., and Servis, K. L. (1968). *J. Chem. Educ.* **45**, 214.
Vollmer, J. J., and Servis, K. L. (1970). *J. Chem. Educ.* **47**, 491.
Waali, E. E., and Jones, W. M. (1973a). *J. Org. Chem.* **38**, 2573.
Waali, E. E., and Jones, W. M. (1973b). *J. Am. Chem. Soc.* **95**, 8114.
Wang, I. S. Y., and Karplus, M. (1973). *J. Am. Chem. Soc.* **95**, 8160.
Weltin, E. E. (1974). *J. Am. Chem. Soc.* **96**, 3049.
Westburg, H. H., Cain, E. N., and Masamune, S. (1969). *J. Am. Chem. Soc.* **91**, 7512.
Wigner, E., and Witmer, E. E. (1928). *Z. Phys.* **51**, 859.
Wilson, E. B., and Wang, P. S. C. (1972). *Chem. Phys. Lett.* **15**, 400.
Wilson, E. B., and Wang, P. S. C. (1973). In "Energy, Structure, and Reactivity: Proceedings of the 1972 Boulder Summer Research Conference on Theoretical Chemistry" (D. W. Smith and W. B. McRae, eds.), p. 42ff. Wiley, New York.
Wollenberg, R. H., and Belloli, R. (1974). *Chem. Br.* **10**, 95.
Woodward, R. B. (1967). *Chem. Soc., Spec. Publ.* **21**, 217.
Woodward, R. B., and Hoffmann, R. (1965). *J. Am. Chem. Soc.* **87**, 395.
Woodward, R. B., and Hoffmann, R. (1969). *Angew. Chem., Int. Ed. Engl.* **8**, 781.
Woodward, R. B., and Hoffmann, R. (1971). "The Conservation of Orbital Symmetry." Academic Press, New York.
Wright, J. S., and Salem, L. (1972). *J. Am. Chem. Soc.* **94**, 322.
Yang, N. C., Carr, R. V., Li, E., McVey, J. K., and Rice, S. A. (1974). *J. Am. Chem. Soc.* **96**, 2297.
Zimmerman, H. E., Robbins, J. D., and Schantl, J. (1969) *J. Am. Chem. Soc.* **91**, 5878.

Author Index

Numbers in *italics* refer to the pages on which the complete references are listed.

A

Abrahamson, E. W., 183, 241, 243, *269*, 276, *300*
Agosta, W., 126, *135*
Ahlberg, P., 67, *71*
Ahlrichs, R., 50, *78*
Akhtar, M., 38, *71*
Alder, K., 102, *135*
Alford, J. A., 173, *178*
Al Holly, M. M., 130, *137*
Allegra, G., 111, *139*
Allen, L. C., 46, 49, 51, *71*, *73*, *77*, *78*
Allred, E. L., 83, *135*, 282, *297*
Almenningen, A., 82, *135*
Alston, P. V., 202, 232, 234, *266*, *297*
Altenbach H.-J., 168, *179*
Altmann, J. A., 159, *178*, 273, *302*
Anastassiou, A. G., 165, *174*, 287, *298*
Anderko, J. A., 151, *177*
André, J. M., 82, *135*
André, M. C., 82, *135*
Andrews, G. D., 285, 290, *298*
Andrews, U. H., 290, 292, *299*
Andrist, A. H., 255, *267*, 274, 285, 287, *298*
Annunziata, R., 173, *174*
Ansmann, A., 88, *140*, 225, *271*
Anzenhofer, K., 111, *137*
Aoyagi, E. I., 58, *72*
Arbuzov, B. A., 169, *174*
Ardis, A. E., 146, *176*
Arendale, W. F., 82, *135*
Arigoni, D., 169, *174*
Armstrong, D. R., 4, *72*
Ashe, A. J., III, 254, *267*
Aso, C., 61, *75*
Ašperger, S., 147, 148, *174*

Atkinson, R. S., 171, *174*
Atovmyan, L. O., 163, *178*
Atwell, W. H., 170, *175*
Ausloos, P., 50, *71*
Avram, M., 91, *137*

B

Bach, R. D., 110, 111, *139*
Bader, R. F. W., 195, *267*
Bahn, C. A., 11, *72*
Baird, M. S., 16, *71*
Baird, N. C., 24, *71*
Baird, W. C., Jr., 253, *268*
Bakal, Y., 130, *137*
Bakuzis, P., 59, *72*
Baldwin, J. E., 80, 94, 99, 100, 102, 106, 109, 111, 114, 115, *135*, 145, 171, 173, *175*, 255, *267*, 274, 280, 282, 285, 287, 289, 290, 291, 294, 295, *298*, *302*
Balquist, J. M., 293, *298*
Bampfield, H. A., 105, 106, 108, 116, *135*
Bandurco, V. T., 164, *178*
Barbaro, G., 226, *267*
Barborak, J. C., 69, *71*
Barket, T. P., 165, *178*
Barner, R., 61, 64, *71*
Barnett, G., 171, *175*
Bartlett, P. D., 96, *135*, *139*, 172, *175*, 218, *267*
Barton, D. H. R., 168, *175*
Barton, T. J., 101, *135*, 168, 170, *175*
Bast, K., 222, *267*
Bastens, L., 101, 102, 123, *137*
Bastiansen, O., 82, *135*
Bastide, J., 222, 234, 237, *267*
Battiste, M. A., 65, 67, *71*, 171, *176*
Bauld, N. L., 112, *135*, *136*, 242, *267*

303

AUTHOR INDEX

Bayer, A. C., 11, 77
Beck, B. R., 282, 297
Beeson, J. H., 171, 178
Beetz, T., 112, 135
Beisiegel, E., 172, 175
Belloli, R., 274, 302
Benary, E., 166, 176
Bennett, M. J., 70, 75
Ben-Shoshan, R., 111, 135
Bente, P. F., III, 50, 74
Benz, J., 282, 298
Benzon, M. S., 280, 299
Bergeron, R., 172, 176
Bergman, R. G., 70, 71, 278, 293, 298
Bernardi, F., 189, 271
Berson, J. A., 20, 54, 59, 63, 67, 71, 77, 250, 251, 267, 282, 286, 293, 294, 298
Berthier, G., 82, 139
Bertrán, J., 231, 267
Bertrand, M., 12, 58, 75, 76, 80, 91, 104, 105, 106, 116, 130, 131, 135, 139, 280, 298
Berwin, H. J., 50, 51, 77
Betz, W., 64, 71
Bezman, S. A., 162, 176
Bhacca, N. S., 237, 269
Bhatnagar, A. K., 173, 175
Bickelhaupt, F., 287, 300
Biefield, C. G., 71
Binamé, R., 90, 136, 137
Bingham, R. C., 292, 298
Binsch, G., 94, 138, 188, 270
Bird, P. H., 162, 176
Bischof, P. K., 50, 71
Bischoff, F., 126, 139
Bladek, R., 23, 71
Block, R., 289, 298
Bly, R. S., 61, 71
Boche, G., 71, 274, 282, 287, 298
Boettcher, R. R., 298
Böhme, H., 118, 135
Boles, D. L., 279, 301
Bollinger, J. M., 10, 15, 71
Bond, A., 165, 175
Borčić, S., 147, 148, 174, 254, 269
Borden, W. T., 295, 298
Bordwell, F. G., 148, 153, 159, 160, 175
Borgulya, J., 61, 64, 71
Borrmann, D., 80, 135
Bowen, M. F., 50, 76
Boyd, R. J., 164, 172, 178

Brady, W. T., 80, 97, 135
Brake, P. F., 60, 71
Bramblett, J. D., 165, 176
Brandt, D. R., 90, 140
Brauman, J. I., 64, 67, 78, 246, 270, 278, 285, 298, 301
Braun, M., 119, 135
Brennan, M. E., 171, 176
Brennan, T. M., 60, 71
Breslow, R., 254, 267, 296, 298, 300
Briegleb, G., 201, 267
Brinich, J. M., 10, 15, 71
Broadhurst, M. J., 59, 76, 168, 175
Brook, P. R., 97, 98, 105, 106, 108, 116, 127, 135, 136
Brookhart, M., 38, 68, 69, 71, 74, 287, 300
Brown, H. C., 51, 71
Brown, R. D., 183, 191, 220, 267
Brown, R. S., 50, 51, 71, 77
Bryan, D. B., 282, 298
Buck, H. M., 37, 69, 74, 76, 77
Buckley, N. C., 38, 76
Buenker, R. J., 4, 75, 76, 82, 136, 274, 278, 288, 289, 298, 300, 301
Bunnett, J. F., 183, 270, 274, 301
Burger, K., 171, 175
Burgess, J. R., 173, 178
Burka, L. T., 283, 299
Burnelle, L., 4, 76
Buss, V., 46, 49, 50, 53, 71, 76, 78
Buter, J., 168, 175
Byers, G. W., 171, 178
Byrd, L. R., 94, 98, 105, 140

C

Cain, E. N., 66, 75, 282, 302
Caine, D., 60, 71
Campbell, P. H., 22, 71
Cantacuzène, D., 4, 71
Caple, G., 246, 270
Caramella, P., 205, 235, 267
Carbó, R., 231, 267
Carless, H. A. J., 103, 136, 139
Carpenter, B. K., 52, 71, 293, 298
Carpino, L. A., 144, 165, 175
Carr, R. V., 296, 302
Čársky, P., 278, 298
Carter, W. L., 293, 298
Casanova, J., 16, 71

Case, R. S., 296, *298*
Caserio, M. C., 94, 98, 105, *140*, 274, *299*
Cassar, L., 162, *175*
Cauquy, G., 170, *177*
Cellerino, G., 237, *267*
Cessac, J., 242, *267*
Chadwick, J. C., 38, *71*
Chadwick, T., 246, *270*
Chan, T. H., 168, *175*
Chang, C.-S., 52, *73*, 112, *135, 136*
Chang, Y. C., 173, *176*
Chang, Y.-M., 208, 236, *269*
Chao, P., 145, *175, 177*
Chapman, O. L., 130, *136*
Chapuisat, X., 182, *269*
Chen, A. F.-T., 12, *76*
Chen, F., 167, *178*
Childs, M. E., 170, *175*
Childs, R. F., 24, 25, 28, 34, 59, *71, 72, 73, 76*
Chiu, N. W. K., 22, 23, *71, 72*
Choi, S. C., 145, 173, *175*
Cholod, M. S., 153, *178,* 223, *270*
Chong, S.-L., 50, *72*
Chow, V. L., 294, *300*
Christensen, P. A., 25, *72*
Christl, M., 222, 236, *267*
Christoffersen, R. E., 82, *140*
Chu, Y., 37, *73*
Ciurdaru, G., 12, *76*
Clarady, J. C., 101, *135*
Clark, D. T., 4, 7, *72*
Clark, G. M., 29, *72*
Clarke, G. M., 20, *71*
Clarke, S. C., 171, *175*
Clemens, K. E., 43, 44, *73*
Clevenger, J. V., 52, *73*
Clifford, P. R., 57, *75*
Clinton, N. A., 50, 51, *77*
Clive, D. L. J., 168, *175*
Coates, R. M., 67, 69, *72*
Cobb, R. L., 142, *178*
Cockcroft, R. D., 251, *270*
Coffey, P., 4, *72*
Coffman, D. D., 126, *138*
Cole, T. M., Jr., 25, 32, *77*
Collins, F. S., 274, 292, *299, 302*
Collins, P. M., *73*, 130, *136, 137*
Commeyras, A., 50, 51, 52, *75*
Conover, W. W., 119, *136*
Cooke, B. J. A., 27, *77*

Cooke, R. S., 290, 295, *299*
Cossement, E., 90, 98, *136, 137*
Coulson, C. A., 183, *267*
Cox, A. P., 82, *136*
Coyle, J. D., 160, *175*
Crandall, J. K., 83, 119, *136*
Crawford, W. C., 102, *139*
Criegee, R., 37, *72,* 279, *299*
Cripps, H. N., 113, *136*
Cristol, S. J., 6, 7, *72*
Crowley, K. J., 161, *175*
Crumrine, D. S., 60, *78*
Csizmadia, I. G., 49, *74,* 82, *138*
Cueille, G., 59, *72*
Cundall, R. B., 160, *175*
Cuthbertson, E., 169, *175*

D

Dai, S. H., 109, 114, *136*
D'Amore, M. B., 285, *298*
Dannenberg, J. J., 289, *299*
Darling, T. R., 44, *77,* 171, *175*
Daub, J., 64, 69, *71*
Dauben, W. G., 58, *72,* 162, *175,* 273, 282, 289, *299*
Daudey, J. P., 277, *299*
Davalt, M., 285, *298*
Davis, R. E., 111, *136*
Dawson, J. B., 60, *71*
DeBardelen, J. F., Jr., 60, *71*
de Boer, C. D., 254, *268*
de Boer, J. J., 111, *137*
de Boer, T. J., 4, *74*
de Brie, M. J. A., 82, 83, *140*
de Bruin, G., 142, *175*
de Haan, J. W., 37, 69, *76, 77,* 83, *140*
Dehmlow, E. V., 16, *72*
Dejonghe, J. P., 90, *137*
Delaunois, M., 92, *136*
Del Bene, J., 82, *136*
Del Re, G., 274, *301*
De Maré, G. R., 282, *299*
de Meijere, A., 283, *299*
DeMember, J. R., 50, 52, 57, *75*
Demole, E., 27, *75*
Deno, N. C., 21, 31, *72*
Denyer, C. V., 168, *175*
de Perez, C., 38, *73,* 117, *137*
De Puy, C. H., 6, 7, 20, *72*

Dervan, P. B., 294, *298*
Dessau, R., 102, *138*
Deugau, K., 22, *71*
Devaquet, A., 53, 54, 66, *72*, 188, 230, *267*, 274, *299*
Deville, C. G., 205, *267*
de Vries, L., 67, *78*
Dewar, M. J. S., 4, 49, 50, 64, 70, *71, 72, 77*, 87, 88, *136*, 153, 171, *175*, 183, 184, 185, 210, *267*, 278, 285, 292, 296, *298, 299*
de Wolf, W. H., 287, *300*
Deyrup, C. L., 65, 67, *71*, 171, *176*
Deyrup, J. A., 171, *175*
Diaz, A. F., 40, 65, *72, 76*
Dieffenbacher, A., 97, 98, *139*
Diels, O., 102, *136*
Dimroth, K., 130, *139*
Disselnkötter, M., 126, *138*
Dixon, D. A., 49, *72*
Dmitrienko, G. I., 285, *300*
Dodson, R. M., 145, *175*
Doering, W. von E., 254, *267*, 279, 280, 283, 295, *299*
Dolbier, W. R., Jr., 109, 114, *136*
Do Minh, T., 94, 98, 100, *136*
Dondoni, A., 226, *267*
Dougherty, R. C., 183, 259, *267*, 274, *299*
Dowd, P., 111, *136*
Draghici, C., 91, *137*
Dreiding, A. S., 97, 98, *139*
Drenth, W., 83, *140*
Driessen, P. B. J., 69, *76*
Duffin, D., 16, *72*
Duke, A. J., 97, 98, *135, 136*, 142, *178*
Duke, J. R. C., 97, *135*
Duke, R. E., Jr., 203, 205, 222, 234, 235, 236, 238, 239, 265, *269*
du Manoir, J. R., 143, *175*
Dumont, W., 91, 92, 94, 96, 100, 123, *136*
Duncan, W. G., 106, *136*
Dunkerton, L. V., 280, *300*
Dunkin, I. R., 68, *76*
Durst, T., 153, *176*
Dusseau, C. H. V., 4, *74*
Dyas, C., 61, *74*
Dygos, D. K., 17, *72*, 245, *268*

E

Eastwood, F. W., 172, *176*
Eaton, D. R., 82, *136*

Eaton, P. E., 162, *175*
Eckelman, W. C., 170, *176*
Edelson, S. S., 44, *72, 77*
Ege, G., 172, *175*
Egger, K. W., 103, *136*
Egorova, T. G., 25, *74*
Ehrenson, S., 273, *299*
Eichelberger, J. L., 166, *178*
Eick, H. A., *71*
Eisch, J. J., 219, *267*
Eisenstein, O., 221, 231, *267*
El Ghandour, N., 234, 237, *267*
Elliott, L. E., 170, *176*
Elser, W., 129, *137*
Emmons, W. D., 173, *175*
Enanoza, R. M., 172, *176*
Enderer, K., 283, *301*
Engelman, C., 61, *74*
England, D. C., 92, 96, 105, 127, *136*
Epiotis, N. D., 61, *72*, 88, 133, *136*, 189, 192, 202, 209, 219, 233, 237, 242, 246, 248, 250, 251, 252, 255, 262, 263, 264, 265, *267, 268*, *271*, 274, *299*
Erker, G., 112, *139*
Erman, W. F., 59, *72*
Eschenmoser, A., 174, *176*
Evans, M. G., 87, *136*, 193, 194, *268*, 273, *299*

F

Fahrenholtz, S. R., 38, *77*
Fahrni, P., 61, 64, *71*
Faler, G. R., 173, *178*
Farneth, W. E., 285, *298*
Farnum, D. G., 31, 52, 58, *72*, 104, *136*
Fedorova, A. V., 80, 130, *136, 139*
Fehn, J., 171, *175*
Feiler, L. A., 90, 92, 94, 99, *136, 138*
Felix, D., 174, *176*
Fellenberger, K., 11, *72, 77*, 245, *270*
Felty, R. E., 253, *268*
Fenoglio, R. A., 20, 58, *78*
Ferreira, R., 275, *299*
Feuer, J., 88, *137*, 220, 230, 260, 262, *268*
Fick, J., 146, *176*
Filipescu, K., 60, *72*
Findlay, D. M., 285, *300*
Finkenbine, J. R., 168, *175*
Firl, J., 83, *137*
Fischer, N. H., 144, 172, *176*
Fisher, H., 80, *137*

AUTHOR INDEX

Fleischhauer, J., 171, *176*
Fleming, I., 219, 253, *268*, 282, *299*
Fleming, R. H., 80, 109, 111, 114, 115, *135*, 280, 295, *298*
Fletcher, V. R., 91, *137*
Fletcher, W. H., 82, *135*
Flowers, M. C., 284, 285, 290, *299*
Flygare, W. H., 83, *140*
Follweiler, D. M., 69, *71*
Fong, F. K., 51, *72*
Foote, C. S., 173, *176*
Ford, P. W., 291, 294, *298*, *302*
Fornasier, R., 173, *174*
Fountain, K. R., 90, *140*
Fowler, F. W., 279, *302*
Franklin J. L., 50, *72*
Franzus, B., 253, *268*
Fraser, A. R., 162, *176*
Fraser, R. R., 143, *175*
Fray, G. I., 38, *71*
Freeburger, M. E., 166, *178*
Freeman, F., 223, *268*
Freeman, J. P., 164, *176*
Freeman, P. K., 289, *300*
Fretz, E. R., 69, *72*
Frey, H. M., 94, 103, *137*, 284, 285, 290, 296, *299*
Friedrich, E. C., 58, 67, *76*, *78*
Friedrich, L. E., 58, *78*, 246, *268*
Fueno, T., 4, *78*, 213, *270*, *271*
Fujimoto, H., 88, *138*, 183, 189, 212, 214, 239, *268*, *269*, 274, *299*
Fujimoto, T. T., 173, *176*
Fujita, K., 53, *77*
Fukui, K., 88, *137*, *138*, 182, 183, 189, 212, 214, 239, *268*, *269*, 273, 274, *299*
Fukutome, H., 213, *271*
Fünfschilling, P., 282, *301*

G

Gajewski, J. J., 13, 58, *72*, 111, 112, 113, *137*, 283, *299*
Gandour, R. W., 205, *267*
Garbisch, E. W., Jr., 220, *268*
Gascoigne, R.M., 39, *72*
Gašić, M., 58, *72*, *77*
Gaspar, P. P., 170, *176*
Gassman, P. G., 17, 18, 67, *72*, 245, *268*
Gasteiger, J., 67, *73*, *74*, 152, *176*
Gaughan, E. J., 90, *140*

Gell, K., 239, *270*
George, J. K., 203, 205, 222, 234, 235, 236, 238, 239, *269*, 292, *299*
George, T. F., 274, *299*
Gheorgiu, M. D., 91, *137*
Ghosez, L., 38, *73*, 80, 83, 90, 92, 94, 101, 102, 117, 118, 119, 120, 122, 123, 124, 132, 133, 134, *136*, *137*, *138*, *139*, *140*
Gianni, F. L., 233, *268*
Gibson, T. A., 278, *301*
Gil, G., 104, 105, 116, *135*, 280, *298*
Gilbert, J. C., 257, *268*, 284, *299*
Gilbert, K. E., 165, *176*
Gilchrist, T. L., 80, *137*
Giles, W. B., 88, *137*, 202, 260, 262, *268*
Gillespie, J. P., 280, *301*
Gimarc, B. M., 241, *268*
Ginsburg, D., 151, *177*
Gleiter, R., 94, *137*, 165, *176*, 295, *299*
Goddard, W. A., III, 274, 289, 291, *299*, *300*
Goe, G. L., 283, *299*
Goering, H. L., 52, *73*
Goetz, D. W., 51, *73*
Gold, H., 126, *138*
Goldhamer, D., 80, *140*
Goldschmidt, Z., 130, *137*, *138*
Goldstein, C. J., 25, *74*
Goldstein, E. J., 170, *178*
Goldstein, M. J., 65, 67, 69, *73*, *74*, 208, *268*, 280, *299*
Gollnick, K., 168, *177*
Gompper, R., 38, *77*, 80, 119, 128, 129, *137*, *140*, 219, *268*
Goodbrand, H. B., 143, *177*
Goodlett, V. W., 89, 90, 94, 104, 127, *139*
Goodlett, W., 83, *135*
Gordon, M., 82, *139*
Goré, J., 105, 106, *135*
Gosnik, T., 246, *270*
Gott, P. G., 89, 90, 92, 94, 104, 127, *137*, *139*
Gouesnard, J. P., 127, *137*
Graefe, J., 16, *73*
Graham, W. H., 164, *176*
Grant, D. M., 83, *135*
Gras, J.-L., 280, *298*
Grashey, R., 80, 94, 100, 101, 102, *138*
Gratz, J. P., 165, *178*
Grayston, M. W., 289, 290, 291, *298*
Greeley, R. H., 25, 32, *77*
Green, G. E., 63, *73*
Green, M., 165, 166, *175*, *176*

Greenberg, A., 4, *74*
Greene, F. D., 165, 171, *176, 178*
Greenwood, G., 42, 43, *73*
Gregory, A. R., 289, *299*
Griesbaum, K., 37, *73,* 80, *137*
Griffin, R. H., 57, *73*
Griffith, R. C., 287, *298*
Griffiths, J., 25, 28, 60, *73,* 130, *137*
Grigg, R., 168, *175*
Grimaldi, J., 130, 131, *135*
Grimme, W., 112, *137,* 280, 287, *299*
Grishin, Y. K., 83, *137*
Grossman, J., 279, *302*
Grubber, M. J., 6, 8, *73*
Gruber, G. W., 280, *300*
Gruber, W., 53, *74*
Grummitt, O., 146, *176*
Günther, H., 59, *73*
Gutman, U., 130, *137*

H

Haake, M., 118, *135*
Haddon, R. C., 70, *72*
Hagedorn, A. A., 31, *72*
Hagen, E. L., 55, *76*
Hall, C. D., 165, *176*
Hall, J. A., 82, 83, *138,* 205, 224, 240, *267, 269*
Hall, L. H., 220, 230, *268*
Halpern, J., 162, *175*
Halpern, Y., 57, *75*
Halton, B., 171, *176*
Haluska, R. J., 254, *270*
Hamersma, J. W., 40, *74*
Hammond, G. S., 254, *268*
Hammond, W. B., 44, *77*
Hamon, D. P. G., 91, *137*
Hanack, M., 58, *73*
Hanas, H. D., 129, *140*
Hancock, K. G., 29, *72*
Handloser, L., 58, *74*
Hanford, W. E., 80, *137*
Hansen, H.-J., 60, 61, 64, *71, 73, 76, 78,* 254, *271,* 281, *300*
Hanstein, W., 50, 51, *77*
Harger, M. J. P., 171, *174*
Hariharan, P. C., 4, 50, 52, 53, 55, *73,* 76
Harnisch, J., 172, *178*
Harris, D. L., 40, 67, *71, 72*
Harrison, J. M., 97, 98, *136*

Hart, H., 25, 28, 60, 70, *71, 73,* 128, 129, 130, *136, 137, 139*
Hartzell, G. E., 145, *176*
Hasek, R. H., 89, 90, 92, 94, 104, 127, *137, 139*
Hassner, A., 91, *137*
Hasty, N. M., 59, *71,* 77, 173, *176*
Hata, Y., 173, *176*
Hatcher, G. B., 102, *138*
Hausser, J. W., 6, 7, 8, *72, 73*
Haveaux, B., 83, *137*
Havinga, E., 268, *300*
Hay, P. J., 289, *300*
Hayakawa, Y., 38, *75*
Hayes, R. R., 289, *301*
Haywood-Farmer, J., 65, 67, *71, 73*
Heap, N., 63, *73*
Hechtl, W., *71*
Hegedić, D., 147, 148, *174*
Hehre, W. J., 48, 49, 50, 53, 54, 59, 63, 66, *72, 73, 74, 78,* 82, *137,* 292, *300*
Heiber, M., 112, *139*
Heimgartner, H., 281, *300*
Heinrich, B., 126, *139*
Hekman, M., 296, *299*
Helder, R., 168, *176, 179*
Heldeweg, R. F., 154, *176*
Hendrickson, J. B., 172, *176*
Henri-Rousseau, O., 234, 237, *267, 268*
Henzel, K. A., 59, *76*
Herndon, W. C., 88, *137,* 208, 220, 230, 260, 262, *268*
Herzberg, G., 273, *300*
Hesbain-Frisque, A. M., 120, 122, *138*
Hess, B. A., 66, *73*
Hess, R. E., 104, *136*
Hewitt, T. G., 111, *130*
Heyman, M. L., 165, *178*
Hiberty, P. C., 48, 59, *73*
Higgins, J., 166, *178*
Higley, D. P., 284, *299*
Hill, A. E., 43, *73*
Hill, R. K., 60, *71*
Hine, J., 159, *176*
Hine, K. E., 34, *73*
Hirano, T., 227, *269*
Hirano, Y., 166, *178*
Hiyama, T., 172, *176*
Hobson, J. D., 130, *137*
Hoff, E. F., Jr., 97, *135*

AUTHOR INDEX 309

Hoffmann, H. M. R., 42, 43, 44, *73*, 258, *268*
Hoffmann, J. M., Jr., 254, *267*
Hoffmann, R., 3, 4, 7, 34, 37, 42, 45, 50, 52, 54, 57, 59, 61, 64, 65, 67, 69, 70, *73*, *74*, *77*, *78*, 80, 85, 86, 94, *137*, *140*, 141, 162, 165, 166, *176*, *179*, 183, 184, 208, 214, 228, 229, 239, 241, 242, 248, 249, 256, 259, 263, *268*, *271*, 273, 274, 279, 280, 288, 289, 295, *300*, *302*
Höfle, G., 145, *175*
Hogeveen, H., 25, 70, *74*, 154, *176*
Hoornaert, C., 120, 122, *138*
Hopf, H., 103, *137*, 290, *299*
Hopkinson, A. C., 82, *138*
Hoppe, I., 253, *270*
Horn, U., 174, *176*
Horsley, J. A., 292, *300*
Houk, K. N., 82, 83, 88, 133, *138*, 193, 195, 202, 203, 205, 208, 212, 216, 222, 223, 224, 226, 227, 228, 231, 232, 233, 234, 235, 236, 237, 238, 239, 240, 254, 255, 256, 265, *267*, *268*, *269*, *270*
Howell, J. M., 166, *176*
Hoyt, E. B., Jr., 148, 153, 160, *175*
Huang, Y. Y., 25, *72*
Huber, H., *71*
Huber, R., 279, *299*
Hudson, R. F., 190, 206, *269*
Hughes, C. T., 15, *76*
Hughes, R. P., 111, *138*
Huisgen, R., 67, *71*, *73*, *74*, 80, 90, 92, 94, 99, 100, 101, 102, *136*, *138*, 152, *176*, 222, 236, 245, *267*, *269*
Humski, K., 254, *269*
Hunt, K., 98, 127, *135*, *136*
Hunt, W. J., 289, *300*
Husk, G. R., 219, *267*
Huybrechts, G., 282, *299*
Hwang, R.-J., 170, *176*

I

Imamura, A., 227, *269*, *300*
Immirzi, A., 111, *139*
Inagaki, S., 88, *138*, 212, 214, *269*
Inamoto, N., 166, *177*
Inukai, T., 227, 231, *269*
Irie, T., 65, 67, *77*
Isaac, P. A. H., 162, 163, *177*

Isaacs, N. S., 94, 99, 100, 102, *137*, *138*, 148, *176*
Isaev, I. S., 5, 25, 59, *74*
Ishiguro, M., 170, *176*
Ishikawa, M., 170, *176*
Ishiyama, H., *74*
Itô, S., 41, *74*
Itoh, I., 41, *74*
Izawa, Y., 166, *178*
Izzo, P. T., 130, *138*

J

Jacobs, T. L., 111, *138*
Jaffé, H. H., 82, *136*
Jakubetz, W., 278, *300*
Janousek, Z., 118, *138*
Jarvis, B. B., 148, 153, 160, *175*
Jean, Y., 182, *269*, 292, *300*
Jenkins, J. A., 63, 67, *71*
Jenkins, R. L., 170, *176*
Jensen, H. P., 169, *174*
Jewett, J. G., 49, 57, *73*, *76*
Johanson, R. G., 57, *75*
Johns, J. W. C., 83, *138*
Johnson, A. W., 168, *175*
Johnson, B., 58, *72*, *77*
Johnson, B. L., 171, *175*
Johnson, D. R., 15, *76*
Johnson, J. R., 82, 83, 104, *136*, *138*
Johnstone, R. A. W., 33, *74*
Jones, E. R. H., 126, 130, *138*
Jones, G., II, 286, 294, *298*, *300*
Jones, H., 58, *77*
Jones, J., 254, *267*
Jones, M., Jr., 161, *176*
Jones, W. J., 82, *139*
Jones, W. M., 285, *302*
Jongejan, E., 4, *74*
Jonkman, H. T., 70, *74*
Joos, R., 174, *176*
Jordan, K. D., 275, *300*
Jorgenson, W. L., 184, 197, *269*
Josan, J. S., 172, *176*
Jotham, R. W., 187, *269*
Jovanovich, A. P., 40, 69, *74*
Joy, D. R., 42, *73*
Jug, K., 4, *72*
Jula, T. F., 16, *77*

AUTHOR INDEX

Jullien, R., 59, 72
Jung, F., 153, 176
Jutz, C., 118, 138
Juvet, M., 170, 175

K

Kaesz, H. D., 67, 78
Kamm, K. S., 160, 179
Kammerer, R. C., 111, 138
Kane, M., 130, 136
Kanematsu, K., 239, 270
Kang, S., 4, 76
Kapecki, J. A., 99, 100, 102, 135, 280, 300
Kaplan, M. S., 282, 298
Karger, M. H., 219, 268
Kari, R. E., 49, 74
Karplus, M., 182, 271, 288, 302
Kashman, Y., 166, 176
Kasuya, M., 70, 73
Kataoka, T., 239, 270
Katô, M., 34, 61, 75
Katz, H., 274, 300
Katz, T. J., 102, 138, 166, 178
Kearns, D. R., 173, 176, 214, 269
Kedrowski, R. A., 170, 176
Keeley, D. E., 58, 77
Kees, F., 172, 178
Kein, W., 111, 138
Kelemen, J., 278, 299
Kellogg, M. S., 294, 299
Kellogg, R. M., 112, 135, 160, 168, 175, 176
Kemp-Jones, A. V., 67, 75
Kende, A. S., 130, 138, 284, 300
Kernaghan, G. F. P., 42, 43, 73
Kiefer, E. F., 114, 138
Kielbania, A. J., Jr., 162, 175
Kim, C. W., 171, 178
Kim, H., 83, 140
Kim, M., 165, 176, 179
Kimling, H., 171, 176
King, J. F., 143, 172, 175, 176
Kinnel, R. B., 289, 300
Kippenhan, R. C., Jr., 175
Kirkley, R. K., 165, 175
Kirkpatrick, J. L., 67, 72
Kirmse, W., 12, 53, 74, 161, 176, 253, 270
Kirschner, S., 4, 72, 153, 171, 175, 296, 298
Kiselev, V. D., 220, 269
Kispert, L. D., 61, 74
Kistiakowsky, G. B., 109, 138

Kitahara, Y., 240, 271
Klabunde, K. J., 168, 176
Klärner, F.-G., 286, 300
Kleveland, K., 154, 176
Klopman, G., 51, 74, 190, 206, 269
Klumpp, G. W., 257, 268
Knoth, W. H., 126, 138
Koenig, F. R., 40, 74
Kohlmer, H., 70, 74
Kojima, T., 227, 231, 269
Kollmar, H., 41, 74, 278, 300
Koltzenburg, G., 86, 138
Kondo, K., 145, 176
Konig, H., 102, 136
Konovalov, A. I., 169, 174, 220, 269
Koock, S. U., 61, 71
Koopmans, T. A., 200, 269
Koppitz, P., 92, 136
Koptyug, V. A., 5, 25, 59, 74
Kostyanovskii, R. G., 163, 178
Kouwenhoven, C. G., 152, 179
Kraft, K., 86, 138
Krauss, H. J., 160, 177
Krebs, A., 171, 176
Kreiter, C. G., 64, 67, 78
Krespan, C. G., 92, 96, 105, 127, 136
Kresze, G., 155, 177
Kretschmer, G., 171, 177
Krist, J., 53, 74
Kropp, P. J., 160, 177
Krow, G. R., 80, 83, 101, 118, 138, 139
Kucherov, V. F., 80, 139
Kuehne, M. E., 92, 138
Kuhn, H. J., 168, 177
Kukla, M. J., 274, 295, 301
Kumada, M., 170, 176
Kumai, S., 53, 77
Kunitake, T., 61, 75
Kurland, D., 27, 74
Kurtz, P., 126, 138
Kusuda, K., 26, 74
Kutzelnigg, W., 4, 50, 74, 78
Kuzubova, L. I., 25, 59, 74
Kwant, P. W., 70, 74

L

Lacey, R. N., 80, 138
Lack, R. E., 27, 77
La Combe, E. M., 172, 177

Lai, K.-H., 61, *74*
Laila, A. A. R., 148, *176*
Lalezari, I., 169, *177*
Lambert, J. B., 40, 69, *74*
Landheer, I. J., 287, *300*
Langlet, J., 277, *299*, *300*
Lankin, D. C., 130, *137*
Lantus, I., 151, *177*
Laroche, P., 101, 102, 123, *137*
Larrabee, R. B., 144, *177*
Lassila, J. D., 130, *136*
Lathan, W. A., 49, 50, *73*, *74*, *78*
Lawrence, R. B., 83, *138*
Lebedev, S. A., 83, *137*
Lechtken, P., 296, *300*, *302*
Ledlie, D. B., 14, *74*
Lee, E. K. C., 103, *136*, *139*
Lee, H. H., 130, *138*
Lee, L., 164, *178*
Leermakers, P. A., 171, *178*
Leforestier, C., 288, *301*
Lefour, J.-M., 231, *267*
Le Gras, J., 80, 104, 105, 106, 116, *135*
Lehr, R., 27, *74*
Leitch, L. C., 62, *75*
Lemal, D. M., 143, 145, 148, 162, 164, *175*, *177*
Leroy, G., 82, *135*
Lesk, A., 4, *74*
Letcher, J. H., 82, *138*
Levin, C. C., 228, 229, *268*
Lewans, E. G., 172, *176*
Lewin, A. L., 65, *78*
Lewis, C. P., 287, *300*
Lewis, M. C., 12, *76*
Leyendecker, F., 289, *298*
Li, E., 296, *302*
Liang, G., 48, 51, 54, 57, 66, *75*, *76*
Liberles, A., 4, *74*
Libit, L., 279, *300*
Lillien, I., 58, *74*
Lin, F. F. S., 165, *176*
Lin, H. C., 48, *75*
Lin, H.-N., 172, *176*
Lindsay, D. G., *71*
Lipscomb, W. N., 49, *72*
Lischka, H., 278, *300*
Litt, M., 118, *140*
Little, J. C., *269*
Liu, K.-C., 60, *77*

Liu, K.-T., 51, *71*
Lodder, A. E., 37, *74*
Loeschen, R. L., 130, *136*
Loew, L. M., 61, *78*
Lombardo, L., 285, *300*
Longone, D. T., 289, *300*
Longuet-Higgins, H. C., 183, 241, 243, *267*, *269*, *276*, *300*
Lossing, F. P., 49, 50, *74*
Love, G. M., 130, *137*
Lowe, J. P., 277, *300*
Lui, C. Y., 50, 51, 52, 57, *75*
Luknitskii, F. I., 80, *138*
Luskus, L. J., 222, 226, 234, 235, 236, 237, 239, *269*
Lustgarten, R. K., 38, 68, 69, *71*, *74*

M

McAdams, L. V., III, 144, *175*
McAdoo, D. J., 50, *74*
McCausland, J. H., 149, 169, *177*
McCay, I. W., 171, *177*
McClure, D. E., 166, 173, *178*
McDermott, J. X., 162, *177*
McDonald, R. N., 19, *74*, *75*
McDonald, W. S., 108, *135*
McElwee, D. E., 208, *269*
McGregor, S. D., 143, 148, 162, 164, *177*
McIntosh, C. L., 143, 170, *175*, *177*
McIver, J. W., Jr., 288, *300*
McLafferty, F. W., 50, *74*, *75*
McLean, S., 285, *300*
MacNicol, D. D., 169, *175*
McVey, J. K., 296, *302*
Mack, W., 222, *267*
Mackenzie, K., 257, *269*
Mader, R. A., 58, *72*
Madija, R., 61, 64, *71*
Magdesieva, N. N., 169, *177*
Mah, T., 233, *268*
Mahler, J. E., 67, *76*
Majorski, Z., 57, 58, *75*
Makino, S., 44, *75*
Malkus, H., 67, *71*
Mallikarjuna Rao, V. N., 26, *74*
Mallory, F. B., 165, *176*
Malloy, T. P., 12, *78*
Malojčič, R., 254, *269*

AUTHOR INDEX

Malpass, J. R., 130, *137*
Malrieu, J.-P., 277, *299*, *300*
Mamatyuk, V. I., 25, 59, *74*
Mango, F. D., 162, *177*, 275, *300*
Mansfield, G. H., 126, *138*
Mantz, I. B., 167, *178*
Manuel, G., 16, *77*, 170, *177*
Marchand-Brynaert, J., 80, 83, 118, 119, 120, 123, 124, *137*, *139*, *140*
Mark, V., 218, *269*
Märkl, G., 126, *139*
Marshall, T. B., 104, *136*
Martin, D., 168, 169, *177*
Martin, H.-D., 278, 289, 296, *299*, *300*, *301*
Martin, J. C., 89, 90, 92, 94, 104, 127, *137*, *139*
Martz, M. D., 168, *175*
Marvell, E. N., 246, *270*
Masamune, S., 66, 67, 70, *75*, 171, *177*, *282*, *302*
Masse, G. M., 143, *177*
Massol, M., 170, *178*
Mateescu, G. D., 51, 52, *75*
Mathieu, J., 274, *300*
Matsumoto, M., 145, *176*
Matsura, T., 60, *75*
Maurin, R., 91, 104, *135*, *139*
Mavrov, M. V., 80, *139*
Mayr, H., 100, *138*
Mazerolles, P., 16, *77*, 170, *177*
Meesters, A., 25, *72*
Mehrotra, I., 151, *177*
Mehta, G., 58, *72*
Meilahn, M. K., 15, *76*
Meinwald, J., 280, 287, *300*
Meisinger, R. H., 152, 161, *177*, *178*
Menachem, Y., 166, *176*
Merkel, P. B., 173, *176*
Merlet, P., 4, *75*, 274, *300*
Metcalfe, J., 103, *139*
Metts, L., 159, *178*
Meyers, A. I., 142, 151, *177*
Meyerson, S., 62, *75*
Michl, J., 247, 259, 264, *270*, 274, 278, *301*
Mielert, A., 223, *270*
Miller, B., 60, 61, *74*, *75*
Miller, I. J., 22, *71*
Miller, W. G., 31, *76*
Minato, T., 88, *138*, 214, *269*
Mitchell, R. H., 168, *177*
Miyagi, M., 274, *299*
Miyashi, T., 286, *298*

Mock, W. L., 143, 144, 147, 148, 149, 150, 151, 152, 155, 156, 157, 162, 163, 169, *177*
Mogoleska, P. D., 109, *139*
Molin, M., 153, *176*
Mollere, P. D., 254, *270*
Mollet, P., 90, 102, 123, *137*
Montaigne, R., 90, 94, 102, 123, *137*, *139*
Montanari, F., 173, *174*
Montgomery, F. C., 171, *178*
Montgomery, L. K., 96, *139*
Monti, H., 12, *75*
Moore, H. W., 94, 98, 105, 106, *136*, *140*
Moore, W. R., 109, 110, 111, *139*
Moret, T., 231, *267*
Mori, A., *74*
Moriconi, E. J., 102, *139*
Morris, M. R., 130, *139*
Moschel, A., 37, *72*
Moser, C., 292, *300*
Moss, R. A., 165, *176*, 228, 229, *268*
Mostashari, A., 31, *72*
Mueller, D. C., 16, *77*
Muetterties, E. L., 166, *176*
Mühlstädt, M., 16, *73*
Mulder, J. J. C., 87, *139*, 274, *301*, *302*
Muller, P., 254, *270*
Müller, R. K., 174, *176*
Mulliken, R. S., 160, *177*, 189, 191, 192, *270*
Munchausen, L. L., 193, 223, *269*
Murata, I., 240, *271*
Murray, R. J., Jr., 128, 129, *139*
Murray, R. W., 173, *177*
Murrell, J. N., 189, *270*
Muszkat, K. A., 274, 289, *301*

N

Naegele, W., 37, *73*
Nagata, C., 182, 183, *268*
Nakahira, T., 57, 58, *78*
Nakamura, A., 111, *139*
Nakashima, T., 67, *75*
Nakatsuka, N., 66, *75*
Nakayama, S., 166, *177*
Nal, F., 254, *270*
Napierski, J., 296, *298*
Naser-Ud-Din, J., 118, *139*
Nebzydoski, J. W., *71*
Neckers, D. C., 171, *178*
Negishi, A., 145, *176*

Nelsen, S. F., 280, *301*
Nelson, A. J., *175*
Nelson, J. P., 145, *175*
Newton, M. D., 289, *301*
Nguyen Trong Anh, 87, *139*, 221, 227, 231, 267
Nickon, A., 53, *75*
Nieuwpoort, W. C., 70, *74*
Nishiguchi, I., 18, *72*
Nohira, H., 278, *301*
Northington, D. J., 265, *269*
Noyori, R., 34, 38, 44, 61, *75*, 162, *177*
Nozaki, H., 172, *176*
Nugent, R. M., 155, 156, 157, *177*

O

Oberdier, J. P., 13, 58, *72*
O'Donnell, M. J., 90, *137*
Ogura, K., 60, *75*
Ohara, O., 61, *75*
Ohloff, G., 27, *75*, 254, *270*
Ohnishi, Y., 34, 61, *75*
Okada, T., 213, *270*
Okamoto, T., 28, *75*, 80, 111, *139*
Okamura, M. Y., 114, *138*
Okazaki, R., 166, *177*
Olah, G. A., 10, 15, 48, 50, 51, 52, 54, 57, 66, *71*, *75*, *76*, 83, *139*
Oliver, S. S., 40, *74*
Ollerenshaw, J., 290, *298*
Olsen, H., 164, *177*
Olsen, J. F., 4, *76*
Ona, H., 70, *75*, 171, *177*
O'Neal, H. E., 171, *178*
Oosterhoff, L. J., 37, *74*, 87, *139*, 263, *270*, 273, 274, *301*, *302*
Orchard, S. W., 282, *301*
Orville-Thomas, W. J., 82, *139*
Osborn, J. A., 162, *176*
Ostlund, N. S., 50, *76*
Östman, B., 33, *76*
Ottenbrite, R. M., 232, 234, *266*
Otteson, D., 88, *137*, 260, 262, *268*
Otto, P., 92, 99, *138*
Ozretich, T. M., 110, 111, *139*

P

Pacifici, J. G., 171, *178*
Paddon-Row, M. N., 171, *177*, 239, *270*, 283, *301*

Paige, J. N., 145, *176*
Panade, K. C., 65, *78*
Pant, B. C., 170, *177*
Paquette, L. A., 12, 59, 68, *76*, 101, *139*, 152, 161, *177*, *178*, 254, *270*, 282, 286, 295, *301*
Parham, W. E., 6, 8, 15, 16, *76*
Parrington, B., 24, 25, 28, *71*, *76*
Parry, F. H., III, 97, *135*
Pasto, D. J., 12, *76*
Patsch, M., 11, *77*
Paul, K. P., 128, 129, *139*
Paust, J., 6, *76*, 245, *270*
Pautini, G., 111, *139*
Pavlik, J. W., 60, *72*
Pavlović, D., 147, 148, *174*
Payne, G. B., 134, *139*
Pazos, J. F., 171, *178*
Pearson, R. G., 4, *76*, 196, *270*
Pearson, S. C., 165, *175*
Pedersen, L. D., 293, *298*
Perlberger, J.-C., 254, *270*
Perrin, C. L., 274, *301*
Person, W. B., 192, *270*
Perst, H., 130, *139*
Peter-Katalinic, J., 59, *76*
Petrov, A. A., 80, 130, *136*, *139*
Pettit, R., 67, *76*, 111, *135*, 296, *298*
Peyerimhoff, S. D., 4, *75*, *76*, 274, 278, 288, 289, *298*, *300*, *301*
Pfeiffer, G. V., 49, *76*
Pfeiffer, J. G., 58, *78*
Pickenhagen, W., 254, *270*
Pierson, G. O., 171, *178*
Pike, W. C., 67, *72*
Pincock, R. E., 65, 67, *71*, *73*
Pinschmidt, R. K., Jr., 274, 282, 287, *298*
Pittman, C. U., Jr., 21, 31, 61, *72*, *74*, *76*
Pitzer, K. S., 83, *139*
Pledger, H., 126, *139*
Polanyi, M., 193, *268*, 273, *299*
Ponomarev, S. V., 83, *137*
Pople, J. A., 4, 49, 50, 52, 53, 55, *73*, *74*, *76*, 82, *137*, *139*, 276, 277, *301*
Porri, L., 111, *139*
Porter, R. D., 48, 52, 57, *75*
Potter, C. D., 58, *76*
Pouliquen, J., 171, *175*
Poulter, C. D., 58, *76*
Powell, J., 111, *138*
Pozzoli, A. P., 274, *301*
Prins, W. L., 160, 168, *176*

Prinzbach, H., 289, *301*
Prociv, T. M., 289, *299*
Pullmann, B., 82, *139*
Purdham, J., 70, *75*

Q

Quast, H., 172, *178*
Quin, L. D., 165, *178*

R

Racanelli, P., 111, *139*
Radlick, P., 173, *176*
Radom, L., 4, 50, 52, 53, 55, *73*, 76
Rajeswari, K., 29, 30, 39, 76
Ramage, R., 281, *301*
Ramirez, R., 279, *299*
Randle, M., 189, *270*
Rastelli, A., 274, *301*
Ray, W. C., 173, *178*
Raymond, K. N., 162, *175*
Razumova, N. A., 166, *179*
Rebbert, R. E., 50, *71*
Reese, C. B., 11, 12, 14, 15, 16, *71*, 76
Rehberg, R., 171, *176*
Reich, D. A., 90, *140*
Reilly, J. L., 83, *139*
Reinhoudt, D. N., 152, *179*
Rey, M., 97, 98, *139*
Rhoads, S. J., 251, *270*
Rice, S. A., 296, *302*
Richardson, W. H., 171, *178*
Richey, H. G., Jr., 38, 58, *76*
Riecke, E. E., 284, *300*
Riedel, K., 13, *77*
Riegl, J., 118, *139*
Riemenschneider, J. L., 51, 52, *75*
Rinehart, J. K., 15, *76*
Ring, M. A., 170, *176*
Rivière, P., 170, *178*
Robbins, J. D., 287, *302*
Roberts, J. D., 80, *139*, 218, *270*
Roberts, S., 97, 98, *139*
Roček, J., *270*
Rodewald, L. B., 12, *76*
Rodgers, T. R., *73*
Roe, R., Jr., 97, *135*
Roedig, A., 126, *139*
Rogido, R., 101, *135*

Roothaan, C. C. J., 160, *177*
Rose, T. L., 282, *301*
Rosenberg, J. L., 67, *76*
Rosenfeld, J. C., 9, 55, *76*
Rosmus, P., 278, *300*
Ross, J., 274, *299*
Roth, W. R., 112, *139*, 283, 284, *301*
Rother, H. J., 112, *137*
Roussel, A., 90, 98, 102, 123, *137*, *139*
Rowland, C., 278, *301*
Roy, U. V., 114, *135*
Ruban, E., 13, *77*
Ruhoff, J. R., 109, *138*
Runge, W., 83, *137*
Rynbrandt, R. H., 144, *175*

S

Sabin, J. R., 83, *140*
Sachdev, K., 279, 283, *299*
St. John, W. M., 289, *301*
Saito, I., *74*
Sajka, S. A., 83, *136*
Sakai, M., 24, 40, 65, 67, 70, *72*, 75, 76, 171, *178*
Sakan, K., *74*
Salem, L., 50, 54, *71*, *76*, 88, *140*, 184, 187, 188, 196, 197, 230, 250, 261, *267*, *269*, *270*, 274, 278, 288, 291, 292, 294, 295, *298*, *300*, *301*, *302*
Saltiel, J., 159, *178*
Santelli, M., 58, *76*
Santiago, C., *269*
Saskai, K., *74*
Saskai, T., 239, *270*
Satgé, J., 170, *178*
Sato, H., 231, *269*
Sattar, A., 281, *301*
Sauer, J., 80, 94, 100, 101, 102, *138*, 223, *270*
Sauer, J. C., 80, *137*
Sauers, R. F., 145, *175*
Saunders, M., 9, 55, 59, 62, *76*, 77, 254, *267*
Schaap, A. P., 173, *178*
Schachtschneider, J. H., 162, *177*
Schakel, M., 257, *268*
Schantl, J., 287, *302*
Scharf, H. D., 171, *176*
Scheidt, F., 12, *74*, 253, *270*
Scheinbaum, M. L., 253, *268*
Schilling, P., 48, *75*

AUTHOR INDEX 315

Schinski, W. L., 167, *178*
Schipper, P., 69, *76*
Schlatmann, J. L. M. A., *268*, *300*
Schless, P., *282*, *301*
Schleyer, P. v. R., 4, 6, 9, 11, 46, 49, 50, 52, 53, 55, 57, 58, 69, 70, *71*, *72*, *73*, *74*, *75*, *76*, *77*, *78*, 245, *270*
Schlyer, D. J., 170, *176*
Schmid, H., 59, 60, 61, 64, *71*, *73*, *76*, *78*, 254, *271*, 281, *300*
Schmid, M., 61, *76*
Schmidbauer, E., 168, *179*
Schmidt, A. H., 296, *298*, *300*
Schmidt, D., 287, *300*
Schmidt, E. A., 44, *73*
Schmidt, E. K. G., 295, *299*
Schmidt, R. R., 36, *77*
Schmidt, T., 284, *301*
Schmidt, W., 288, 289, *301*
Schnack, L. G., 6, *72*
Schneider, G., 287, *298*
Schneider, H. J., 58, *73*
Schöllkopf, U., 6, 11, 13, *72*, *76*, *77*, 245, 253, *270*
Schönefeld, J., 16, *72*
Schönleber, D., 279, *299*
Schreiber, J., 174, *176*
Schubert, R., 222, *270*
Schueller, K., 96, *139*, 218, *267*
Schulman, J. M., 289, *301*
Schulte-Elte, K. H., 27, *75*
Schumacher, H., 25, 32, *77*
Schuster, D. I., 60, *77*, 171, *178*
Schuster, G. B., 246, *268*
Schuster, P., 278, *300*
Schweickhardt, C., 279, *299*
Sclove, D. B., 171, *178*
Seebach, D., 119, *135*, 274, *301*
Segal, G., 288, *301*
Seidl, P., 67, *71*
Semeliuk, G. P., 49, 50, *74*
Sequeira, R. M., 6, 7, *72*
Servis, K. L., 274, *302*
Seyden-Penne, J., 227, *267*
Seyferth, D., 16, *77*
Shafiee, A., 169, *177*
Shanshal, M., 59, *77*
Shapley, J. R., 162, *176*
Sharafi-Ozeri, S., 274, *301*
Sharkey, W. H., Jr., 113, *136*

Sharpless, K. B., 169, *174*
Sharts, C. M., 80, *139*, 218, *270*
Shatavsky, M., 65, *78*
Shaw, A., 11, 12, 15, 16, *76*
Sheehan, J. C., 171, *178*
Sheeran, P. J., 92, *138*
Sheridan, J., 82, *136*
Shibaeva, R. P., 163, *178*
Shih, C. N., 111, 112, 113, *137*
Shih, S., 4, *75*, 274, *300*
Shillady, D. D., 202, 232, 234, *266*, *297*
Shingu, H., 182, *268*
Shleider, I. A., 5, *74*
Shono, T., 53, *77*
Shoppee, C. W., 27, *77*
Shortridge, T. J., 173, *175*
Shudo, K., 28, *75*
Shuler, K. E., 273, *301*
Sidani, A., 118, 120, *139*
Sieck, L. W., 50, *71*
Silber, E., 88, *137*, 260, 262, *268*
Silbey, R., 275, *300*
Silver, D. M., 273, *301*
Simmons, H. E., 183, *270*, 274, *301*
Sinanoğlu, O., 59, *77*
Sims, J., 203, 222, 226, 234, 235, 236, *269*, *270*
Siu, A. K. Q., 289, *301*
Sjöberg, S., 33, *76*
Skattebøl, L., *139*, 154, *176*, 188
Skell, P. S., 153, 168, 170, *176*, *178*, 223, *270*
Slater, J. C., 278, *301*
Slegir, W., 296, *298*
Smael, P., 152, *179*
Smale, G., 4, 7, *72*
Smirnov-Zamkov, I. V., 37, *77*
Smith, H. A., 109, *138*
Smith, H. O., 41, 70, *74*
Smith, J. C., 253, *268*
Smith, K. R., 257, *268*
Smithers, R. H., 43, 44, *73*
Snyder, J. P., 164, 165, 172, *177*, *178*
Sonnenberg, J., 67, *78*
Sonveaux, E., 132, 133, 134, *140*
Sorensen, T. S., 21, 22, 23, 25, 29, 33, 39, *71*, *72*, *76*, *77*
Sousa, L. R., 172, *179*
Spadaro, J. J., Jr., 11, *77*
Spangler, C. W., 287, *301*
Spiewak, J. W., 144, *175*
Sprecher, R. F., 220, *268*

Srinivasan, R., 161, *178*
Staley, S. W., 287, *301*
Stanbury, P., 99, 100, *138*
Stanescu, L., 91, *137*
Stanton, R. E., 288, *300*
Staral, J. S., 48, 54, 57, 66, *75*, *76*
Stark, B. P., 142, *178*
Staudinger, H., 71, 89, *140*
Stehouwer, D. M., 289, *300*
Stein, G., 102, *135*
Steinberg, H., 4, *74*
Steiner, G., 101, *138*
Steiner, R. P., 287, *302*
Steinheimer, T. R., 172, *178*
Stephenson, L. M., 166, 173, *178*, 246, *270*, 278, *301*
Steppel, R. N., 19, *74*
Steur, R., 82, 83, *140*
Stevens, H. C., 90, *140*
Stevens, R. M., 292, *300*
Steward, R. F., 82, *137*
Stewart, B., 172, *177*
Stewart, C. A., Jr., 218, 219, *270*
Stille, J. K., 166, *178*
Stofko, J. J., Jr., 55, 62, *76*
Stohrer, W.-D., 69, 70, *74*, *77*, 279, 289, *302*
Stoicheff, B. P., 82, *140*
Stone, J. M. R., 83, *138*
Stoos, F., *270*
Storr, R. C., 80, *137*
Story, P. R., 38, *77*, 173, *178*
Stowell, J. C., 274, *301*
Strandberg, M. W. P., 82, 83, *138*
Strausz, O. P., 94, 98, 100, *136*
Strebles, M. R. D., 14, *76*
Streklov, T., 254, *269*
Strickler, S. J., 83, *139*
Strozier, R. W., 82, 83, *138*, 203, 205, 208, 222, 224, 227, 234, 235, 236, 240, 255, *269*
Studeneer, A., 129, *137*
Su, T. M., 9, 11, *76*, *77*
Suciu, E. N., 165, *178*
Sunko, D. E., 254, *269*
Sustmann, R., 49, *77*, 88, *140*, 188, 221, 222, 225, 226, *270*, *271*
Sutherland, J. K., 16, *72*
Sutter, B., 64, *73*
Suzui, A., 173, *177*
Suzuki, T., 162, *177*
Swenton, J. S., 295, *302*
Swilinski, W. F., 245, *270*

Sysak, P. K., 173, *178*
Szeimies, G., 57, 58, 59, *78*, 172, *178*, 274, *298*

T

Tabor, T. E., 19, *75*
Taguchi, V., 25, *71*
Takada, S., 66, *75*
Takatsuki, K., 240, *271*
Takaya, H., 44, *75*, 162, *177*
Takaya, T., 151, *177*
Tanida, H., 65, 67, *77*
Tanny, S. R., 279, *302*
Tappen, D. C., 170, *176*
Taylor, D. R., 80, 114, *140*
Tee, O. S., 4, 7, *77*, 159, *178*, 273, *302*
Thomas, L. F., 82, *136*
Thompson, G. L., 282, 286, *301*
Thompson, H. W., 82, *136*
Thoumas, F., 16, *77*
Thrush, B. A., 282, *301*
Tiernan, T. O., 50, *71*
Tomioka, H., 166, *178*
Tonne, P., 13, *77*
Tordeux, M., 4, *71*
Toshima, N., 289, *298*
Toth, M., 282, *299*
Traetteberg, M., 82, *135*
Traficante, D. D., 109, *139*
Traylor, T. G., 50, 51, *71*, *77*
Trayrelis, V. J., 168, *178*
Trent, J. E., 17, *72*, 245, *268*
Treptow, R. S., 59, *72*
Trill, H., 221, *270*
Trindle, C., 59, *77*, 274, 292, *297*, *299*, *302*
Trost, B. M., 58, *77*, 119, *140*, 166, 167, *178*
Tsuji, T., 65, 67, *77*
Tsuruta, H., 287, *300*
Tsushima, T., 65, *77*
Tufariello, J. J., 11, *77*
Turbitt, T. D., 45, *77*
Turk, S. D., 142, *178*
Turnblom, E. W., 166, *178*
Turner, J. O., 31, *72*
Turro, N. J., 44, *72*, *77*, 171, *175*, *178*, 296, *302*

U

Ubersax, R. W., 58, *78*

AUTHOR INDEX 317

Uchic, J. T., 6, 7, 8, *73*
Ullenius, C., 294, *298*, *302*
Ulrich, H., 80, *140*
Unland, M. L., 82, *138*

V

Vahrenholt, F., 88, *140*, 225, *271*
Van Cauwelaert, F. H., 225, *271*
Van Den Elzen, R., 153, *176*
van der Hart, W. J., 274, *302*
van der Hout-Lodder, A. E., 37, *77*
van der Lugt, W. T. A. M., 162, *178*, 274, *302*
van de Ven, L. J. M., 37, *77*, 83, *140*
Van Dine, G. W., 6, 11, *76*, *77*, 245, *270*
van Dongen, J. P. C. M., 82, 83, *140*
Vanlierde, H., 90, 102, 123, *137*
van Tamelen, E. E., 25, 32, *77*
Van Tilborg, W. J. M., 4, *74*, 152, *179*
Van Wazer, J. R., 82, *138*
Vasella, A., 169, *174*
Vaughan, W. E., 109, *138*
Vedejs, E., 282, 287, *302*
Vesley, G. F., 171, *178*
Viehe, H. G., 83, 118, *137*, *138*
Vigdorovich, O. A., 220, *269*
Visser, J. P., 152, *179*
Vogel, E., 168, *179*
Vogel, P., 55, 59, *76*, *77*
Vollmer, J. J., 274, *298*, *302*
Vovsi, B. A., 80, *138*
Vukov, R., 66, 70, *75*

W

Waali, E. E., 285, *302*
Waegell, B., 16, *71*, 130, 131, *135*
Wagner, H.-U., 38, *77*, 119, *140*
Wang, I. S. Y., 182, *271*, 288, *302*
Wang, P. S. C., 274, *302*
Wanless, G. G., 37, *73*
Warburton, M. R., 114, *140*
Ward, S. D., 33, *74*
Warhurst, E., 87, *136*, 193, 268, 273, *299*
Waring, A. J., *73*, 130, *137*, *139*
Warrener, R. N., 171, *177*, 239, *270*, 283, *301*
Wassenaar, S., 168, *175*
Watanabe, M., 173, *176*
Watson, P. L., 239, *270*, 283, *301*
Watts, C. R., 222, 226, 234, 235, 236, *269*
Watts, W. E., 45, *77*

Weber, H., 282, *298*
Weber, W. P., 170, *175*
Webster, B., 104, *136*
Wege, D., 20, *71*, 285, *300*
Weglein, R. C., 53, *75*
Weidler-Kubanek, A., 118, *140*
Weiler, J., 82, *135*
Weimann, L. J., 82, *140*
Weingarten, H., 118, *140*
Weiss, R. G., 18, *77*
Weiss, V. W., 83, *140*
Weltin, E. E., 274, *302*
Wenkert, E., 59, *72*
Werthemann, D. P., 160, *179*
West, R., 26, *74*
Westburg, H. H., 282, *302*
Westerman, P. W., 48, *75*, 83, *139*
Wetmore, R., 288, *301*
Wexler, A., 295, *302*
Weyenberg, D. R., 170, *175*
Weyler, W., Jr., 94, 98, 105, 106, *136*, *140*
Whalen, D., 58, *72*, *77*
Whalen, W. L., 282, *299*
White, A. M., 50, 52, *75*
White, J. D., 165, *176*, *179*
White, J. F., 162, *177*
White, R., 162, *176*
Whitehead, M. A., 172, *178*
Whitesides, G. M., 162, *177*
Whitham, G. H., 11, 16, *77*
Whitlam, G. H., 63, *73*
Whitney, M. C., 126, 130, *138*
Wiberg, K. B., 20, 57, 58, 59, *78*
Widmer, U., 61, 64, *78*, 254, *271*
Wiest, H., 223, *270*
Wigner, E., 273, *302*
Wilcox, C. F., 61, *78*
Wildsmith, E., 282, *299*
Willcott, M. R., III, 292, *300*
Willey, F. G., 273, *299*
Williams, D. R., 189, *270*
Williams, J. E., Jr., 49, *77*, *78*
Williams, J. K., 113, *136*
Williams, J. M., Jr., 148, 153, 160, *175*
Williams, J. R., 44, *77*
Williams, V. Z., Jr., 58, *78*
Willis, B. J., 168, *175*
Wilson, E. B., 274, *302*
Wilson, G., 80, *140*
Wilson, H. R., 171, *178*
Wilt, J. W., 12, *78*

Wingard, R. E., Jr., 152, 161, *177, 178*
Winkler, J., *75*
Winnerwisser, G., 83, *138*
Winstein, S., 24, 25, 38, 40, 57, 58, 59, 64, 65, 67, 68, 69, *71, 72, 74, 76, 77, 78*
Witiak, J. L., 170, *175*
Witmer, E. E., 273, *302*
Wolf, A. D., 52, *72*
Wolf, R., 279, *299*
Wollenberg, R. H., 274, *302*
Wollweber, H., 219, *271*
Woodward, R. B., 3, 7, 34, 37, 42, 45, 52, 54, 59, 61, 64, *73, 78,* 80, 85, 86, *140,* 141, 162, *179,* 183, 208, 214, 228, 239, 241, 242, 248, 249, 256, 259, 263, *268, 271,* 273, 274, 280, *300, 302*
Worchel, A., 130, *137*
Wright, D. B., 114, *140*
Wright, H. E., 130, *136*
Wright, J. J., 292, *300*
Wright, J. S., 196, *270,* 291, *302*
Wright, M., 11, 16, *77*
Wu, E. S. C., 287, *302*
Wulfman, D. S., 172, *178*
Wynberg, H., 168, *176, 179*

Y

Yakovleva, T. V., 166, *179*
Yalpani, M., 169, *177*
Yamabe, S., 214, *269,* 274, *299*

Yamaguchi, H., 171, *177*
Yamaguchi, K., 4, *78,* 213, *270, 271*
Yamak, S., 88, *138*
Yamamoto, H., 18, *72,* 165, *174*
Yang, N. C., 296, *302*
Yasuji, I., 166, *178*
Yates, K., 4, 7, *77,* 159, *178,* 273, *302*
Yates, R. L., 189, 262, *268, 271*
Yelvingon, M. B., 171, *178*
Yokoyama, K., 38, *75*
Yonezawa, T., 182, 183, *268*
Yong, K. S., 6, 8, 15, *76*
Yoshifuji, M., 166, *177*
Yoshikawa, Y., 168, *178*
Yoshimasa, H., 166, *178*
Yu, C. Y., 164, *178*

Z

Zahradník, R., 278, *298*
Ziegler, E., 129, *140*
Zika, R. G., 168, *175*
Ziman, S. D., 166, 167, *178*
Zimmer, G., 246, *270*
Zimmerman, H. E., 60, *78,* 87, *140,* 160, 172, *178,* 199, *271,* 287, *302*
Zsindely, J., 59, 61, 64, *76, 78,* 254, *271*
Zubtsova, L. I., 166, *179*
Zuman, P., 225, *271*
Zurawski, B., 50, *78*
Zweifel, G., 29, *72*

Subject Index

A

ab initio, calculations, 291, 292
Activated complex, 273, 278, 283
Addition
 electrophilic, theory of, 207-208
 [$_\sigma 2 + _\sigma 2$], 280, 285, 286
Aldoketenes, reaction with cyclopentadiene, 31-32
Allene(s), *see also* specific allenes
 dimerization, 60-68
 electronic structure, 3-10
 frontier molecular orbitals, 8, 9, 59
 geometry, 6
 p orbital populations, 6-7
 reaction with dienes, 93-94
  with ketenes, 49-58
  with olefins, 69-72
Allene cycloadditions
 [2 + 2], 59-76
  allenes + olefins, 69-72
  degenerate rearrangements, 73-76
  dimerization, 60-68
  diradical pathway, 59-60
 [2 + 4], 93-94
Allene + olefin cycloadditions, [2 + 2], 69-72
 isotope effects, 70-71
 stereospecificity, 70
 transition state, 71-72
Allyl cation, cycloaddition of, 36, 39, 42-44
 internal, 39, 40
 stereochemistry of, 43
Allylenecyclopropanes, 284
Allyl radical, 278
1-Aminonaphtho[1,8-*de*]triazine, 281
Anthracene, 275, 296
Asymmetrical reactions, 274

B

Basketene, 282
Benzenium cation, photochemical electrocyclic closure, 21, 24, 25
Benzene, 275, 282, 296
8,9-Benzobicyclo[5.2.0]nona-1,3,5,8-tetraene, 285
4,5-Benzotetracyclo[4.4.0.0^{2,10}.0^{3,9}]deca-4,7-diene, 287
8,9-Benzotricyclo[5.2.0.0^{1,6}]nona-2,4,8-triene, 285
Benzvalene, 296
Bicyclo[3.1.1]hept-2-ene, 290
Bicyclo[2.2.0]hexa-2,5-diene, 296
Bicyclo[2.2.0]hexane, 280
Bicyclo[3.1.0]hex-2-ene(s), 282, 290, 291, 294, 295
Bicyclo[3.1.0]hexenyl cations
 from benzenium cations, 24, 25
 sigmatropic rearrangements in, 59, 60
Bicyclo[5.2.0]nona-1,3,5,8-tetraenes, 285
Bicyclo[4.2.1]nona-2,4,7-triene, 281
cis-Bicyclo[4.3.0]nona-2,4,7-triene, 287
Bicyclo[5.2.0]nona-2,5,8-triene, 287
Bicyclo[6.1.0]nona-2,4,6-triene, 287
Bicyclo[5.1.0]octa-2,4-diene, 280
Bicyclo[4.2.0]octane, 291
Bicyclo[2.1.0]pentane(s), 290-292, 294
Bicyclo[2.1.0]pent-2-ene, 284, 285
Bicyclopropyl, 283, 284
Bis(trifluoromethyl) ketene
 reaction with butadiene, 96-97
  with 1,3-dimethylallene, 50
  with enol ethers, 30-31
  with ethyl vinyl ether, 25-26
  with norbornadiene, 97
  with *cis*-propenylpropyl ether, 30-31
Bonding
 antisymmetric, 289, 295
 conjugated single, 295
 maintenance of, 273
Bridge flipping, in bicyclic cations, 68, 69
Butadiene, 276, 278
t-Butylcyanoketene
 reaction with *S*-(+)-1,3-dimethylallene, 55-57
  with 1,3-diphenylallene, 57

319

320   SUBJECT INDEX

  with 1-methyl-1-*t*-butylallene, 50–51
  with norbornadiene, 97
  with phenylacetylene, 21
Butylethylketene
  reaction with butadiene, 19
  with 2-methoxybutadiene, 96

C

Carbene(s), 223, 229
  "pull-push" character, 16
Carbocations, *see also* specific cations
  boron analogs, 29
  bridged, 46–54
  $(CH)_5^+$ type, 70
  $(CH)_6^{+2}$ type, 70
  cycloaddition reactions of, 34–45
  pericyclic stabilization in, 64–69
  electrocyclic reactions of, 1–34
  laticyclic stabilization in, 68
  longicyclic stabilization in, 69
  molecular orbital calculations of, 3, 4, 5, 18, 24, 38, 49–52, 55, 59, 70
  photochemistry of, 6, 24, 25, 26, 28, 32, 34
  sigmatropic reactions of, 45–70
Carbodiimides, electronic structure, 4–5
Carbon chain lengthening, *see* Vicinal carbofunctionalization
Carbon dioxide, electronic structure, 4–5
Carbon-13 labeling, 285, 291
Carbon monoxide extrusion, *see* Decarbonylation
Carbon shift, 280, 281, 284, 286, 287, 292, 294, 295
Catalysts, metallic and enzymic, 275
Cheletropic reactions, 255–257, 265, *see also* specific substances
Chemiluminescence, 296, 297
Chloroketene, reaction with *trans*- and *cis*-cyclooctene, 34–35
Chlorosulfonylisocyanate
  reaction with 2,2-dimethyl-1-methylenecyclopropane, 22–23
  with olefins, stepwise, 41
Closed-shell repulsion, 188–190
Collisional deactivation, 285
Configuration, 275
  doubly excited, 277, 288
  dominant, 291
  ground state, 277, 288
  electronic, 275–277

  interaction, 275–279, 287–289, 291, 292, 297
  perturbative treatment using localized orbitals, 277
Configurational composite, 278
Configurational mixing, 289
Cope rearrangement, 253–255, 282
Core energy, 292
Correlation diagrams
  electronic configuration, 276
  electronic state, 276
  molecular orbital, 274, 276, 277
Correlation rules
  electronic states, 273
  orbital symmetry, 273
Cumulene(s), 239–240
  carbon-13 magnetic resonance, 8
  chemistry of, 2–3
  cycloadditions, [2 + 4], 93–103
    dienophilic reactivity, 93–95
    intermolecular, 95–102
    intramolecular, 102–103
  dipole moments, 8
  electronic structure, 3–10
    valence bond theory, 5
  frontier molecular orbitals, 8–10
  geometry, 6
  infrared spectra, 8
  p orbital populations, 6–7
  proton magnetic resonance, 8
Cyanoalkenes, 221, 223
Cycloadditions, 11–14, 206–240, 260–263, *see also* specific compounds, Diels-Alder reaction
  of allyl cations, 36, 39, 42–44
  catalysis of, 226–228, 235
  cumulenes, 224–225, 229
  1,3-dipolar, 222
  miscellaneous [4 + 2] processes, 44–45
  $\pi 2 + \pi 2$ processes, 36–39
  $\pi 2 + \pi 4$ processes, 41–44
  $\pi 4 + \pi 6$ processes, 45
  $\pi 2 + \sigma 2$ processes, 39–41
  rules for, 35–36
  theory of, 208–220
  of tropylium ions, 41–42, 45
  [2 + 2], 212–215
  of vinyl cations, 36–38
Cyclobutadiene dimers, 296
Cyclobutane(s), 278, 291, 294
Cyclobutene, 276, 278

SUBJECT INDEX 321

Cyclobutenyl cations, rearrangements in, 5, 6
Cyclobutyl bonds, 295
Cyclohepta-1,4-dienes, 294
Cyclohexa-1,3-diene, 282
Cyclohexa-1,4-diene, 282, 290
Cyclohexene, 283, 284, 291
Cyclohexenyl cations
  from closure of homodienylic cations, 30, 31
  from cycloaddition reactions, 39–40
1,2-Cyclononadiene, dimerization, 61–64
cccc-Cyclononatetraene, 287
Cyclooctatetraene, 296
Cyclopentadiene(s), 281
  vibrationally excited, 285
Cyclopentenyl cations, stereospecific electrocyclic formation of, 22–24
Cyclopropane(s), 278, 290, 294
  isomerization of, 292, 293
Cyclopropyl allyl cations, internal cycloadditions in electrocyclic closures, 29–30
Cyclopropyl bonds, 290, 295
Cyclopropylcarbinyl cations
  interchange with cyclobutyl and homoallyl systems, 56–59
  stereochemistry of migrating center in degenerate rearrangements, 57
  stabilization of, 56
  bicyclic analogs, 10–12, 15, 16
  calculated energy of, 5
  chloro-substituted, 14–17
  cyclopropanone as model for, 4
  electrocyclic opening, 3–16
  stereochemistry of opening, 6–10
Cyclopropyl radical, 278
Cyclopropyl rings, 289
Cycloreversions
  ketene + diene, 99
  + olefin, 45–47

D

Decarbonylation, cheletropic CO extrusion, 171
2,4-Dehydrohomoadamantane, 291
1,8-Dehydronaphthalene, 281
Deuterium labeling, 282, 286, 287, 290–293, 295, 296
1,3-Diadamantylallene, dimerization, attempted. 64–65

Diatomic molecules, fragmentations of, 278
Dichlorodicyanobenzoquinone, 257
Dichloroketene
  reaction with 4-t-butylcyclohexene, 21–22
  with 2-cholestene, 21–22
  with cyclopentadiene, 19
  with *cis*- and *trans*-cyclooctene, 27
Diels-Alder reaction, 218–219, 221, 281, *see also* Cycloaddition
  photochemical, 262–263
  retro, 282, 287
  theory of, 192–195, 215–220, 229, 231–233
Dienes, 216–217, 219, 223
Dienylic cations
  conformations of, 22, 24
  cycloadditions involving, 41–42
  electrocyclic closure, 21–25
Dihydrogen fluoride cation, 278
cis-3a,7a-Dihydroindene, 287
trans-4a,4b-Dihydrophenanthrene, 288, 289
Dihydro-*m*-xylenes, 279, 280
Dimerization
  allenes, 60–68
  ketenes, 10, 48, 49
$R-(-)$-1,3-Dimethylallene, reaction with acrylonitrile, 71–72
6,7-Dimethylbicyclo[3.2.0]hept-1-enes, 283
6,7-Dimethylbicyclo[2.1.0]pentane-5-spirocyclopropane, *cis* and *trans* isomers, 283
5,5-Dimethylcyclohexa-1,3-diene, 279, 280
trans-3,4-Dimethylcyclopentene, 281
trans-15,16-Dimethyldihydropyrene, 288
trans-8,9-Dimethyl-8,9-dihydrotetralin, 281, 282
Dimethylenecyclobutane rearrangement, 64, 67, 68
2,3-Dimethylenecyclohexadiene, intramolecular cycloaddition reactions, 66–67
1,4-Dimethylenecyclohexane, 289
Dimethylketene
  dimerization, 48–49
  kinetics of cycloadditions, 37
  reaction with *trans*-butene, 28
  with $R-(-)$-1,3-dimethylallene, 52–54
  with 2,2-dimethyl-1-methylenecyclopropane, 22–23
  with *cis*- and *trans*-enol ethers, 27
  with *N*-isobutenyl-pyrrolidine, 23–24
Dimethyl[2.2]metacyclophane, 288

3,7-Dimethyl-7-methoxycarbonylcyclohepta-1,3,5-triene, 286
1,4-Dimethyl-7-methoxy-2-oxabicyclo[3.2.2]nona-3,6,8-triene, 286
1,2-Diphenylcyclobutanes, *cis* and *trans* isomers, 294
Diphenylketene
  kinetics of cycloadditions, 35–39
  reaction with conjugated vinyl ether, 100
    with *cis*- and *trans*-cyclooctene, 27
    with cyclopentadiene, 1
    with 2,2-dimethyl-1-methylenecyclopropane, 22–23
    with *cis*- and *trans*-enol ethers, 27
    with 1-ethoxybutadiene, 95–96
    with ethylene, 20
    with ethyl vinyl ether, 21
    with *cis, trans*-2,4-hexadiene, 31
    with 2-methoxybutadiene, 96
    with norbornene, 25
    with propene, 21
    with styrene, 21
    with tetramethylallene, 50
    with tropone, 100–101
Diphenylketene-*N*-methylimine, reaction with 1-diethylaminopropyne, 77
Diphenylketene-*N*-phenylimine, reaction with 1-diethylaminopropyne, 78
Diphenylketene-*p*-tolylimine
  reaction with bis(trifluoromethyl) ketone, 78–79
    with phenylisocyanate, 78–79
1,3-Dipoles, 206, 222
trans-1,2-Dipropenylcyclobutanes, 294
Diradical, 278, 291, 293, 294
Dispiro[2.0.2.4]deca-7,9-diene, 283
trans,trans-1,8-Distyrylnaphthalene, 280
Dithioketenes, reaction with cyclopentadiene, 20
1,2-Divinylcyclopropanes, *cis* and *trans* isomers, 294

E

Electrocyclic closure
  azapentadienyl cations, 28
  benzenium cations, 24–25
  bishomoallyl cations, 20–21
  diphenylmethyl cations, 32–33
  3,5-homodienyl cations, 29–31
  4,6-homodienyl cations, 29, 30
  homotropylium cations, 25–26
  pentadienyl cations, 21–25
  1-phenylallyl cations, 31–32
  polyenylic cations, 33–34
  protonated dienones, 26–28, 34
  rules for, in cations, 3
  tropylium cation, 25
Electrocyclic openings
  bicyclic cyclobutyl cations, 20
  bicyclic cyclopropyl cations, 10–16
  cyclopropyl cations, 3–17
  α-haloepoxides, 19
  inner and outer rotations, 7, 8, 10, 11, 19
  nitrenium ions, 17, 18
  rules for, in cations, 3
Electrocyclizations, 241–248, 263–265
  cyclobutenes, 197–199, 263–264
Electron affinities, 201, 203, 220, 222, 225
  of alkenes, 203–205
  cumulenes, 205
  of 1,3-dipoles, 206
Electron repulsion, 276
Electronic configuration, predominant, 279
Ene reaction, 280, 281
Energy relations, in cation reactions, 5, 21, 32, 58–59
Epimerization
  one-center, 293, 294
  two-center, 292–294
Ethane, 278
Ethoxyketene, reaction with *cis*-butene, 35
Ethyl cation, 49–50
Ethylene, 286–288, 291
Ethylene–tetralin Diels–Alder adduct, 287
1-(Z)-Ethylidene-2-methylcyclobutane, 281
8-(Z)-Ethylidenespiro[2.5]octa-4,6-diene, 284
o-Ethylstyrene, 283, 284

F

Fluorene, 285
Fluorine atom, 278
Fulvenes, 238–239

G

Geometrical distortions, 288
Germylene additions cheletropic reactions, 169, 170
Group theory, applications of, 273, 288

H

Half-electron method, 278
α-Haloenamines from keteniminium salts, 9, 80
Halonium ions, 48
Hexachlorocyclopentadiene, 218
1,5-Hexadiene, 280
Hexamethylbenzene, 279, 280
Homoadamantene, 291
Homo ene reaction, 283
Homotropylium cations
  photochemical electrocyclic closure in, 25, 26
  possible sigmatropic rearrangements in, 63
Hot-molecule effects, 285
Hydride shifts
  1,2-, 52–54
  1,3-, 54
  1,4-, 59
  1,5-, 62
  1,6-, 63
Hydrogen
  atom, 278
  ion, 278
  molecule, 278
  shift, 280–282, 285, 292
Hydrogen fluoride, 278

I

Indenes, 285
Interaction terms, 277
Intermediates, 273, 275, 278, 284, 286, 292, 293
Inversion, one-center and two-center, 292–294
Interstate crossing, 279
Intersystem crossing, 275, 296, 297
Ionization potentials, 200–201, 220, 222, 225
  alkenes, 203–205, 221
  cumulenes, 205
  1,3-dipoles, 206
Isocyanates
  electronic structure, 4–5
  reaction with olefins, 22–23
Isomer(s), "bond" and "stretch," 289
Isomerization
  least motion, 286
  [$_\sigma 2 + _\sigma 2$], 283, 286, 290, 291

3-Isopropylidenetetracyclo[3.3.0.0^{2,8}.0^{4,6}]octane, 289, 290
3-Isopropylidenetricyclo[3.2.1.0^{2,4}]oct-6-ene, 289, 290

K

Ketene(s), *see also* specific compounds
  anion-stabilizing substituents, 25–26
  dimerization, 10, 48, 49
  electronic structure, 4–10
  frontier molecular orbitals, 8–10
  geometry, 6
  p orbital populations, 6–7
  "pull-push" character, 16
  reaction with cyclopentadiene, 31–34
    with dicyclopentadiene, 41–43
    with dienes, 19–20, 31–34
    with enamines, 23–24
    with enol ethers, 25–27, 30, 31
    with olefins, 20–23, 27, 28, 35, 41–43
    with ynamines, 24–25
Ketene + allene cycloadditions, [2 + 2], 49–58
  activation parameters, 53
  solvent effects, 53
  stereochemistry, 51–57
  transition state, 53–57
Ketene cycloadditions, [2 + 4], 95–103
  activation parameters, 99–100
  dipolar intermediate, 97–99
  intermolecular, 95–102
  intramolecular, 102–103
  retro-Diels–Alder reaction, 99
  solvent effects, 99–100
Ketene + olefin cycloadditions, [2 + 2], 11–47
  activation parameters, 38
  aromatic transition states, 14
  cycloreversions, 45–47
  dienes, 19–20
  dipolar mechanism, 28–31
  1,4-dipole formation, 17
  endo substituent preference, 32–33
  HOMO-LUMO interaction, 15–17, 21
  kinetics, 35–39
  olefins, simple, 20–22
  orthogonal approach, 11–17, 33
  oxetane formation, 29–30
  periselectivity, 19–26

324 SUBJECT INDEX

reactivity of *cis*- vs *trans*-olefins, 38–41
regioselectivity, 19–26
secondary deuterium isotope effects, 44–45
skew approach, 33
solvent dependence, 30–31, 35–43
stepwise pathway, 23–26
stereochemistry, 26–35
cis-stereospecificity, 27–28
steric effects, 18, 35–43
substituent effects, 17, 35–43
theoretical analysis, 11–18
theoretical calculations, 16–17
thermodynamic product preference, 33–34
transition states, 12–18
Ketenimine(s), *see also* specific compounds
electronic structure, 4–10
frontier molecular orbitals, 8–10, 76
geometry, 6
p orbital populations, 6–7
reaction with dienes, 76–77
with isocyanates, 79
with ketones, 79
with olefins, 76–77
with ynamines, 77–78
Ketenimine cycloadditions, [2 + 2], 76–79
dipolar intermediates, 77–78
Keteniminium salt cycloadditions, [2 + 2], 79–92
[2 + 2] and [4 + 2] adducts, 90, 94, 95
concerted pathway, 89
reaction with acetylenes, 83, 86
with dienes, 88–92
with olefins, 83–88
transition state, 82–83, 88, 91, 92
Keteniminium salts, *see also* specific salts
electronic structure, 3–10
frontier molecular orbitals, 8–9, 81, 82
geometry, 6
methods of preparation, 80
p orbital populations, 6–7
reaction with acetylenes, 83, 86
with allenes, 91–92
with dienes, 88–92
with halide ions, 80
with olefins, 83–88
Ketones protonated, as models for carbocations, 2, 4, 26, 27, 28, 34, 42

L

LCAO MO theory, 278
Lead tetraacetate, 281
Least energy path, 289
Least motion principle, 273
  nitrous oxide elimination and, 164
  sulfolene reaction and, 159
Linear free energy relationships, 225–226
Lumibullvalene, 295

M

Matrix element, configuration interaction, 289
Maleic anhydride, 221, 225
Methane, elimination of, 279
2-Methylbicyclo[2.1.0]pent-2-ene, 285
Methylcyclopentadienes, 285
Methylene, 288
Methylenecyclobutane(s), 295, 296
  rearrangement, 73–74
Methylenecyclobutanone rearrangement, 75–76
Methylenecyclopropane, 292
1-Methylenecyclopropanes, 284
cis,*cis*-1-Methylene-2-(pent-2-enylidene)-cyclohexane, 281, 282
2-Methyl-1-(penta-1,3-dienyl)cyclohexene, *cis*,*cis* and *cis*,*trans* isomers, 281, 282
cis-3-Methyl-4-(trans-propenyl)cyclohexene, 294
Methyl radical, 278
MINDO/CI, 292
MINDO/2, 292
Multistep reaction mechanisms, 274, 275
Multi-dimensional hypersurface, 293

N

Naphthalene, 296
peri-Naphthobicyclo[3.1.1]heptene, 280
Naphtho[*b*-2,3]bicyclo[2.2.0]hexa-2,5-diene, 296
peri-Naphthotricyclo[4.2.0.0^{2,8}]octene, 281
Naphthvalene, 296
Nazarov cyclization, 27
Nenitzescu's hydrocarbon, 282

SUBJECT INDEX

Nitrogen eliminations, cheletropic reactions, 162–165
Nitrosoaziridines, *see* Nitrous oxide eliminations
Nitrous oxide eliminations, cheletropic reactions, 162–165
Nondynamic reaction path, 288
Noncrossing rule, 277
Norbornene, 290
Norcaradiene, 286
Norpinene, 290
Nortricyclene, 290

O

Open-shell molecules, 274
Orbital
  complex molecular, 276
  highest occupied molecular, 277, 278
  lowest unoccupied, 277
  nonbonding, 294, 295
  subjacent, 249–252, 295
    control, 54, 55
  superjacent, 249–252
  virtual, 277
Orbital momentum correlation rules, 273
Orbital occupancy, 276
Oxyallyl cation, 38–39
  cycloadditions of, 43–44

P

Pauli exclusion principle, 276
1,2,3,5,6-Pentamethylhomofulvene, *exo* and *endo* isomers, 279, 280
Pericyclic selectivity, 283
Phenyl azide, 221, 222, 225
8-Phenylbicyclo[5.2.0]nona-1,3,5,8-tetraene, 285
2-Phenylindene, 285
8-Phenyltricyclo[5.2.0.0^{1,6}]nona-2,4,8-triene, 285
Phosphine or phosphite additions, *see* Phospholenes
Phospholenes, cheletropic reactions, 165 166
Photochemical reactions, 274

Photochemistry, sulfur dioxide extrusions, 159–161
cis-Piperylene, 292
Polyenylic cations
  electrocyclic closure, 33, 34
  from electrocyclic cyclization of phenylallyl cations, 31, 32
  possible formation from diphenylmethyl cations, 32
[4.4.2]Propella-2,4-diene, 286
[2.2.2]Propellane, 289
cis-Propenylallene, reaction with methyl vinyl ketone, 105
trans-Propenylallene, reaction with methyl vinyl ketone, 106
cis-Propenylcyclopropane, 283, 284
trans,trans-1-Propenyl-2-methylcyclopropane, 281
1-Propyl cation, 50–51

Q

o-Quinoid ketene, reaction with ketones, 101–102

R

Radical-chain mechanisms, 279
Reaction dynamics, 288
Reactions, favored versus unfavored, 274
Rearrangements, *see* specific types
Relief of strain, 292
Reversion, [$_\sigma$2 + $_\sigma$2], 280, 289–291

S

SCF calculations, 292
SCF configurations, 288
Selenium extrusions, cheletropic reactions, 168, 169
Semibullvalene, 287
Sigmatropic rearrangements
  in bicyclo[3.1.0]hexenyl cations, 59–60
  in cyclopropenylcarbinyl cations, 54
  in cyclopropylcarbinyl cations, 56–59
  definition of, in cations, 44–45

shifts, 248–255, 265
  1,2-, 52–54
  1,3-, 54
  1,4-, 59
  1,5-, 62
  1,6-, 63
  3,4-, 64
Silylene additions, cheletropic reactions, 169, 170
Skeletal inversion, 292
Slater determinant, 276
Spin correlation rules, 273
Spin forbidden, 275, 296
Spin-orbit interaction, 275
State(s)
  electronic, 275, 277
  singlet, 278, 289, 292
    low-lying excited, 288
  triplet, 278, 279, 289
  vibrational, 275
State crossing, 296
State symmetry, 276
Stereochemical paths, 274
cis-Stilbene, 288, 289
Styrenes, 227
Substitution
  electrophilic, 182, 183, 223
    theory of, 182, 183
  nucleophilic, 182
    theory of, 182, 206–207
  radical, 182
    theory of, 182
Sulfolene reactions, 142–161
  compared with Diels–Alder (1,2-addition), 153–159
  energy hypersurface for, 158
  entropic considerations, 158, 159
  kinetics and mechanism, 146–153
    five-membered ring, 146, 148
    homoconjugate elimination, 150
    seven-membered ring, 149, 150
    steric constraints, 151–153
    sulfolane fragmentation, 151
    three-membered ring, 148, 149
  stereochemistry and mechanism, 143, 144
  thiazine oxides and, 155–157
Sulfuranes, cheletropic reactions, 166, 167
Sulfur dioxide, cheletropic reactions, see Sulfolene reactions
Sulfur extrusions, see also Sulfuranes
  cheletropic reactions, 168, 169

Sulfur monoxide extrusions and additions, cheletropic reactions, 144, 145
Symmetry element, 274, 277

T

Tellurium extrusions, cheletropic reactions, 168, 169
Tetralin, 283, 284, 286, 287
Tetrachloroallene, reaction with cyclopentadiene, 94
Tetracyclo[4.2.1.0^{2,8}.0^{5,7}]non-3-ene, 281
Tetracyclo[3.3.0.0^{2,8}0^{4,6}] octane, 289, 290
Tetracyclo[4.2.0^{1,3}.0^{5,7}]octane, 289, 290
1,2-Tetramethylenebicyclo[2.2.2]octa-2,5-diene, 287
1,2-Tetramethylene Dewar benzene, 287
Tetramethyleneethane, 63–68
  ESR, 64
  geometry, 65–67
  organometallic derivatives, 64
  triplet species, 66
Tetramethylketeniminium tetrafluoroborate (TMK)
  reaction with acetylene, 83
    with allene, 91
    with butadiene, 88–89
    with cis- and trans-butenes, 86–87
    with t-butylacetylene, 86
    with cyclopentadiene, 89–90
    with cyclopentene, 84
    with dicyclopentadiene, 87–88
    with 1,1-dimethylallene, 91
    with 2,3-dimethylbutadiene, 90
    with 2,2-dimethyl-1-methylenecyclopropane, 85–86
    with ethylene, 83
    with cis,trans-2,4-hexadiene, 88–89
    with isobutene, 85
    with methylacetylene, 86
    with propene, 84
    with styrene, 84
    with tetramethylallene, 91
Thiazine oxides, see Sulfolene reactions
Through-bond coupling, 295
Through-bond interactions, 295
TME, see Tetramethyleneethane
TMK, see Tetramethylketeniminium tetrafluoroborate

Toluene, 279
Trajectory calculations, 288
Triatomic molecules, fragmentation of, 278
Tricyclo[4.4.0.0^{1,4}]deca-2,5-diene, 287
Tricyclo[3.3.2.0^{2,6}]deca-3,7,9-triene, 295
endo,endo-Tricyclo[4.3.1.0^{9,10}]dec-7-ene, 294
1-exo, 6-endo-Tricyclo[4.3.1.0^{9,10}]dec-7-ene, 294
Tricyclo[3.1.0.0^{2,4}]hexane, 290
Tricyclo[4.3.0.0^{2,9}]nona-4,7-diene, 287
Tricyclo[3.3.0.0^{2,6}]octa-3,7-diene, 287
Tricyclo[3.3.0.0^{2,8}]octa-3,6-diene, 287
Tricyclo[3.2.1.0^{2,4}]oct-6-ene, 289, 290
endo, endo-Tricyclo[5.3.1.0^{10,11}]undec-8-ene, 294
1-exo,7-endo-Tricyclo[5.3.1.0^{10,11}]undec-8-ene, 294
Trimethylene, 292, 293
2,3,4-Triphenyltetracyclo[3.3.0.0^{2,8}.0^{4,6}]octane, 289, 290
2,3,4-Triphenyltricyclo[3.2.1.0^{2,4}]oct-6-ene, 289, 290
Tropilidene, 286
Tropylium cation
  analogs of, 25, 26, 63
  cycloadditions involving, 41–42, 45
  photochemical electrocyclic closure in, 25
Two-electron effects, 295

V

Valence isomerization, 285, 296
Variational mixing, 291
Vibrational interaction, 275

"Vicinal carbofunctionalization," 20
Vinylallenes
  reaction with dimethylacetylene-dicarboxylate, 104
  with methyl vinyl ketone, 104–105
  with tetracyanoethylene, 104
Vinyl cation, cycloadditions of, 36–38
Vinylcumulene cycloadditions, 104–111
Vinylketenes
  frontier molecular orbitals, 106–107
  reaction with ethyl vinyl ether, 110–111
Vinylketenimines
  frontier molecular orbitals, 106–110
  reaction with 1,1-dicyanostyrene, 109
    with N-diethylamino-phenylacetylene, 109–110
    with dimethylacetylene-dicarboxylate, 108
    with tetracyanoethylene, 108
2-Vinyl-1-methylenecyclopropanes, 284
5-Vinylpentamethylcyclopentadiene, 279, 280

W

Wave functions, 275
  determinantal, 276, 277
  single determinantal, 278
  single product, 275
  spin-density, 276
Wigner–Witmer rules, 273

X

m-Xylene, 279

ORGANIC CHEMISTRY
A SERIES OF MONOGRAPHS

EDITORS

ALFRED T. BLOMQUIST*   HARRY H. WASSERMAN
Department of Chemistry   *Department of Chemistry*
Cornell University   *Yale University*
Ithaca, New York   *New Haven, Connecticut*

1. Wolfgang Kirmse. CARBENE CHEMISTRY, 1964; 2nd Edition, 1971
2. Brandes H. Smith. BRIDGED AROMATIC COMPOUNDS, 1964
3. Michael Hanack. CONFORMATION THEORY, 1965
4. Donald J. Cram. FUNDAMENTALS OF CARBANION CHEMISTRY, 1965
5. Kenneth B. Wiberg (Editor). OXIDATION IN ORGANIC CHEMISTRY, PART A, 1965; Walter S. Trahanovsky (Editor). OXIDATION IN ORGANIC CHEMISTRY, PART B, 1973
6. R. F. Hudson. STRUCTURE AND MECHANISM IN ORGANO-PHOSPHORUS CHEMISTRY, 1965
7. A. William Johnson. YLID CHEMISTRY, 1966
8. Jan Hamer (Editor). 1,4-CYCLOADDITION REACTIONS, 1967
9. Henri Ulrich. CYCLOADDITION REACTIONS OF HETEROCUMULENES, 1967
10. M. P. Cava and M. J. Mitchell. CYCLOBUTADIENE AND RELATED COMPOUNDS, 1967
11. Reinhard W. Hoffman. DEHYDROBENZENE AND CYCLOALKYNES, 1967
12. Stanley R. Sandler and Wolf Karo. ORGANIC FUNCTIONAL GROUP PREPARATIONS, VOLUME I, 1968; VOLUME II, 1971; VOLUME III, 1972
13. Robert J. Cotter and Markus Matzner. RING-FORMING POLYMERIZATIONS, PART A, 1969; PART B, 1; B, 2, 1972
14. R. H. DeWolfe. CARBOXYLIC ORTHO ACID DERIVATIVES, 1970
15. R. Foster. ORGANIC CHARGE-TRANSFER COMPLEXES, 1969
16. James P. Snyder (Editor). NONBENZENOID AROMATICS, VOLUME I, 1969; VOLUME II, 1971

*Deceased.

17. C. H. Rochester. ACIDITY FUNCTIONS, 1970
18. Richard J. Sundberg. THE CHEMISTRY OF INDOLES, 1970
19. A. R. Katritzky and J. M. Lagowski. CHEMISTRY OF THE HETEROCYCLIC N-OXIDES, 1970
20. Ivar Ugi (Editor). ISONITRILE CHEMISTRY, 1971
21. G. Chiurdoglu (Editor). CONFORMATIONAL ANALYSIS, 1971
22. Gottfried Schill. CATENANES, ROTAXANES, AND KNOTS, 1971
23. M. Liler. REACTION MECHANISMS IN SULPHURIC ACID AND OTHER STRONG ACID SOLUTIONS, 1971
24. J. B. Stothers. CARBON-13 NMR SPECTROSCOPY, 1972
25. Maurice Shamma. THE ISOQUINOLINE ALKALOIDS: CHEMISTRY AND PHARMACOLOGY, 1972
26. Samuel P. McManus (Editor). ORGANIC REACTIVE INTERMEDIATES, 1973
27. H.C. Van der Plas. RING TRANSFORMATIONS OF HETEROCYCLES, VOLUMES 1 AND 2, 1973
28. Paul N. Rylander. ORGANIC SYNTHESES WITH NOBLE METAL CATALYSTS, 1973
29. Stanley R. Sandler and Wolf Karo. POLYMER SYNTHESES, VOLUME I, 1974; VOLUME II, 1977
30. Robert T. Blickenstaff, Anil C. Ghosh, and Gordon C. Wolf. TOTAL SYNTHESIS OF STEROIDS, 1974
31. Barry M. Trost and Lawrence S. Melvin, Jr. SULFUR YLIDES: EMERGING SYNTHETIC INTERMEDIATES, 1975
32. Sidney D. Ross, Manuel Finkelstein, and Eric J. Rudd. ANODIC OXIDATION, 1975
33. Howard Alper (Editor). TRANSITION METAL ORGANOMETALLICS IN ORGANIC SYNTHESIS, VOLUME 1, 1976
34. R. A. Jones and G. P. Bean. THE CHEMISTRY OF PYRROLES, 1976
35. Alan P. Marchand and Roland E. Lehr (Editors). PERICYCLIC REACTIONS, VOLUME I, 1977; VOLUME II, 1977
36. Pierre Crabbé (Editor). PROSTAGLANDIN RESEARCH, 1977